Java 程序设计

项目化教程

主　编　范凌云　兰　伟　杨　东

副主编　夏　科　邓　柳

復旦大學 出版社

内容提要

  本书通过系列项目全面、系统地介绍了 Java 语言程序设计的基础知识，包括基本语法、编程方法、面向对象的特征，以及 Java 语言的异常处理、多线程、输入输出和文件操作等内容。在此基础上，本书阐述了 Java 语言在图形界面设计、数据库操作和网络编程等方面的应用。本书每个项目均配有大量练习题和上机题，可帮助读者掌握书中的主要内容，检验学习效果。

  本书以"教、学、做"一体化的教学模式来体现教学内容和单元结构，做到"讲练结合、讲中练、练中学"，易于学习者消化和吸收所学内容，并锻炼实操能力，达到学以致用的效果。

  本书适合作为高职高专学校相关专业学习 Java 编程的入门级教材，也适合计算机爱好者、软件开发人员学习 Java 语言时使用。

扫码获取相关程序代码

前言

Java 语言是当前计算机程序设计中使用最广泛的语言之一,具有安全、跨平台、可移植、健壮等显著特点。因此,自诞生以来,Java 语言迅速被业界认可并广泛应用于桌面应用程序、Web 应用程序、分布式系统和嵌入式系统应用程序等的开发中。在此形势下,国内高校在计算机及相关专业广泛开设了 Java 程序设计相关课程,旨在培养学生的编程能力,提高学生使用 Java 语言解决实际问题的能力,使学生建立良好的程序设计思想和编程习惯。

本书以软件开发工程师的岗位需求为主线,以培养学生的综合编程素质为目标,以强化理论学习与实际应用相结合为宗旨,系统介绍了 Java 语言的相关知识,内容翔实,图文并茂,具有鲜明的高等教育特色。

本书的主要特点如下。

一、注重理论知识的实用性

本书遵循"理论够用,重在实践"的原则,根据 Java 的学科特点构建知识体系。全书共两大项目:学生成绩管理系统、简单网络聊天软件。按照"基础知识→编程技能→核心应用"的结构和"由浅入深,由深到精"的学习模式,将学生成绩管理系统分解为 7 个子项目:Java 程序设计初识、面向过程的学生成绩信息处理、面向对象的学生成绩信息处理、学生成绩信息的异常处理、学生成绩信息保存到文件、创建学生成绩管理系统的图形界面、学生成绩管理系统的数据库编程。每个项目又分为若干相对独立的任务,通过各种典型工作任务,将知识点融入各个案例中。

二、教学设计体现"用、学、仿、创"的教学模式

在学生明确学习目标的前提下,可以先让学生使用本书案例,以便直观地理解体会工作任务的结果,接下来再学习相关知识点,然后再模仿完成案例,最后在充分理解原理和使用方法的基础上,通过拓展训练完成创新的过程,体现了创新型教育。

三、注重理论与实践的统一

本书每个项目和任务都选择了典型案例并进行深入浅出的分析,融入情境实训,以行为示范引导学生对专业知识的学习和掌握。突出教材的专业性、应用性和实践性,有利于学生

动手、动脑、增强能力。

  本书由范凌云、兰伟、杨东担任主编，夏科、邓柳担任副主编。兰伟编写项目1；范凌云编写项目2、项目4、项目5、项目7、项目8；杨东、夏科、邓柳编写项目3、项目6。全书最后由范凌云统稿和审定。

  本书编写过程中参考了大量国内外书刊和业界的研究成果，也得到企业一线开发人员的大力支持，还参考借鉴了网络中百度文库和博文的讲解。在此，谨向各方表示衷心的感谢。

  虽然作者在编写本书时尽了最大努力，但书中难免有疏漏和错误之处，敬请广大读者、专家批评指正。

编 者

2020年6月

目 录

项目 1   Java 程序设计初识 ······ 1
   1.1   任务 1   认知 Java ······ 1
       1.1.1   Java 语言概述 ······ 1
       1.1.2   Java 语言的应用 ······ 2
       1.1.3   Java 语言发展简史 ······ 2
       1.1.4   Java 语言的特点 ······ 3
       1.1.5   Java 程序的运行机制 ······ 4
   1.2   任务 2   搭建 Java 开发环境 ······ 4
       1.2.1   Java 开发环境 ······ 5
       1.2.2   JDK 的安装与配置 ······ 6
       1.2.3   MyEclipse 的安装 ······ 21
   1.3   任务 3   编写第一个 Java 程序 ······ 30
       1.3.1   "Hello World"程序解释 ······ 31
       1.3.2   Java 编码规范 ······ 32
       1.3.3   使用记事本完成"Hello World"程序设计 ······ 33
       1.3.4   使用 MyEclipse 完成"Hello World"程序设计 ······ 38
       1.3.5   常见错误 ······ 43
   1.4   习题 ······ 46

项目 2   面向过程的学生成绩信息处理 ······ 48
   2.1   任务 1   学生成绩信息的表示 ······ 48
       2.1.1   关键字和标识符 ······ 49
       2.1.2   数据类型 ······ 51
       2.1.3   常量和变量 ······ 56
       2.1.4   数据表示的实现 ······ 58
   2.2   任务 2   学生课程的综合成绩计算 ······ 60
       2.2.1   运算符 ······ 61
       2.2.2   表达式和语句 ······ 67
       2.2.3   综合成绩计算 ······ 71

2.3 任务3 学生课程的综合成绩等级判断 ································· 72
　　2.3.1 if-else 语句 ·· 72
　　2.3.2 switch 语句 ··· 76
　　2.3.3 使用 if 语句实现综合成绩等级判断 ···························· 80
　　2.3.4 使用 switch 语句实现综合成绩等级判断 ····················· 81
2.4 任务4 学生课程的综合成绩等级分布统计 ··························· 83
　　2.4.1 Scanner 类 ··· 83
　　2.4.2 循环语句 ·· 85
　　2.4.3 综合成绩等级分布统计实现 ······································ 92
2.5 任务5 学生课程的综合成绩的保存、排序(数组) ··················· 95
　　2.5.1 数组 ··· 95
　　2.5.2 数组的常用方法 ·· 103
　　2.5.3 综合成绩的保存、排序实现 ······································ 105
2.6 习题 ·· 107

项目3　面向对象的学生成绩信息处理 ································· 114

3.1 任务1 学生信息、课程信息和学生成绩的表示(类) ··············· 114
　　3.1.1 类 ·· 115
　　3.1.2 类的对象 ·· 116
　　3.1.3 方法成员 ·· 118
　　3.1.4 类和类成员的修饰符 ··· 124
　　3.1.5 类的使用 ·· 125
　　3.1.6 学生成绩信息的表示实现(类) ··································· 129
3.2 任务2 学生成绩查询(类) ··· 132
　　3.2.1 子类 ··· 132
　　3.2.2 抽象类和抽象方法 ·· 135
　　3.2.3 接口 ··· 136
　　3.2.4 包 ·· 141
　　3.2.5 类及类成员的访问权限 ·· 142
　　3.2.6 Java 的应用程序接口(API) ····································· 143
　　3.2.7 包装类 ·· 144
　　3.2.8 常用的字符串类 ··· 145
　　3.2.9 Vector 类 ··· 159
　　3.2.10 学生成绩查询实现(类) ·· 161
3.3 习题 ·· 166

项目4　学生成绩信息的异常处理 ·· 174

4.1 任务 学生成绩输入异常的处理(类) ·································· 174

		4.1.1 异常概述	175
		4.1.2 异常处理	178
		4.1.3 学生成绩输入异常处理实现(类)	185
4.2	习题		187

项目 5　学生成绩信息保存到文件 …… 189

5.1	任务　学生成绩信息保存到文件 …… 189
	5.1.1 输入/输出流概念 …… 190
	5.1.2 输入/输出类 …… 190
	5.1.3 目录和文件管理 …… 191
	5.1.4 文件的顺序访问 …… 197
	5.1.5 文件的随机访问 …… 222
	5.1.6 其他常用的流 …… 226
	5.1.7 学生成绩信息保存到文件实现(类) …… 234
5.2	习题 …… 239

项目 6　创建学生成绩管理系统的图形界面 …… 241

6.1	任务 1　学生成绩管理系统功能分析 …… 242
	6.1.1 系统功能分析 …… 242
	6.1.2 图形用户界面设计概述 …… 243
6.2	任务 2　学生成绩管理系统主界面设计 …… 245
	6.2.1 JFrame 类 …… 246
	6.2.2 菜单 …… 248
	6.2.3 学生成绩管理系统主界面设计实现 …… 251
6.3	任务 3　学生成绩管理系统登录界面设计 …… 254
	6.3.1 标签 JLabel …… 254
	6.3.2 文本框 …… 258
	6.3.3 按钮 JButton …… 262
	6.3.4 布局管理器 …… 264
	6.3.5 事件处理机制 …… 270
	6.3.6 学生成绩管理系统登录界面设计实现 …… 291
6.4	任务 4　学生成绩管理系统信息管理窗口设计 …… 295
	6.4.1 复选框 JCheckBox …… 295
	6.4.2 单选按钮 JRadioButton …… 299
	6.4.3 下拉列表框 JComboBox …… 305
	6.4.4 面板 JPanel …… 311
	6.4.5 对话框 …… 313
	6.4.6 学生成绩管理系统信息管理窗口设计实现 …… 326

6.5 任务 5 学生成绩管理系统信息查询窗口设计 ……………………………… 361
6.5.1 滚动条 JScrollPane …………………………………………… 362
6.5.2 表格 JTable …………………………………………………… 364
6.5.3 JTabbedPane …………………………………………………… 380
6.5.4 JSplitPane ……………………………………………………… 381
6.5.5 列表框 …………………………………………………………… 383
6.5.6 文件选择框 ……………………………………………………… 390
6.5.7 学生成绩管理系统信息查询窗口设计实现 …………………… 394
6.6 习题 ………………………………………………………………………… 401

项目 7 学生成绩管理系统的数据库编程 ……………………………………… 405
7.1 任务 1 创建学生成绩数据库 S_Score ……………………………………… 405
7.1.1 MySQL 数据库 ………………………………………………… 406
7.1.2 SQL 基础语法 …………………………………………………… 424
7.1.3 创建学生成绩数据库 S_Score(包含 user 表、StudentScore 表) …… 430
7.2 任务 2 实现学生成绩管理系统界面功能 …………………………………… 432
7.2.1 JDBC 概述 ……………………………………………………… 433
7.2.2 JDBC 访问数据库的操作步骤 ………………………………… 434
7.2.3 学生成绩管理系统功能实现 …………………………………… 457
7.3 习题 ………………………………………………………………………… 504

项目 8 简单网络聊天软件 ……………………………………………………… 505
8.1 任务 1 一对一网络聊天软件的单机模拟实现 ……………………………… 505
8.1.1 IP 地址和 InetAddress 类 ……………………………………… 506
8.1.2 URL 类和 URLConnection 类 ………………………………… 509
8.1.3 TCP 和 UDP 程序设计 ………………………………………… 513
8.2 任务 2 一对一网络聊天软件的多线程实现 ………………………………… 532
8.2.1 进程与线程 ……………………………………………………… 532
8.2.2 线程的创建 ……………………………………………………… 533
8.2.3 线程的生命周期及调度 ………………………………………… 539
8.2.4 线程的同步与死锁 ……………………………………………… 548
8.3 习题 ………………………………………………………………………… 563

Java 程序设计初识

项目 1 首先介绍了 Java 语言的应用、发展简史、特点、运行机制，然后讲解了 Java 开发环境的搭建，最后介绍了如何编写 Java 程序，为 Java 面向过程的程序设计提供开发环境，奠定开发环境应用基础。

工作 任务

(1) 认知 Java。
(2) 搭建 Java 开发环境。
(3) 编写第一个 Java 程序。

学习 目标

(1) 掌握 Java 语言的特点及运行机制，理解 Java 语言的工作原理。
(2) 掌握 Java 开发环境的搭建，能够独立完成开发环境的安装和配置。
(3) 掌握 Java 开发环境的应用，能够使用开发环境独立编写 Java 程序。

1.1 任务 1 认知 Java

任务描述 及分析

小明是一位 Java 语言的初学者，希望对 Java 语言能够有总体的认知。要完成这个任务，我们首先要了解 Java 语言是什么，可以做什么，其次我们需要知道 Java 语言的发展历程，然后我们需要理解 Java 语言的特点，最后我们需要掌握 Java 语言的工作原理。

相关 知识

1.1.1 Java 语言概述

Java 语言是一门面向对象编程语言，不仅吸收了 C++ 语言的各种优点，还摒弃了

C++里难以理解的多继承、指针等概念，因此Java语言具有功能强大和简单易用两个特征。目前，Java语言已发展成为人类计算机史上影响深远的编程语言，在某种程度上，它甚至超出了编程语言的范畴，成为了一种开发平台，一种开发规范，甚至是一种信仰。Java语言所崇尚的开源、自由等精神，吸引了全世界无数优秀的程序员。事实上，人类有史以来，从没有一门编程语言能够吸引如此多的程序员，也没有一门编程语言能衍生出如此之多的开源框架。

1.1.2　Java语言的应用

Java语言涉及编程领域的各个方面，主要分为4大类：

（1）桌面级应用，尤其是需要跨平台的桌面级应用程序。

（2）企业级应用，这是目前Java应用最广泛的一个领域，包括各种行业应用、企业信息化、电子政务等，如办公自动化（OA）、客户关系管理（CRM）、人力资源（HR）、供应链管理（SCM）、企业设备管理系统（EAM）、财务管理，等等。

（3）嵌入式设备及消费类电子产品，包括无线手持设备、智能卡、通信终端、医疗设备、信息家电（如数字电视、机顶盒、电冰箱）、汽车电子设备等，尤其是手机上的Java应用程序和Java游戏。

（4）其他功能，如进行数学运算、显示图形界面、进行网络操作、进行数据库操作、进行文件操作等。

上述应用，主要通过3个版本的Java实现，分别是Java标准版（JSE）、Java微缩版（JME）和Java企业版（JEE）。

（1）Java标准版：JSE（Java Standard Edition）。JSE是针对桌面开发和低端商务计算解决方案而开发的版本，JSE包含构成Java语言的核心类，体现了Java的主要技术。

（2）Java微缩版：JME（Java Micro Edition）。JME包含JSE中的一部分类，是对标准版JSE进行功能缩减后的版本，用于嵌入式设备和消费类电子产品的软件开发，如手机、PDA、机顶盒等。

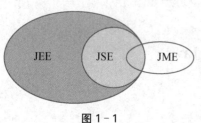

图1-1

（3）Java企业版：JEE（Java Enterprise Edition）。JEE不仅包含了JSE中所有的类，还包含企业级开发应用的类，如EJB、Servlets、JSP、XML等。

虽然Java有3个版本，但最核心的是JSE，JME和JEE是在JSE的基础上发展起来的，三者之间的关系如图1-1所示。

1.1.3　Java语言发展简史

Java语言发展到今天的3个版本，经历了一系列过程：

1991年，Sun公司的詹姆斯·高斯林（James Gosling）领导的绿色计划（Green Project）开始研发一种分布式系统结构，使其能够在各种消费性电子产品上运行。在开发过程中，高斯林等人基于C++开发出一种新的语言Oak（Java的前身）。

1995年，Oak被改名为Java，Sun公司在5月23日的Sun World会议上正式发布Java和HotJava浏览器，这标志着Java正式进军互联网，Java也由此走上繁荣之路。

1996 年 1 月，Sun 公司发布了 Java 的第一个开发工具包(JDK 1.0)，这是 Java 发展历程中的重要里程碑，标志着 Java 成为一种独立的开发工具。

1998 年 12 月 4 日，Sun 发布了 Java 新版本 JDK 1.2，并将 JDK 改名为 Java 2 Software Development Kit，简称 J2SDK，这个版本的发布标志着 Java 已经进入 Java 2 时代。

1999 年 6 月，Sun 公司发布了第二代 Java 平台(简称为 Java 2)的 3 个版本：J2ME(Java 2 Micro Edition，Java 2 平台的微型版)，应用于移动、无线及有限资源的环境；J2SE(Java 2 Standard Edition，Java 2 平台的标准版)，应用于桌面环境；J2EE(Java 2 Enterprise Edition，Java 2 平台的企业版)，应用于基于 Java 的应用服务器。Java 2 平台的发布，是 Java 发展过程中最重要的一个里程碑，标志着 Java 的应用开始普及。

2004 年 9 月 30 日，J2SE 1.5 发布，成为 Java 语言发展史上的又一里程碑。为了表示该版本的重要性，J2SE 1.5 更名为 Java SE 5.0。

2005 年 6 月，在 Java One 大会上，Sun 公司发布了 Java SE 6。此时，Java 的各种版本已经更名，已取消其中的数字 2，如 J2EE 更名为 Java EE，J2SE 更名为 Java SE，J2ME 更名为 Java ME。

目前最新的版本是 2018 年版本 Java 10.0。

1.1.4 Java 语言的特点

Java 语言的广泛应用与它的有效特性密切相关。Java 语言的典型特征如下：

(1) 简洁易学。Java 语言的语法与 C 语言和 C++ 语言很接近，容易学习和使用；Java 还去掉了 C++ 中难以理解的、令人迷惑的特性，如指针、操作符重载、多继承等，更加简洁有效。

(2) 跨平台性。所谓的跨平台性，是指软件可以不受计算机硬件和操作系统的约束而在任意计算机环境下正常运行。这是软件发展的趋势和编程人员追求的目标。之所以这样说，是因为计算机硬件的种类繁多，操作系统也各不相同，不同的用户和公司有自己不同的计算机环境偏好，而软件为了能在这些不同的环境里正常运行，就需要独立于这些平台。在 Java 语言中，其自带的虚拟机很好地实现了跨平台性。

(3) 面向对象。面向对象是指以对象为基本粒度，其下包含属性和方法。对象的说明用属性表达，通过使用方法来操作这个对象。面向对象技术使得应用程序的开发变得简单易用，节省代码。Java 是一种面向对象的语言，也继承了面向对象的诸多优点，如代码扩展、代码复用等。

(4) 安全性。安全性可分为 4 个层面，即语言级安全性、编译时安全性、运行时安全性、可执行代码安全性。语言级安全性指 Java 的数据结构是完整的对象，这些封装过的数据类型具有安全性。编译时要进行 Java 语言和语义的检查，保证每个变量对应一个相应的值，编译后生成 Java 类。运行时 Java 类需要类加载器载入，并经由字节码校验器校验之后才可以运行。Java 类在网络上使用时，它的权限被设置，保证了被访问用户的安全性。

(5) 多线程。多线程在操作系统中已得到了最成功的应用。多线程是指允许一个应用程序同时存在两个或两个以上的线程，用于支持事务并发和多任务处理。Java 除了内置的多线程技术之外，还定义了一些类、方法等来建立和管理用户定义的多线程。

1.1.5 Java 程序的运行机制

计算机高级编程语言分为编译型和解释型,Java 语言是两种类型的集合,其工作流程如图 1-2 所示。

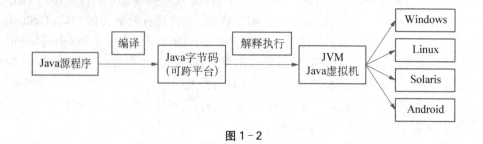

图 1-2

从图 1-2 可以看出,开发 Java 程序的工作流程如下:

(1) 编写源程序。将我们想要计算机完成的事情,通过 Java 表达出来,写成 Java 文件,这个过程就是编写源程序的过程,源程序的后缀为.java。(Java 源程序可以用任何文本编辑器创建与编辑)

(2) 编译。完成 Java 源程序后,机器并不认识我们写的 Java 代码,需要使用 Java 编译器读取 Java 源程序并翻译成 Java 虚拟机能够明白的指令集合,即 Java 字节码文件,Java 字节码文件的后缀是.class。(编译命令是 javac.exe)

(3) 解释。编译生成 Java 字节码文件后,由 Java 虚拟机(JVM)解释运行。(解释命令是 java.exe)

Java 虚拟机是在一台计算机上由软件或硬件模拟的计算机,JVM 读取并运行已经编译生成的字节码 class 文件。由图 1-2 可知所有的 Java 程序都必须在 JVM 上运行,即 *.class 文件只需要认识 JVM,再由 JVM 去适应各个操作系统。由此,不同的操作系统只要安装了符合其要求的 JVM,同一个 Java 程序就可以通过 JVM 在不同的操作系统上被正确执行,从而实现 Java 程序的跨平台运行。

1.2 任务 2 搭建 Java 开发环境

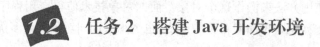

学习 Java 开发的第一步就是构建开发环境,Java 初学者小明想搭建自己的 Java 开发环境。要完成这个工作任务,我们首先要了解有哪些 Java 开发环境并根据特点选取适合个人需求的开发环境,其次根据个人需求选择下载、安装、配置开发环境。

相关知识

1.2.1 Java 开发环境

Java 语言自诞生起经历了多次的磨砺与蜕变,已成为当今最流行、最通用的软件开发语言之一,Java 开发环境也随之经历了多次变革。Java 开发环境发展历程中的典型应用如下:

(1) 1995 年,Sun 公司推出 Java 语言,同时还提供了一个免费的 Java 开发工具集 JDK(Java Development Kit)。JDK 非常适合 Java 初学者,因为初学者使用 JDK 开发 Java 程序能够很快理解程序中各部分代码之间的关系,有利于理解 Java 面向对象的设计思想。但是 JDK 不利于复杂 Java 软件开发,难以开发大规模企业级 Java 应用程序,团体协同开发困难。

(2) 1996 年,Sun 公司推出了 Java WorkShop。Java WorkShop 是第一个供互联网使用的多平台开发工具,也是第一个完全的 Java 开发环境(Java WorkShop 完全用 Java 语言编写)。Java WorkShop 结构易于创建,支持可视化编程,还支持 JDK 1.1.3 及 JavaBeans 组件模型,其 API 和语言特征增加了编译 Java 应用程序的灵活性。但 Java Workshop 中的每一个可视化对象都要用到网格布局,不符合用户的使用习惯。此外,Java Workshop 的调色板性能较弱,无法满足用户需求。

(3) 1997 年,IBM 发布 Visual Age for Java。Visual Age for Java 提供对可视化编程的广泛支持,支持利用 CICS 连接大型机应用,支持 EJB 的开发应用,支持与 WebSphere 的集成开发,支持快速应用开发(RAD),支持无文件式的文件处理,还有方便的 Bean 创建。但 Visual Age for Java 独特的管理文件方式使其集成外部工具非常困难,无法让 Visual Age for Java 与其他工具一起联合开发应用。

(4) 1998 年到 2000 年 JBuilder 诞生。JBuilder 支持最新的 Java、CORBA 技术及各种应用服务器,能自动生成基于后端数据库表的 EJB Java 类,并能用 Servlet 和 JSP 开发和调试动态 Web 应用,还拥有专业化的图形调试界面,支持远程调试和多线程。但 JBuilder 难以把握整个程序各部分之间的关系,对机器硬件要求较高,尤其是内存资源。

(5) 2001 年,业界厂商合作创建了 Eclipse 平台。Eclipse 具有可扩展的开放源代码 IDE,实现了工具之间的互操作性,显著改变了项目工作流程,能够与其他工具完美集成,还能接受 Java 开发者自己编写的开放源代码插件。

(6) MyEclipse,是在 Eclipse 基础上加上自己的插件开发而成的功能强大的企业级集成开发环境,主要用于 Java、Java EE 以及移动应用的开发。MyEclipse 在数据库和 Java EE 的开发、发布以及应用程序服务器的整合方面极大地提高了工作效率。它是功能丰富的 Java EE 集成开发环境,包括完备的编码、调试、测试和发布功能,完整支持 HTML、Struts、JSP、CSS、Javascript、Spring、SQL、Hibernate。

以上开发环境只是众多 Java 开发环境中较为典型的代表,根据学习者小明的初学者特征,我们建议选择 JDK 作为开发环境。此外,为了后续的专业核心课程"Java 企业项目开发"和"动态网页设计 JSP"的学习,我们建议在熟悉 JDK 应用的基础上,应进一步熟练使用 MyEclipse 开发工具。

任务 实施

1.2.2 JDK 的安装与配置

JDK 是 Java 开发工具包,JDK 中包括 Java 编译器(javac)、打包工具(jar)、文档生成器(javadoc)、查错工具(jdb),以及完整的 JRE(Java Runtime Environment,Java 运行环境)。目前,JDK 的最新版本是 Java SE 10(2018 年 3 月 14 日发布)。不同的计算机平台,应该根据自己的操作系统选择相应的 JDK 版本安装。

1. JDK 的下载

(1) 根据操作系统,确定要安装的 JDK 版本(以 XP 系统为例)。

在计算机桌面上右击"我的电脑"图标,在弹出的快捷菜单中选择"属性"。然后,在弹出的"系统属性"窗口中,点击"常规"选项卡,查看"系统"项。如果您的电脑是 64 位,则会明确标明"x64";如果没有标明"x64",则说明您电脑的操作系统是 32 位的。如图 1-3 所示,由于没有标明"x64",则说明本机操作系统是 Windows 32 位系统,因此我们可以选择 jdk-6u45-windows-i586 版。

图 1-3

(2) 进入 oracle 主页。

打开浏览器,在地址栏输入网址 https://www.oracle.com/index.html,进入 oracle 主页,如图 1-4 所示。

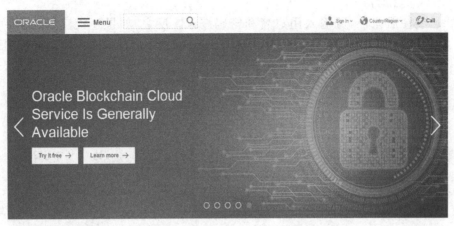

图 1-4

(3) 登录个人帐户。

在 oracle 的主页点击【Sign In】,点击"Oracle Account"对应的【Sign In】,如图 1-5 所示。此时,网页会自动跳转到登录页面,如图 1-6 所示。

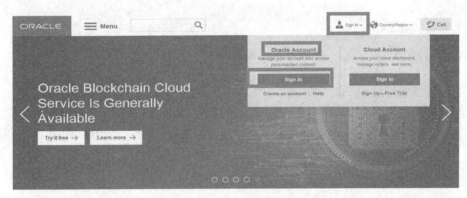

图 1-5

图 1-6

(4) 登录或注册帐户。

如果已经注册帐户,可输入用户名和密码登录。反之,如果还没注册帐户,可点击图1-6中的【创建帐户】按钮,进行免费注册。点击【创建帐户】后,网页自动跳转到创建帐户页面,如图1-7所示。

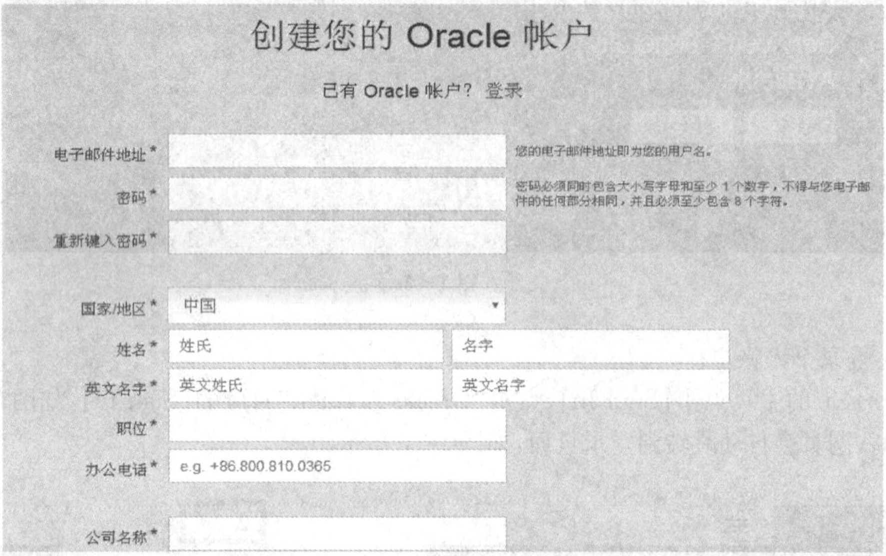

图 1-7

(5) 完善注册信息。

在创建帐户页面完善个人信息后,点击【创建】按钮,如图1-8所示。此时网页会提示你,已向你发送了电子邮件,其中附有确认电子邮件地址的按钮,请你进行确认,如图1-9所示。

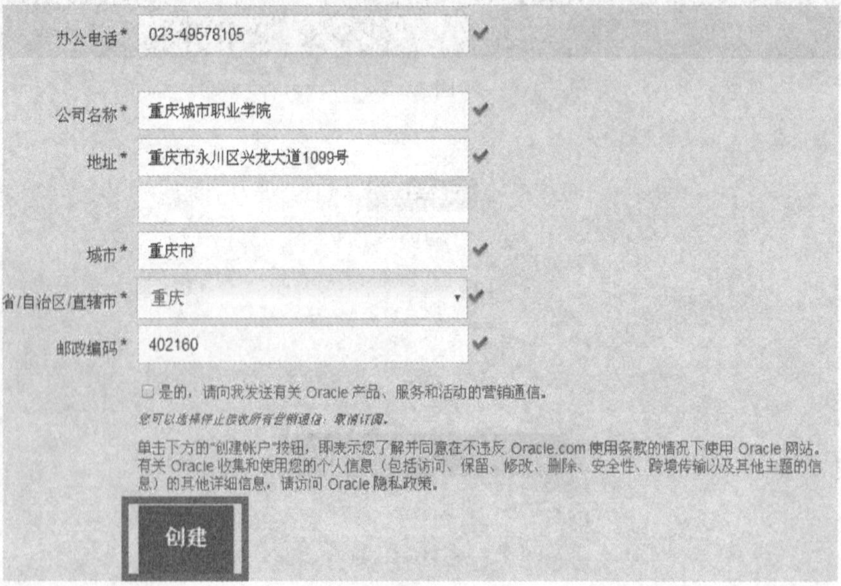

图 1-8

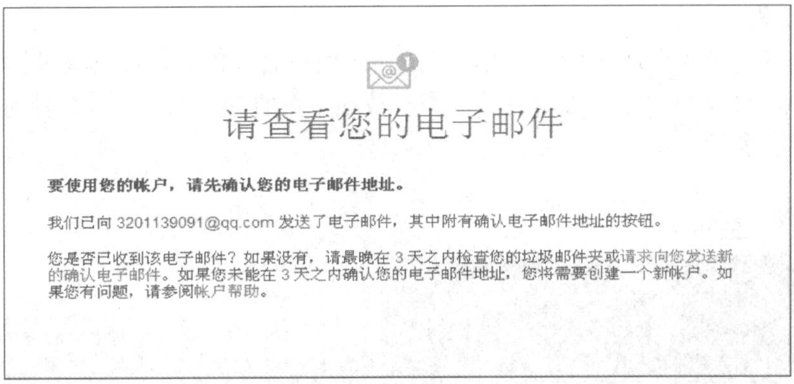

图 1-9

(6) 确认电子邮件地址。

进入注册时填写的电子邮箱,打开收到的确认邮件,点击【确认电子邮件地址】按钮,如图 1-10 所示。此时,网页提示你"创建成功",如图 1-11 所示,点击图 1-11 中的【继续】按钮,网页会自动跳转到登录页面,如图 1-6 所示。

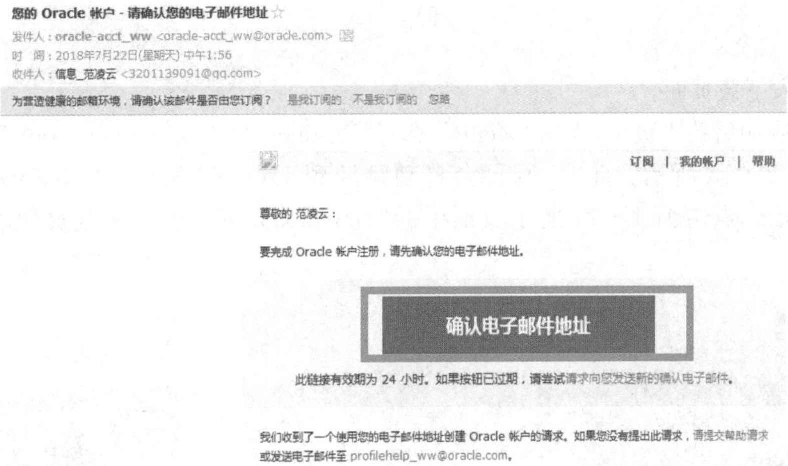

图 1-10

图 1-11

(7) 登录帐户。

在图1-6所示的登录页面中输入注册成功的用户名和密码,点击【登录】按钮,网页会自动跳转到登录成功的主页。如果用户将鼠标放在主页的"Account"上,会看到自己的用户名,如图1-12所示。

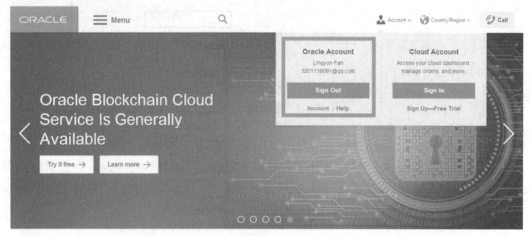

图1-12

(8) 进入下载页面。

在登录成功后的主页上,点击"Menu",选择"Products"—"Developer Tools"—"Java SE SDK",如图1-13所示。此时,网页会自动跳转到Java SE SDK的Overview页面,如图1-14所示。点击图1-14所示页面中的"Downloads"项,进入下载页面,如图1-15所示。

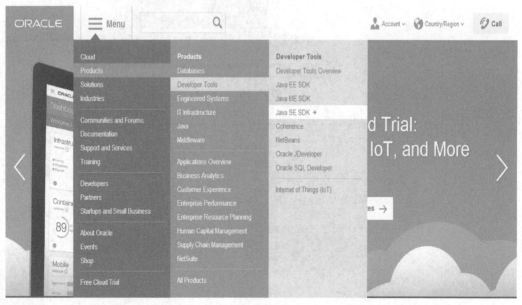

图1-13

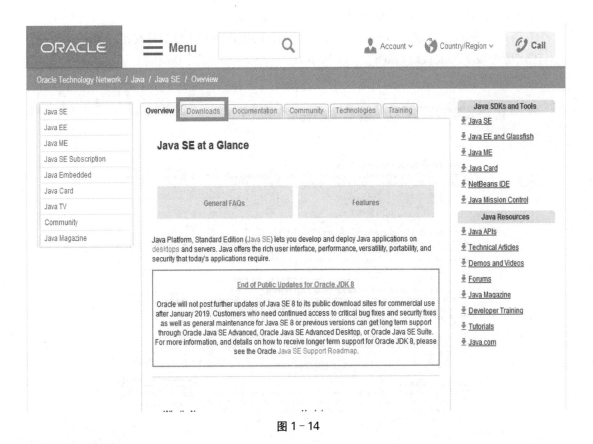

图 1-14

图 1-15

(9)进入 Java Archive(Java 存档)页面。

在下载页面的底部,点击 Java Archive 右侧对应的【DOWNLOAD】按钮,如图 1-16 所示,进入历史版本列表页面,如图 1-17 所示。

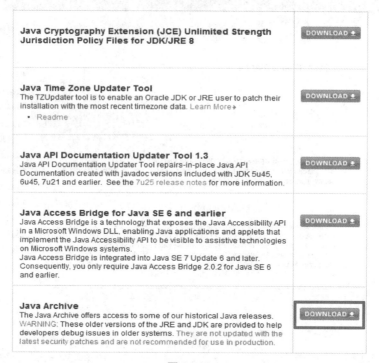

图 1-16

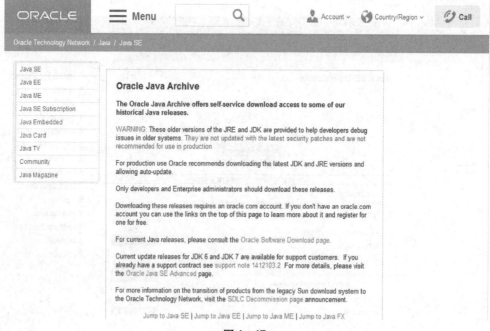

图 1-17

(10) 选择对应版本 JDK。

在历史版本列表页面点击 Java SE 6,如图 1-18 所示,页面会自动跳转到 Java SE 6 的下载页面,如图 1-19 所示。

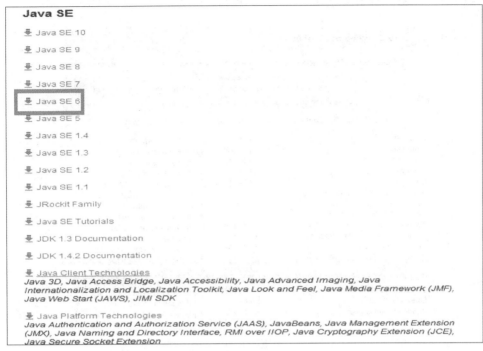

图 1-18

图 1-19

(11) 下载 jdk-6u45-windows-i586.exe。

在 Java SE 6 下载页面中,首先选中"Java SE Development Kit 6u45"中的 Accept

License Agreement(接受许可协议),其次根据我们的操作系统是 Windows 32 位系统,确定选择 JDK 的版本是 jdk-6u45-windows-i586.exe,然后点击对应的下载按钮,如图 1-20 所示。此时浏览器会提示我们选择"jdk-6u45-windows-i586.exe"的保存位置(请注意本例使用的是 360 浏览器,不同浏览器的提示方式可能不一致),如图 1-21 所示。

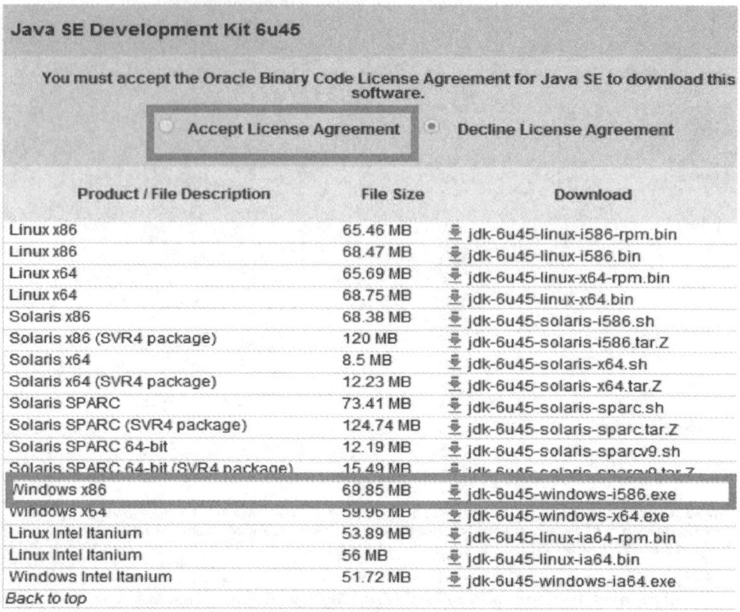

图 1-20

图 1-21

2. JDK 的安装

(1) 双击打开 jdk-6u45-windows-i586.exe 文件,出现如图 1-22 所示的安装界面,点击【下一步】,进入自定义安装界面,如图 1-23 所示。

(2) 在如图 1-23 所示的自定义安装界面,我们可以自定义"开发工具""源代码""公共 JRE""Java DB"的安装位置,具体做法如下(以开发工具为例):在图 1-23 中,选中"开发工

图 1-22

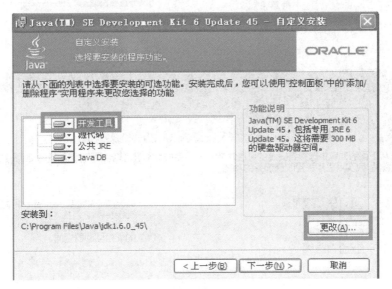

图 1-23

具",点击【更改】按钮,在如图 1-24 所示的"更改当前目标文件夹"窗口中设置需要安装到的路径为"D:\common\Java\jdk1.6.0_45\",点击【确定】,返回上一个界面。此时,显示"开发工具"的安装路径发生变化,如图 1-25 所示。

(3) 图 1-25 所示自定义安装窗口中的相关设置定义好后,点击【下一步】按钮,就进行 JDK 的安装,直到出现 JRE 的安装界面,如图 1-26 所示。

(4) 如果需要更改 JRE 的安装路径,可在图 1-26 所示的窗口中点击【更改】按钮进行

图 1-24

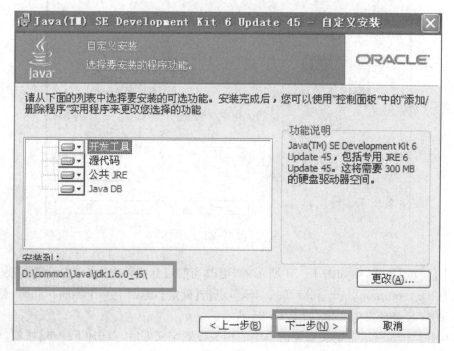

图 1-25

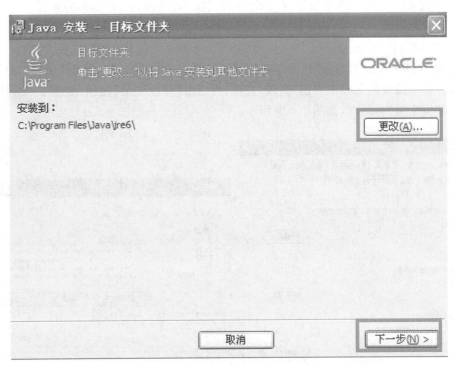

图 1-26

设置。如不需要更改安装路径,可直接点击【下一步】按钮,进行安装,直到出现如图 1-27 所示的界面,表示安装完成。

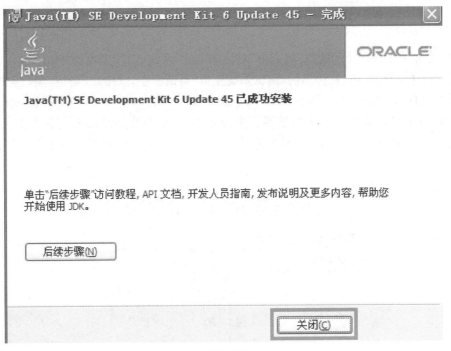

图 1-27

(5) 点击如图 1-27 所示安装成功界面中的【关闭】按钮，完成 JDK 的安装。

3. JDK 的配置

JDK 安装完成后，还要进行 Java 环境的配置，才能正常使用，步骤如下：

(1) 在计算机桌面上右击"我的电脑"图标，在弹出的快捷菜单中选择"属性"，打开系统属性窗口，在系统属性窗口中点击"高级"选项，再点击【环境变量】按钮，如图 1-28 所示，此时打开"环境变量"对话框，如图 1-29 所示。

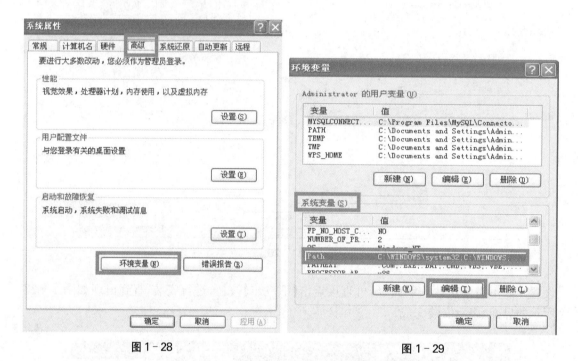

图 1-28   图 1-29

(2) 在如图 1-29 所示的"环境变量"对话框中的"系统变量"里找到"Path"这一项，选中它，然后单击【编辑】按钮，在弹出的"编辑系统变量"对话框中，选中"变量值"项，在"变量值"项的开头添加如下语句"C:\Program Files\Java\jdk1.6.0_45\bin;"（C:\Program Files\Java\jdk1.6.0_45\是在自定义安装界面中设置的 JDK 安装路径，bin 文件夹里存放的是 Java 的各种可执行文件，请注意不要忘了 bin 后面的分号），然后点击【确定】按钮，如图 1-30 所示。

图 1-30

(3) 点击如图 1-29 所示的"环境变量"对话框中"系统变量"下方的【新建】按钮,在弹出的"新建系统变量"对话框中,将"变量名"项设置为:JAVA_HOME,将"变量值"项设置为:C:\Program Files\Java\jdk1.6.0_45,然后点击【确定】按钮,如图 1-31 所示。

图 1-31

(4) 再次点击如图 1-29 所示的"环境变量"对话框中"系统变量"下方的【新建】按钮,在弹出的"新建系统变量"对话框中,将"变量名"项设置为:classpath,将"变量值"项设置为.;,请注意变量值是点和分号,然后点击【确定】按钮,如图 1-32 所示。

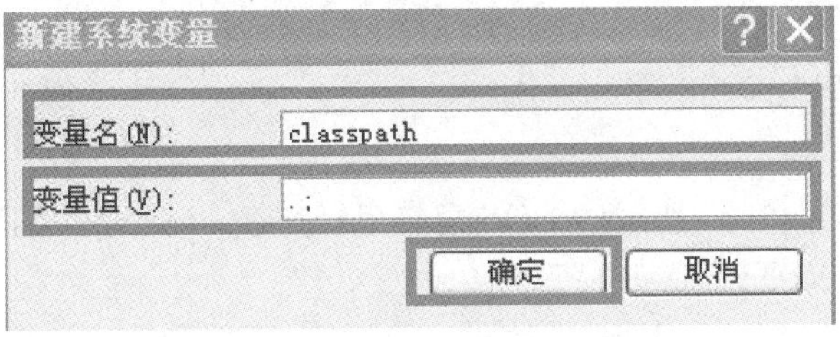

图 1-32

(5) 依次点击如图 1-29 所示的"环境变量"对话框中的【确定】按钮和如图 1-28 所示的系统属性窗口中的【确定】按钮,完成 JDK 的配置。

提示:

• Path 系统变量用于给操作系统提供寻找到 Java 命令工具的路径,通常是配置到 JDK 安装路径\bin。

• JAVA_HOME 系统变量用于提供给其他基于 Java 的程序使用,让它们能够找到 JDK 的位置。通常配置到 JDK 安装路径。注意:这个必须书写正确,全部大写,中间用下划线。

• classpath 系统变量用于提供程序在运行期间寻找所需资源的路径,如类、文件、图片等。此外,强烈建议在 windows 操作系统里,最好在 classpath 的配置里面,始终在前面保持".;"的配置,在 windows 里面"."表示当前路径。

4. JDK 检测

JDK 安装、配置完后,检测安装配置是否成功的步骤如下:

(1) 点击"开始"—"运行",如图 1-33 所示,在弹出的运行对话框中输入"cmd",然后点击【确定】按钮,如图 1-34 所示。

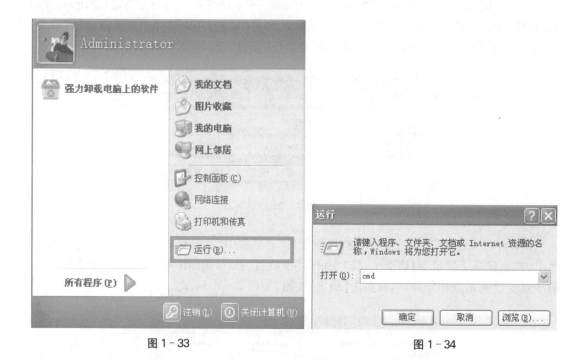

图 1-33　　　　　　　　　　　　　　　　图 1-34

（2）在弹出的如图 1-35 所示的 DOS 窗口里，输入"javac"，然后回车，出现如图 1-36 所示的界面，则表示安装配置成功。

图 1-35

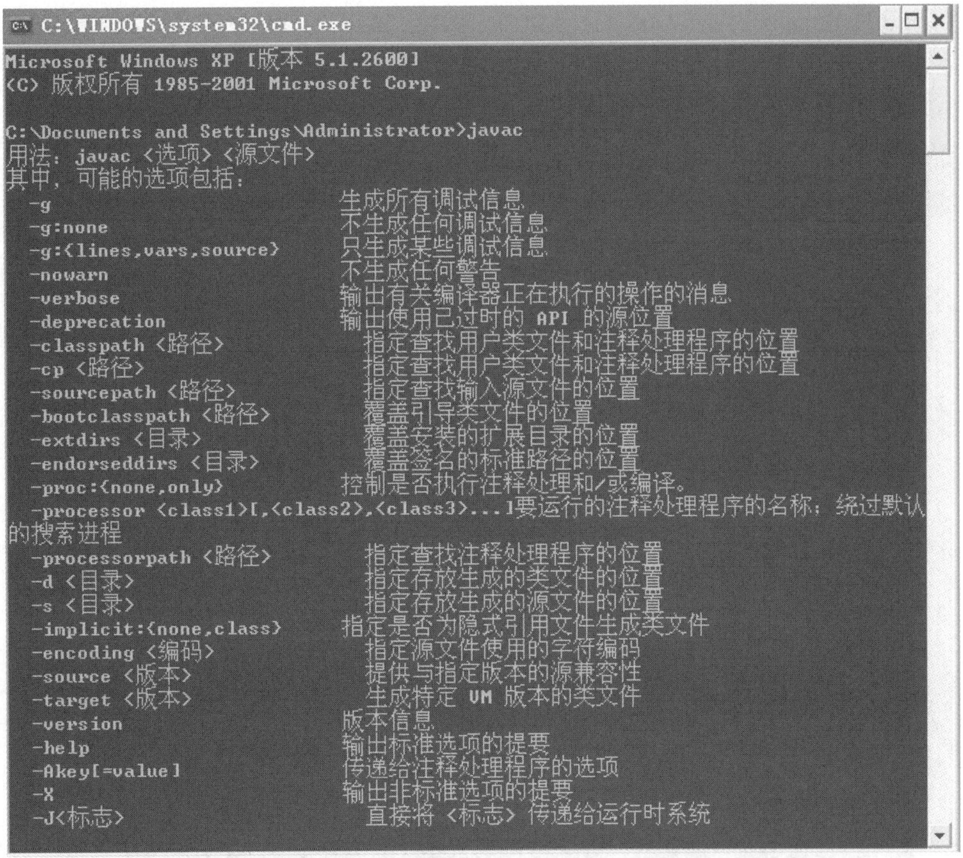

图 1-36

1.2.3 MyEclipse 的安装

MyEclipse 是功能最全面的 Java IDE,被广泛应用于企业级开发、Web 开发、云开发、移动开发。MyEclipse 开发平台同样有多个版本,目前的最新版是 MyEclipse 2017 CI 10 (2017 年 12 月 28 日更新)。本例中我们使用的版本是 MyEclipse 2014。

1. MyEclipse 2014 的安装

(1) 双击打开 MyEclipse 2014 安装文件,出现如图 1-37 所示的安装界面,点击【Next】,进入"License Agreement"界面,如图 1-38 所示。

(2) 在如图 1-38 所示的"License Agreement"界面中,选中"I accept the terms of the license agreement",点击【Next】按钮,进入"Choose Installation Location"窗口,如图 1-39 所示。在该界面中,我们可以点击【Change...】按钮,在弹出的"浏览文件夹"窗口中选择 MyEclipse 2014 的安装位置(本例选的是 D 盘的 MyEclipse 文件夹),如图 1-40 所示。然后点击图 1-40 中的【确定】按钮,此时,你会发现 MyEclipse 2014 的安装位置发生变化,如图 1-41 所示。点击图 1-41 中的【Next】按钮,进入"Choose Optional Software"窗口,如图 1-42 所示。

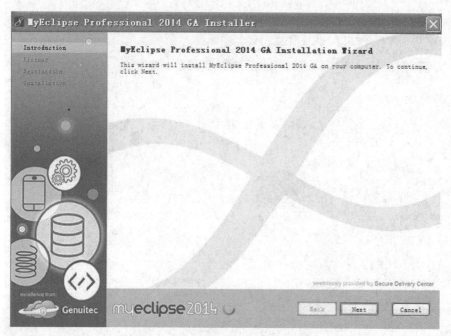

图 1-37

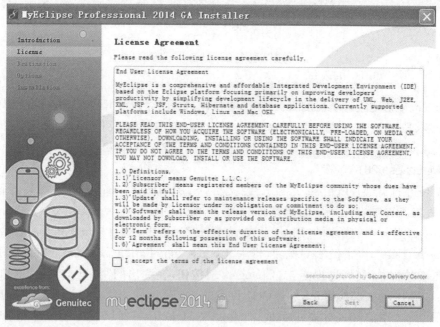

图 1-38

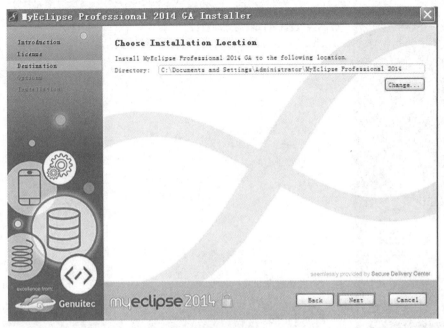

图 1-39

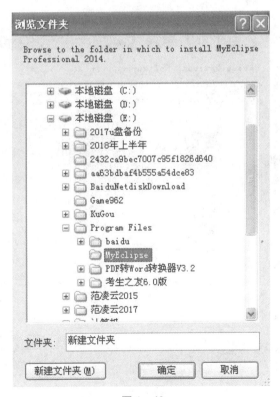

图 1-40

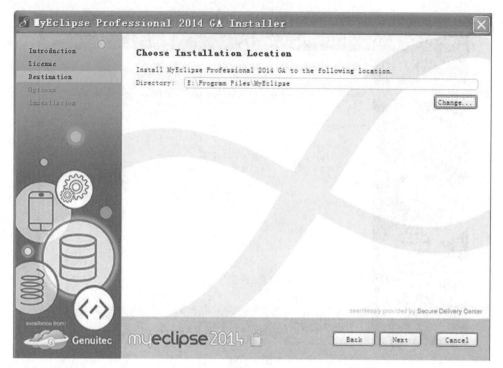

图 1-41

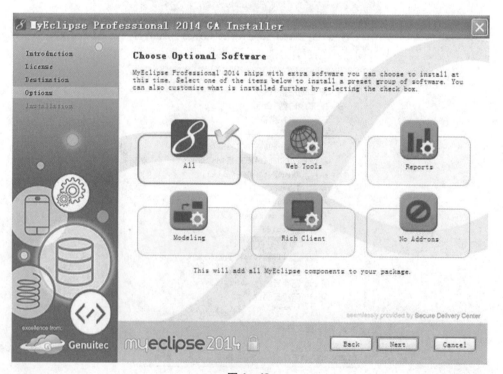

图 1-42

(3) 保持图 1-42 所示的"Choose Optional Software"窗口默认选项"All"不变,然后点击【Next】按钮,进入"Installing Software"窗口,开始安装软件,如图 1-43 所示。软件安装完毕后,进入"Installation Complete"窗口,如图 1-44 所示,点击【Finish】按钮,完成软件安装。

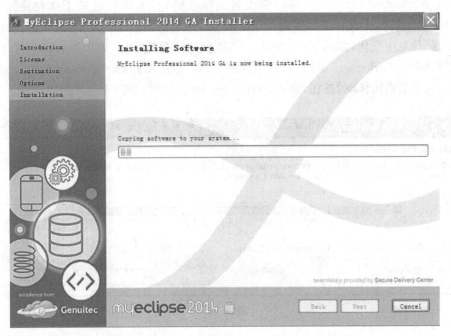

图 1-43

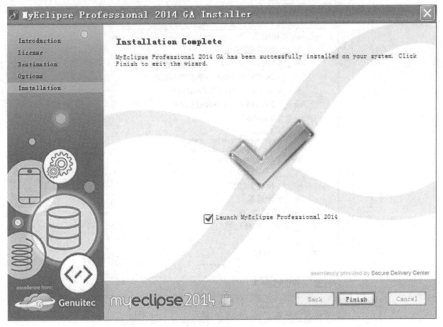

图 1-44

2. MyEclipse 2014 注册

（1）点击"开始"—"所有程序"—"MyEclipse"—"MyEclipse 2014"—"MyEclipse Professional 2014"，打开 MyEclipse Professional 2014 软件，此时会出现"Select a workspace"窗口，如图 1-45 所示。我们可以点击图 1-45 中的【Browse...】按钮，在弹出的"Select Workspace Directory"窗口中设置工作台路径（本例设置的工作台路径是 E 盘 MyEclipse Workspace 文件夹），如图 1-46 所示。然后点击图 1-46 中的【确定】按钮，会发现工作台路径发生变化，如图 1-47 所示。工作台路径设置好后，如果不想每次打开软件时都出现"Select a workspace"窗口，可以勾选图 1-47 中的"Use this as the default and do not ask again"，然后点击【OK】按钮，进入 MyEclipse 2014 软件，如图 1-48 所示。

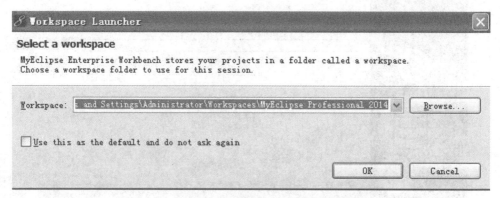

图 1-45

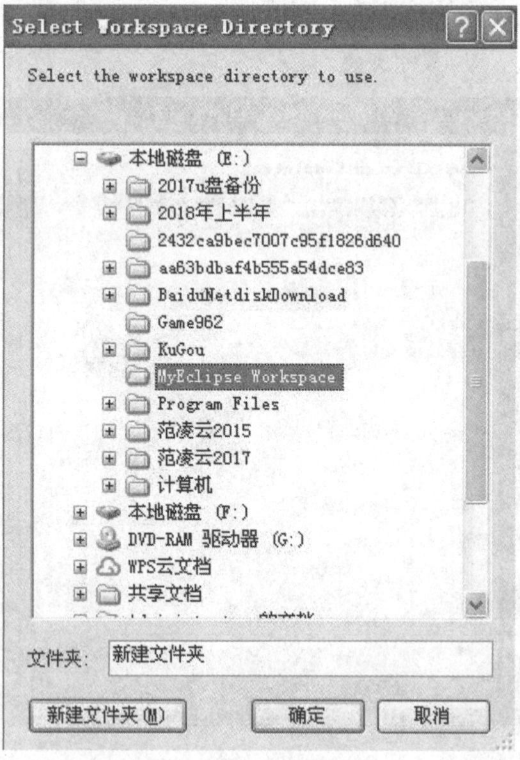

图 1-46

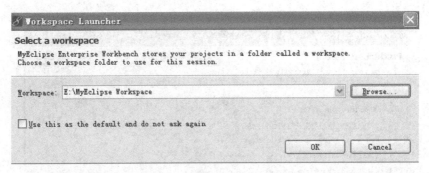

图 1-47

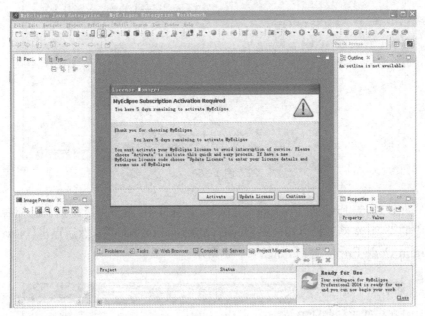

图 1-48

(2) 在如图 1-48 所示的 MyEclipse 2014 软件界面上,会自动出现如图 1-49 所示的"License Manager"窗口。点击图 1-49 中的【Activate】按钮,进入"Product Activation"窗口,如图 1-50 所示。

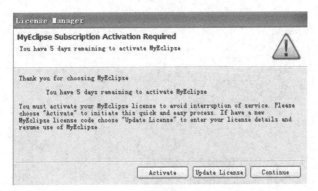

图 1-49

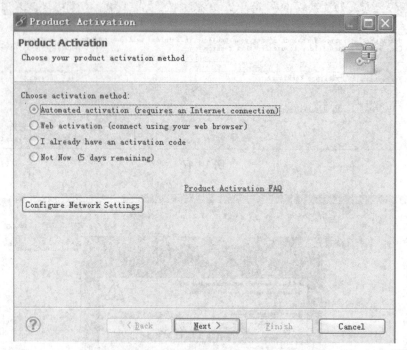

图 1-50

（3）在如图 1-50 所示的"Product Activation"窗口中，用户可以根据自己拥有的激活方式选择对应的类型（本例选择"I already have an activation code"），然后点击【Next】按钮，进入"Activation Code Entry"窗口，如图 1-51 所示。

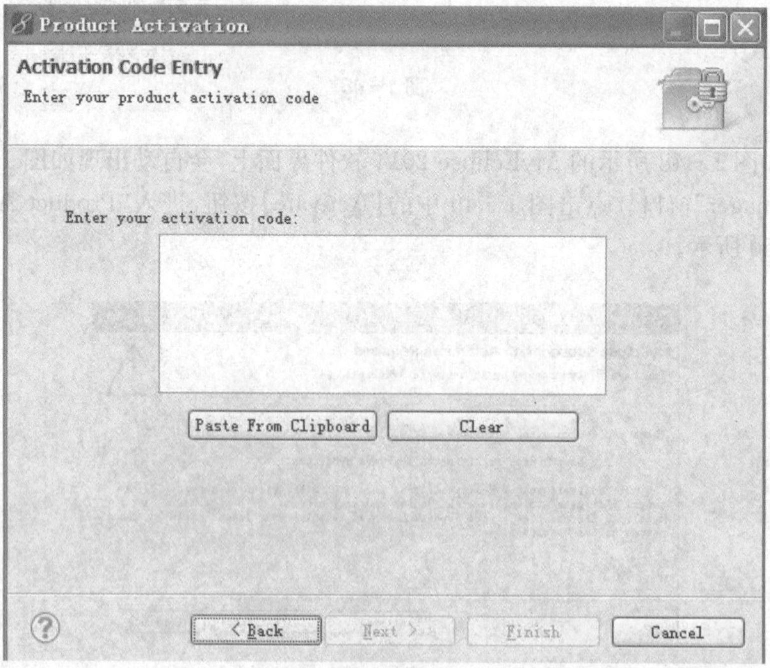

图 1-51

(4) 在如图 1-51 所示的"Activation Code Entry"窗口的"Enter your activation code:"下方输入注册码,然后点击【Finish】按钮,完成 MyEclipse 2014 软件注册。此时,才成功打开软件,如图 1-52 所示。

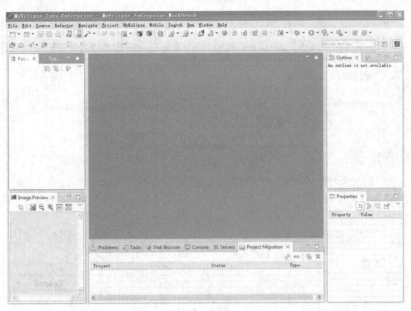

图 1-52

(5) 在如图 1-52 所示的 MyEclipse 2014 软件窗口中点击"MyEclipse"菜单,选中"Subscription Information...",如图 1-53 所示。进入"Update Subscription"窗口,如图 1-54 所示。此时该窗口的"Activation status"提示产品已激活(Product activated),"Subscription Detail"的"Subscription expiration date(YYYYMMDD)"项提示订阅到期日期是 20210722。至此,MyEclipse 2014 安装、注册成功。

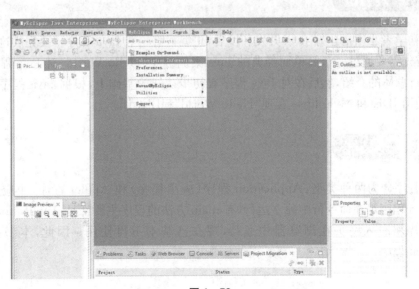

图 1-53

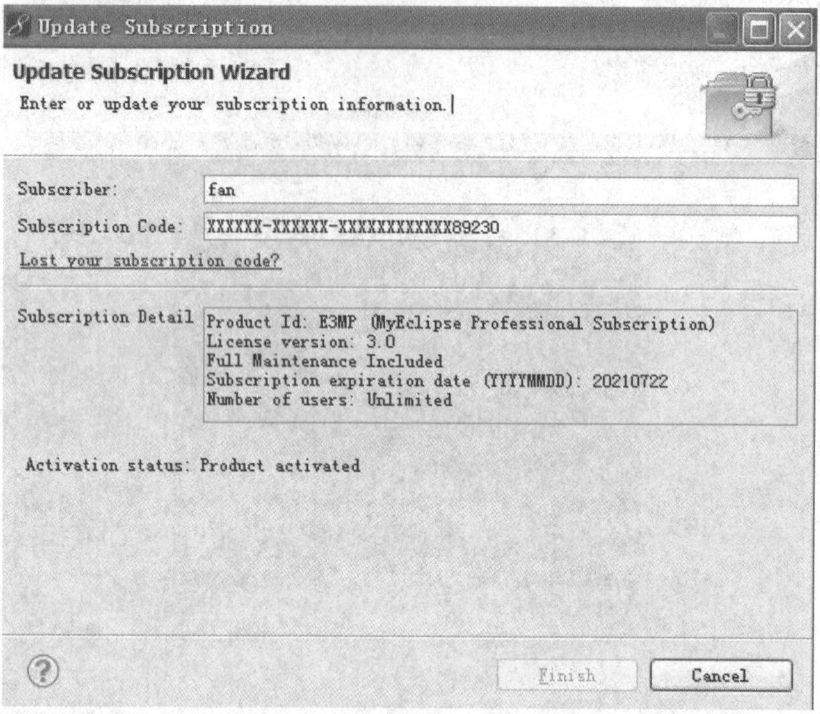

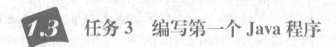

图 1-54

1.3 任务 3 编写第一个 Java 程序

任务描述 及分析

Java 初学者小明搭建好 Java 开发环境后,想编写第一个 Java 程序。要完成这个工作任务,我们首先需要通过一个简单而经典的 Java 案例程序"Hello World"对 Java 程序的基本结构等进行总体的介绍,然后在对 Java 程序有初步认知的基础上,根据 Java 程序的工作流程,分别使用 JDK 和 MyEclipse 实现"Hello World"程序设计。

相关知识 与技能

Java 程序分为两种类型:Application 程序(应用程序)和 Applet 程序(小应用程序)。Application 程序可在命令行中独立运行,有 main 方法的程序主要都是 Application 程序;而 Applet 程序须嵌入网页在浏览器中运行,现在已基本不再使用。因此,本书主要讲解 Application 程序。

1.3.1 "Hello World"程序解释

Java经典案例程序HelloWorld.java的源程序清单如下：

```
1.          //文件名:HelloWorld.java
2.          public class HelloWorld{    //类名要和文件名一致
3.              public static void main(String args[]){
4.                  System.out.println("Hello World!");
5.              }//结束main方法的定义
6.          }//结束类HelloWorld的定义
```

程序说明：

（1）程序的第1行、第2行、第5行、第6行使用"//"声明的部分是Java的注释部分,注释有助于程序的阅读,不会被编译,可以写任何内容。

注意：

在Java程序中,根据功能不同,注释分为单行注释、多行注释、文档注释。

① 单行注释,就是在注释内容前面加双斜线(//),Java编译器会忽略这部分信息。如下所示：

```
int x;//定义一个整型变量x
```

② 多行注释,就是在注释内容前面以单斜线加一个星形标记开头(/*),并在注释内容末尾以一个星形标记加单斜线结束(*/)。当注释内容超过一行时一般使用这种方法。如下所示：

```
/*
int x;
x = 3;
*/
```

③ 文档注释,是以单斜线加两个星形标记开头(/**),并以一个星形标记加单斜线结束(*/)。用这种方法注释的内容会被解释成程序的正式文档,并能包含在如Javadoc之类工具生成的文档中,用以说明该程序的层次结构及方法。

（2）public class是Java中的关键字,表示定义一个公共的类,类的名字是HelloWorld,"{"表示类的开始,与之对应的"}"表示类的结束。

注意：

① Java程序是由类构成的,一个程序可以有多个类,但是只能有一个由public class定义的类,而且由public class定义的类的名字必须与程序的文件名相同。例如,该案例的程序文件名是HelloWorld.java(在程序清单的第1行注释中进行了说明),所以程序中由public class定义的类的名字也是HelloWorld(在程序清单的第2行)。

② 在Java程序中声明类的关键字有两种：public class关键字、class关键字。使用class

关键字声明的类的名字可以与文件名不一致，一个 Java 程序中可以有多个由 class 定义的类，但只能有一个由 public class 定义的类。

③ Java 程序中类的名字中的每个单词的首字母必须大写，这是 Java 的命名规范中规定的。例如，该案例第 2 行中的类名 HelloWorld，它的两个单词 Hello 和 World 的首字母都是大写。

（3）public static void main(String args[])定义 Java 程序的主方法，它是程序的入口点，即 Java 程序以此方法作为起点，开始执行程序。"{"表示主方法的开始，与之对应的"}"表示主方法的结束。

注意：

main()表明该方法为主方法，包含 main()方法的类为主类，例如该程序中的 HelloWorld 即为主类。一个 Java 应用程序只能包含一个 main()方法、一个主类，且程序从主类的 main()方法的第 1 行开始执行，直到执行完 main()方法中的所有代码。

在书写 main()方法时，要求必须按照上面的书写格式和内容进行书写。main()方法前面必须使用 public static void 进行修饰且顺序不变，中间使用空格进行分开，main 小括号中的"String[]args"也是必须的，其中的 args 可以书写成其他的内容，而没有任何的影响。这是一个固定的书写形式，读者现在可能不能理解，可先进行记忆，我们后续再描述各部分的含义。这里 public、static、void 及小括号中的 String[]args 并称为 main()方法的四要素。

（4）System.out.println("Hello World!");用于直接将"()"中的"HelloWorld!"输出。System.out 是标准输出，通常与计算机的接口设备有关，如显示器、打印机等。其后的 println 是指将"()"里的内容打印在标准输出设备显示器上，然后把光标移到下一行的开始位置。

1.3.2 Java 编码规范

在日常生活中大家都要说普通话，目的是让不同地区的人能够通畅交流。编码规范也是程序界的"普通话"，它对于程序员来说非常重要。为什么这么说呢？一个软件 80% 的时间都花在系统的维护上，一般系统的维护不会是最开始编写程序的人进行的，编码规范可以增强代码的可读性，使软件的开发和维护更方便。编码规范其实就是行业人士的一个编码标准，是所有从事这个行业的人都要遵循的一种编码方式。

Java 的编码规范具体如下：

（1）要求类名必须使用 public 修饰。

（2）建议一行只写一条语句。

（3）注意大括号"{}"的位置，"{"方向的大括号在一行代码的末尾，"}"在一行代码的开头，并独占一行。

（4）第一层的语句或注释内容应该比高一层次的语句或注释内容缩进一定的位置。一个制表（相当于按一下键盘上的 Tab 键，8 个空格）的间隙，看起来更有层次感。

任务实施

1.3.3 使用记事本完成"Hello World"程序设计

1. 创建 HelloWorld.java 源程序

（1）在计算机 E 盘根目录下创建一个"新建文本文档.txt"文本文件，如图 1-55 所示。然后修改文件名为"HelloWorld.java"，此时会弹出"重命名"对话框，如图 1-56 所示。点击【是】按钮，完成文件的重命名，如图 1-57 所示。

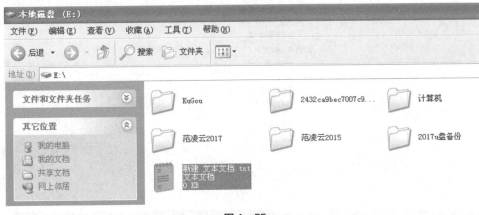

图 1-55

图 1-56

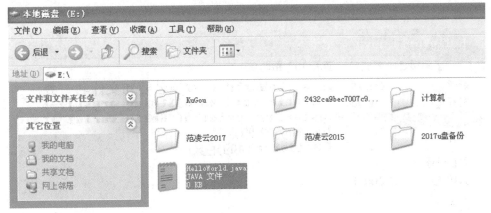

图 1-57

(2) 右击"HelloWorld.java",选择"打开方式"—"记事本",如图 1-58 所示。在打开后的"HelloWorld.java"文件中输入 1.3.1 中的案例代码,如图 1-59 所示。然后点击"文件"—"保存",如图 1-60 所示。最后点击"文件"—"退出",如图 1-61 所示,完成 HelloWorld.java 源程序的创建。

图 1-58

图 1-59

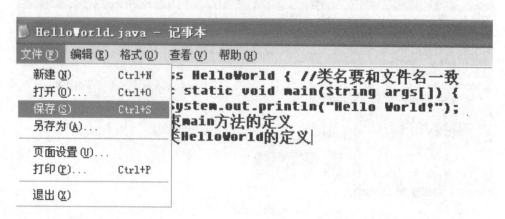

图 1-60

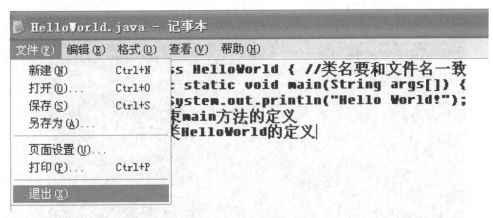

图 1-61

2. 编译 HelloWorld.java 源程序

(1) 点击"开始"—"运行",如图 1-62 所示。打开运行对话框,在运行对话框中输入 "cmd",如图 1-63 所示。然后点击【确定】按钮,进入 cmd.exe 窗口,如图 1-64 所示。接下来输入"e:"(因为 HelloWorld.java 源程序在 E 盘根目录),如图 1-65 所示。最后点击回车键,将路径切换为"E:\>",如图 1-66 所示。

图 1-62

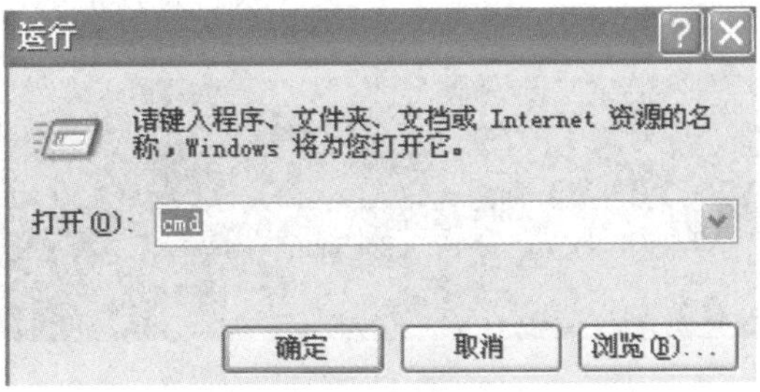

图 1-63

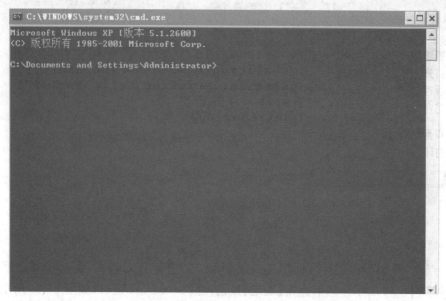

图 1-64

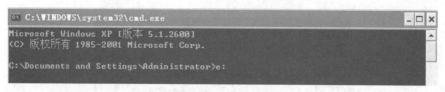

图 1-65

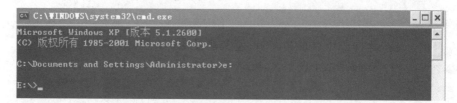

图 1-66

(2) 在如图 1-66 所示的窗口中输入"javac HelloWorld.java",如图 1-67 所示。然后点击回车键。如果编译器未返回任何提示信息,如图 1-68 所示,则编译生成的字节码文件 HelloWorld.class 被存储在与源文件相同的目录中(除非另有指定),如图 1-69 所示。

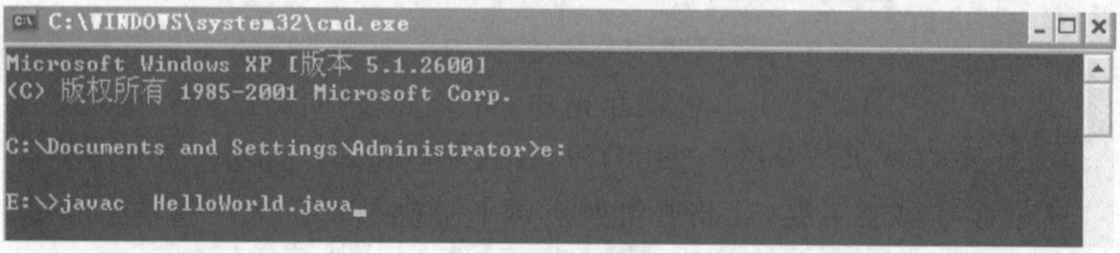

图 1-67

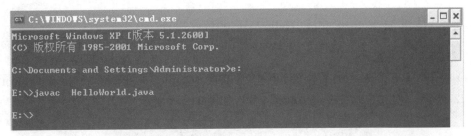

图 1-68

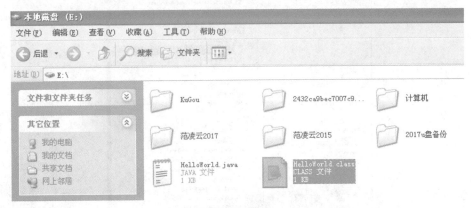

图 1-69

3. 执行 HelloWorld.class 文件

在如图 1-68 所示的窗口中输入"java HelloWorld",如图 1-70 所示。然后点击回车键,运行结果如图 1-71 所示。

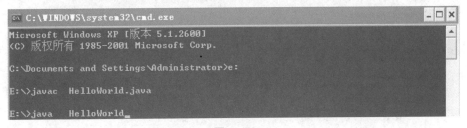

图 1-70

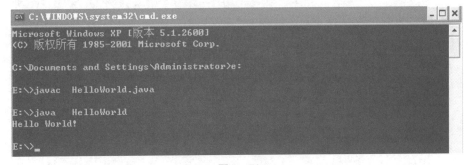

图 1-71

1.3.4 使用MyEclipse完成"Hello World"程序设计

1. 新建Java项目

(1) 打开MyEclipse 2014软件,点击菜单栏的"File"—"New"—"Java Project",如图1-72所示。弹出"New Java Project"窗口,如图1-73所示。

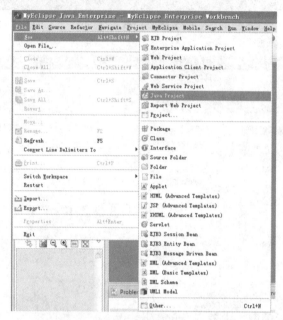

图1-72

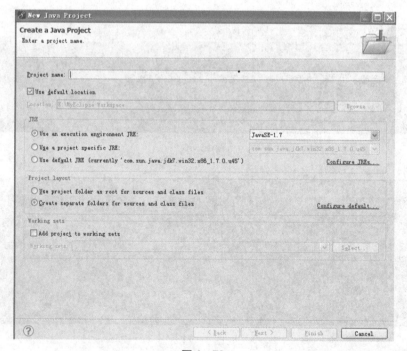

图1-73

（2）在如图1-73所示的窗口中输入Project Name为"HelloWorld"，如图1-74所示。然后点击【Finish】按钮，就完成了"HelloWorld"Java项目的创建。此时，会发现"Package Explorer"窗口出现了"HelloWorld"项目的图标，如图1-75所示。

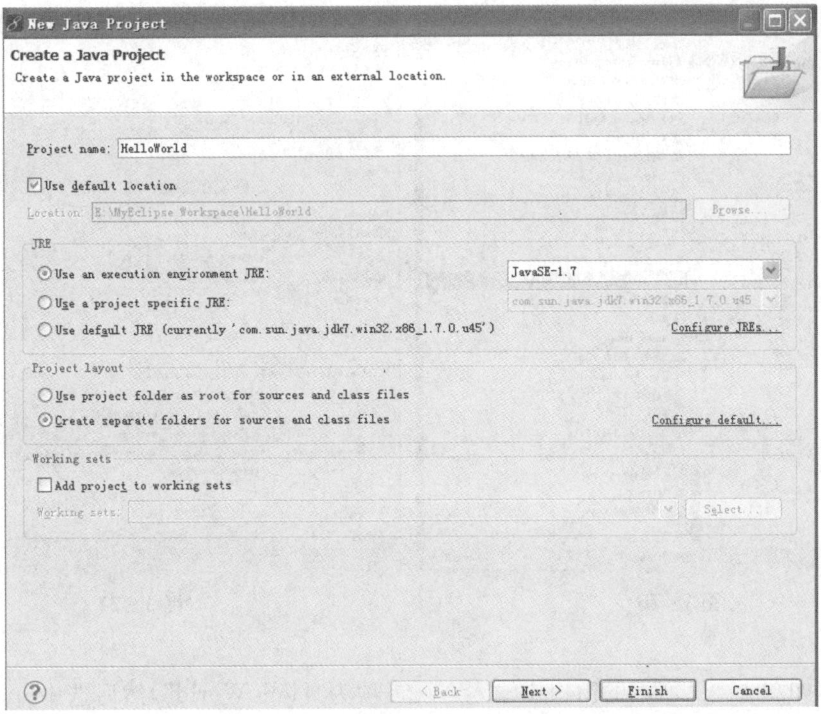

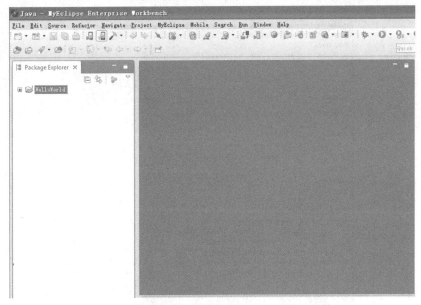

图1-74

图1-75

2. 编写源程序 HelloWorld.java

（1）在如图 1-75 所示的窗口中选中"HelloWorld"项目，然后点击菜单栏的"File"—"New"—"Class"，如图 1-76 所示。接下来"New Java Class"窗口弹出，如图 1-77 所示。

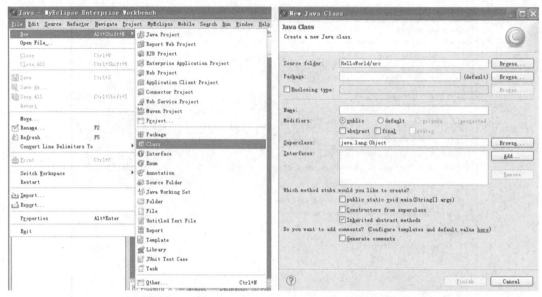

图 1-76　　　　　　　　　　　　　图 1-77

（2）在如图 1-77 所示的窗口中输入 Name 项为"HelloWorld"，然后选中"public static void main(String[]args)"项，不勾选"Inherited abstract methods"，如图 1-78 所示。然后点击【Finish】按钮，完成源程序 HelloWorld.java 的创建，如图 1-79 所示。

图 1-78

图 1-79

（3）在如图 1-79 所示窗口的 HelloWorld.java 代码编辑区中添加代码"System.out.println("Hello World!");"，如图 1-80 所示。然后点击菜单栏的"File"—"Save"，如图 1-81 所示，完成源程序 HelloWorld.java 的编写。

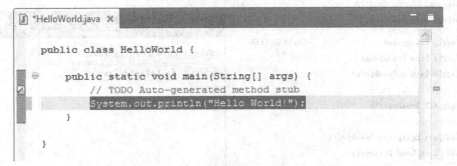

图 1-80

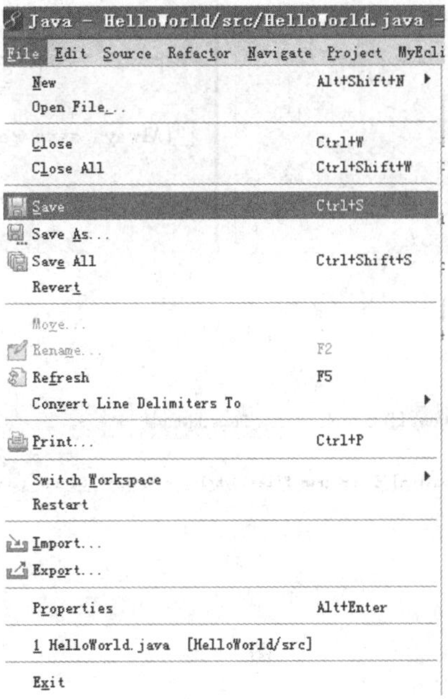

图 1-81

3. 编译、运行 HelloWorld

点击菜单栏的"Run"—"Run",如图 1-82 所示。在弹出的"Save and Launch"窗口中点击【OK】按钮完成"Hello World"程序的编译、运行,如图 1-83 所示。运行结果如图 1-84 所示。

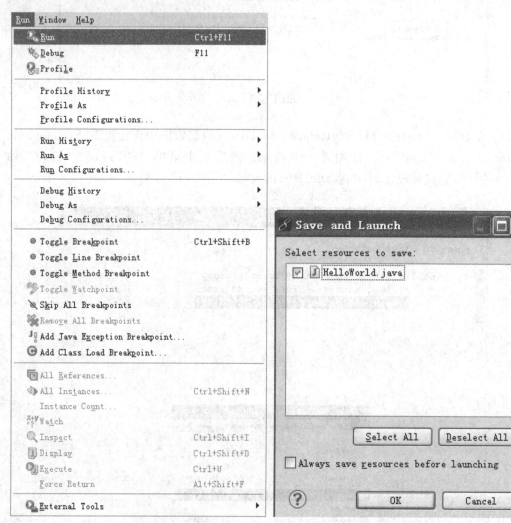

图 1-82   图 1-83

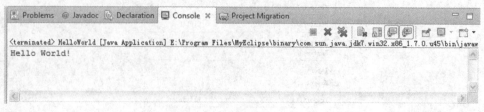

图 1-84

1.3.5 常见错误

程序开发存在一条定律,即"一定会出错"。有时我们会不经意地犯一些错误,或者为了了解代码而故意去制造一些错误。因此,我们要具备一定的查看和排除错误的能力。

下面我们通过 5 个案例讲解常见的错误及其对应的提示,方便理解程序的常见错误。

1. 类名不能随意取

在 1.3.1 中介绍 HelloWorld.java 程序的框架时曾讲过,HelloWorld 是类名,它是由程序开发人员定义的,那么它的名字能够随便定义吗? 我们将"HelloWorld.java"文件中的类名修改成 helloworld,修改后代码如下。

【例 1.1】常见错误 1。

```java
public class helloworld{    //将类名修改成 helloworld
    public static void main(String args[]){
        System.out.println("Hello World!");
    }
}
```

修改保存后,进行编译,会有如图 1-85 提示:

```
E:\>javac HelloWorld.java
HelloWorld.java:2: 类 helloworld 是公共的,应在名为 helloworld.java 的文件中声明
    public class helloworld {    //将类名修改成 helloworld
           ^
1 错误
```

图 1-85

意思是公用类型 helloworld 必须在它自己的定义文件中存在,也就是说没有找到 helloworld 的类。由此我们得出第一个结论:由 public 修饰的类,它的名称必须和类名完全一致!

2. main 方法前的 void 不能少

在 main()方法框架中,void 是告诉编译器 main()方法没有返回结果。既然没有返回结果,那 void 可以省略吗? 去掉 void 后代码示例如下。

【例 1.2】常见错误 2。

```java
public class HelloWorld{
    public static main(String args[]){//去掉 void
        System.out.println("Hello World!");
    }
}
```

修改保存后,进行编译,会有如图 1-86 提示:

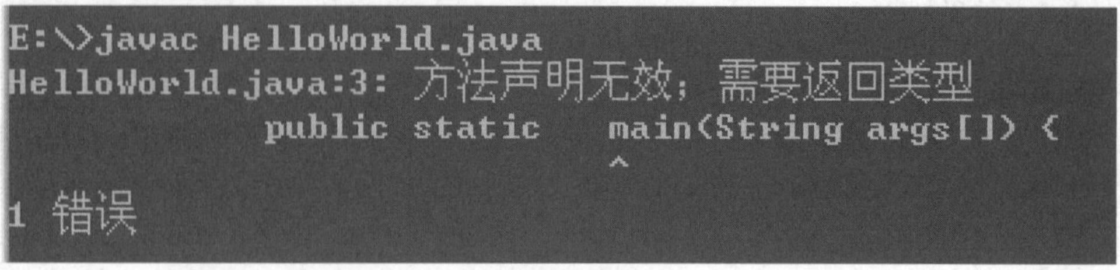

```
E:\>javac HelloWorld.java
HelloWorld.java:3: 方法声明无效；需要返回类型
        public static   main(String args[]) {
                        ^
1 错误
```

图 1-86

通过查看效果，得出第二个结论：main()方法前的 void 不能缺失，且顺序也不能更改！

3. Java 对大小写敏感

我们知道英文都有大小写之分，那 Java 程序中有那么多的英文，是否也有大小写之分呢？现在我们将代码进行修改查看一下效果，修改后的示例如下。

【例1.3】常见错误 3。

```
public class HelloWorld{
    public static void main(String args[]){
        system.out.println("Hello World!");//将 System 的第一个 s 变为小写
    }
}
```

对上例进行编译，会有如图 1-87 提示：

```
E:\>javac HelloWorld.java
HelloWorld.java:4: 软件包 system 不存在
        system.out.println("Hello World!");
              ^
1 错误
```

图 1-87

通过效果查看，得出第三个结论：Java 对大小写敏感！

4. ";"分号不能省略

现在我们将输出语句后的";"分号删除，看看效果。示例如下。

【例1.4】常见错误 4。

```
public class HelloWorld{
    public static void main(String args[]){
        System.out.println("Hello World!")//删除分号
    }
}
```

对上例进行编译,出现如图 1-88 提示:

```
E:\>javac HelloWorld.java
HelloWorld.java:3: 需要 ';'
                System.out.println("Hello World!")//删除分号
                                                  ^
1 错误
```

图 1-88

通过效果获得第四个总结:一个完整的语句后,都要以";"分号结尾,且必须是在英文状态下!

5. 引号的省略

另一个我们常犯的错误就是不小心漏掉一些东西,比如说忘记小括号,或只写一半的小括号;忘记引号或引号只写一半。如下所示的示例就是将引号少写了一半。

【例 1.5】常见错误 5。

```
public class HelloWorld{
        public static void main(String args[]){
            System.out.println("Hello World!);//少了后一半引号
        }
    }
```

对上例编译后,出现如图 1-89 提示:

```
E:\>javac HelloWorld.java
HelloWorld.java:3: 错误: 未结束的字符串文字
            System.out.println("Hello World!);//少了后一半引号
                               ^
HelloWorld.java:3: 错误: 需要';'
            System.out.println("Hello World!);//少了后一半引号
                                             ^
HelloWorld.java:5: 错误: 解析时已到达文件结尾
    }
     ^
3 个错误

E:\>
```

图 1-89

通过展示效果得出第五个结论:输出的字符串内容必须使用引号引起来,且必须是在英文状态下!

到此为止,我们认识了 5 种常见的编写 Java 程序的错误,并且知道如何进行修改和排错。可能有些错误目前我们还不能理解其含义,没有关系。我们当下的任务是在编写的过程中如果出现这些类似的问题,能够知道错在哪里,如何进行修改即可。

任务拓展

1. 编写 Java 程序:输出信息"我是×××!"。
2. 完成自助银行服务系统显示用户信息内容,根据效果图实现代码的编写,注意类名的命名规范。

```
用户名      账号             密码      身份证号码              余额
王丽丽      179708064356     1234      210050619890808185     1000.0
张颖颖      179708064359     4321      210050619891231127     2000.0
刘华        179708064368     4567      410207198904051271     3000.0
```

3. 完成自助银行服务系统主界面的制作,根据效果图实现,注意类名的命名规范。

```
==========================欢迎使用自动银行服务==========================
1. 开户   2. 存款   3. 取款   4. 转账   5. 查询余额      6. 修改密码      0. 退出
======================================================================
请输入选择:
```

1.4 习题

一、选择题

1. 选出下列选项中 Java 注释的正确声明(　　)。
  A. //这里是注释
  B. /*这里是注释*/
  C. /这里是注释
  D. */这里是注释*/

2. 在控制台编译一个 Java 程序,以下选项中正确的是(　　)。
  A. javac Test.java
  B. java Test.java
  C. javac Test
  D. Java Test

3. 在控制台显示消息的语句正确的是(　　)。
  A. System.out.println(我是一个Java程序员了!);
  B. System.Out.println("我是一个Java程序员了!");
  C. system.out.println("我是一个Java程序员了!");
  D. System.out.println("我是一个Java程序员了!");

4. 下列说法正确的是(　　)。
  A. Java 程序中可以有多个 main()方法。

B. Java 程序的类名必须与文件名相同。
C. Java 程序中的 main 方法如果只有一条语句,可以不用大括号括起来。
D. Java 程序的 main() 方法必须写在类中。

二、简答题

1. 请描述 Java 的特点。
2. 请描述开发 Java 程序的步骤,并说出自己对于每一个步骤的理解。
3. 请描述编写 Java 程序的常见错误及解决方案。
4. 书写一个 Java 程序,显示个人的档案信息。

例如,在控制台显示的结果如下:

姓名:张三
性别:男
年龄:23
身高:1.75
爱好:音乐、篮球、足球、游戏
联系方式:13982787382

面向过程的学生成绩信息处理

项目 2 首先介绍了 Java 的标识符、关键字、数据类型、常量、变量,其次讲解了 Java 运算符、表达式与语句,然后讲解了条件与循环语句,最后介绍了数组的定义及使用,为 Java 面向对象程序设计奠定语法基础。

工作 任务

(1) 学生成绩信息的表示。
(2) 学生课程的综合成绩计算。
(3) 学生课程的综合成绩等级判断。
(4) 学生课程的综合成绩等级分布统计。
(5) 学生课程的综合成绩的保存、排序(数组)。

学习 目标

(1) 了解 Java 中的关键字,掌握 Java 中标识符的命名规则,掌握 Java 的基本数据类型及数据类型转换,能够根据任务需求使用 Java 定义变量或声明常量。
(2) 掌握 Java 运算符、表达式的应用,能够根据任务需要编写简单的 Java 语句。
(3) 掌握 if、switch 条件判断语句的使用方法,能够独立编写简单的 Java 判断程序。
(4) 掌握 while、for、do-while 循环语句的使用方法,能够独立编写简单的 Java 循环程序。
(5) 掌握数组的定义和使用方法,掌握 sort 排序,能够根据任务需要独立完成数组的定义和应用。

2.1 任务 1 学生成绩信息的表示

任务描述 及分析

小明要完成学生成绩管理系统,首先要把学生成绩信息保存到计算机中。要完成这个工作任务,第一步先要明确学生成绩信息要保存哪些数据;第二步要掌握 Java 语言中与数据存储有关的基本语法知识(标识符、关键字、数据类型、常量、变量);第三步根据要保存的数据的特点(如数据类型是整数还是实数等),选取合适的数据类型等,实现数据在计算机中的表示。

学生成绩信息要保存的数据如下:

学号,姓名,课程名称,平时成绩,过程性考核成绩,终结性考核成绩,综合成绩,综合成绩等级等。

相关 知识

2.1.1 关键字和标识符

1. 关键字

语文课程中学习文章写作的过程如下:

(1) 学会单个汉字;

(2) 使用学会的单个字组成词;

(3) 使用字词造句;

(4) 句子按照一定格式、含义组织成文章。

同理,学习 Java 语言编写程序的过程也是如此,而关键字和标识符就相当于 Java 程序中的单个汉字。

例如,一个汉语的句子是:整数类型变量 x 的值等于1。它所对应的 Java 语句为:

int x=1;

上述 Java 语句中的 int 就是一个 Java 语言和 Java 的开发及运行平台提前约定好的、代表特定含义(即代表整数类型)的单个字,即只要 Java 的开发及运行平台看到 int 就知道是整数类型的变量。像这种 Java 语言和 Java 的开发及运行平台提前约定好的、赋以特定的含义并用作专门用途的单词,就是关键字,也称保留字。

Java 语言中的关键字如表 2-1 所示:

表 2-1

abstract	boolean	break	byte	case	catch	char	class
const	continue	default	do	double	else	extends	false
final	finally	float	for	goto	if	implements	import
instanceof	int	interface	long	native	new	null	package
private	protected	public	return	short	static	super	switch
synchronized	this	throw	throws	transient	true	try	void
volatile	while						

注意:

(1) 关键字是 Java 语言中自带的(或者说已存在的,因为已经提前约定好了含义,也可以理解为这些单词是保留给 Java 语言专用的),用户(即人)只能按照系统的规定来使用它们,不允许对它们进行修改或自行定义。

(2) 关键字全部为小写字母表示的完整的英文单词或简写。

(3) goto 和 const 不是 Java 语言中使用的关键字,而是 Java 的保留字,也就是说 Java 保留了它们,但没有使用它们。

(4) Java 的关键字是随着新的版本发布不断变化的。

(5) 关键字的具体含义和使用方法,会在后续的具体应用中讲述,请大家不要强行记忆。

2. 标识符

上例的 Java 语句:int x=1;中的 x 代表变量的名字,这个变量名字是用户根据需要自定义的,我们也可以将该变量名字改为 xyz,对应的 Java 语句变为:int xyz=1;像这种用户根据需要自己定义的变量名、类名或方法名称等,叫作标识符。

我们人取名字,要按照一定规则(如吉利、喜庆、辈分等),Java 语言中标识符的命名也有自己的规则:

(1) 首字符只能以字母、下划线、$ 开头,其后可以跟字母、下划线、$ 和数字的组合。如 $abc、_ab、ab123 等都是有效的。

(2) 标识符区分大小写。Java 中,Count 与 count 是两个不同的标识符。

(3) 命名的标识符不允许与关键字相同。如:int int=1;(×)

该例中使用关键字 int 作为变量名,这是不允许的。

(4) Java 语言对标识符的长度未加限制,但实际命名时不宜过长。

(5) 命名的标识符最好能够反映其意义,即能够"见名知义",从而提高程序的可读性。

(6) 如果标识符由多个单词构成,建议从第二个单词开始,首字母大写。如:isText、canRunTheCar 等。

(7) 在命名时,最好遵循 Java 推荐的命名规范。如符号常量全部采用大写字母,类名中每个单词的首字母大写,属性与方法名称首字母小写,等等。

(8) 标识符必须"先定义后使用"。

【例 2.1】下列哪些是错误的标识符?

My book1

Mybook2

_ab3

$a3_

b3_$

4foots

you

a+b

int

错误的标识符及其原因分析如下:

My book1//含有空格,空格不是字母,也不是数字或下划线

4foots//首字符为数字

a+b//加号不是字母,也不是数字或下划线
int//int 是关键字

2.1.2 数据类型

现实生活中,不同的数据信息具有各自的类型,如价钱为实数,姓名为一串字符。数据类型简单地说就是对数据分类,即对数据根据其特点进行类别的划分,划分的每种数据类型都具有区别于其他数据类型的特征,每一种类型的数据都有相应的特点和操作功能,如数字类型的就能够进行加减乘除的操作。Java 语言是一种强类型的语言,每个变量都必须声明其数据类型。例如 Java 语句:int x=1;中 int 声明的变量 x 是整数类型的数据。

1. Java 数据类型的分类

Java 语言的数据类型分为两大类:基本数据类型(Primitive Type)和引用类型(Reference Type)。基本数据类型规定了 8 种简单的数据类型来存储整数、浮点数、字符和布尔值,引用类型又分为数组、类、接口。Java 语言中的数据类型层次如图 2-1 所示:

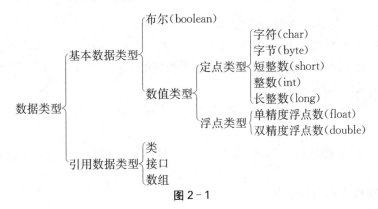

图 2-1

根据学习层次的安排,本任务先介绍基本数据类型,引用数据类型会在后续的任务中介绍。

2. 基本数据类型

Java 的基本数据类型如表 2-2 所示:

表 2-2

分类	数据类型	关键字	长度(位)	取值范围
整数类型	字节型	byte	8	$-2^7 \sim 2^7-1$
	短整型	short	16	$-2^{15} \sim 2^{15}-1$
	整型	int	32	$-2^{31} \sim 2^{31}-1$
	长整型	long	64	$-2^{63} \sim 2^{63}-1$
浮点类型	浮点型	float	32	—
	双精度型	double	64	—

续 表

分类	数据类型	关键字	长度(位)	取值范围
文本类型	字符型	char	16	'\u 0000'~'\u FFFF'
逻辑类型	逻辑型	boolean	8	true、false

(1) 整数类型。

当数字数据不带有小数时,可以声明为整数类型变量,如 5、-23 等都为整数。在 Java 语言中,整数类型数据又分为 byte(字节型)、short(短整型)、int(整型)、long(长整型)。如果数据的取值在 $-2^7 \sim 2^7-1(-128 \sim 127)$ 之间,可以声明为 byte 类型;若超出 byte 类型的范围,但取值在 $-2^{15} \sim 2^{15}-1(-32768 \sim 32767)$ 之间,可声明为 short 类型;若超出 short 类型的范围,但取值在 $-2^{31} \sim 2^{31}-1$ 之间,可声明为 int 类型;若超出 int 类型的范围,但取值在 $-2^{63} \sim 2^{63}-1$ 之间,可声明为 long 类型。例如声明一个字节型变量 sum 时,可以在程序中作出如下声明:

byte sum;//声明 sum 为字节型变量

在 Java 中,整数型的值都是带符号的数字,可以用十进制、八进制和十六进制来表示。例如整数值 22 的十进制、八进制、十六进制表示形式如下:

① 22:十进制的 22;

② 022:八进制的 22(在数值的前面加数字 0 表示八进制),相当于十进制的 18,计算公式:2*8+2=18;

③ 0x22:十六进制的 22(在数值的前面加 0x 表示十六进制),相当于十进制的 34,计算公式:2*16+2=34。

【例 2.2】整数类型变量定义。

```
byte     sno1=22;//定义 byte 类型变量 sno1,其值为十进制的 22
short    sno2=022;//定义 short 类型变量 sno2,其值为十进制的 18
int      sno3=0x22;//定义 int 类型变量 sno3,其值为十进制的 34
long     sno4=5;//定义 long 类型变量 sno4,其值为十进制的 5
```

注意:

① 如果要明确表示是 long 型的值,可以在后面直接跟一个字母"L"或"l",如:long sno4=5L;但由于小写字母"l"与数字"1"容易混淆,因而,尽量不要使用小写字母"l"。

② 整数类型数据的取值范围(见表 2-2)是按 Java 编程语言规范定义的且不依赖于平台。

(2) 浮点类型(float、double)。

在日常生活中经常遇到带小数的数值数据,如身高等。在数学中,这种带小数的数值称为实数;在 Java 语言中,这种数据类型称为浮点数类型。Java 语言中浮点类型分为 float 型和 double 型。float 型数据的长度为 32 位,取值范围为 3.402823E+38~1.401298E-45;double 型数据的长度为 64 位,取值范围为 1.797693E+308~4.9000000E-324。

在 Java 语言中,一个数字如果包括小数点或指数部分,或者在数字后带有字母 F 或 f(float)、D 或 d(double),则该数字为浮点型数据。例如:

1.2//带小数点的浮点型数据
1.2E3//带指数的浮点型数据

【例2.3】浮点型变量定义。

float score1 = 5.6F;//定义float类型变量score1,其值为5.6
float score2 = 5.6f;//定义float类型变量score2,其值为5.6
float score3 = 5.6E8f;//定义float类型变量score3,其值为$5.6*10^8$
float score4 = 5.6E8F;//定义float类型变量score4,其值为$5.6*10^8$
double score5 = 5.6D;//定义double类型变量score5,其值为5.6
double score6 = 5.6d;//定义double类型变量score6,其值为5.6
double score7 = 5.6;//定义double类型变量score7,其值为5.6
double score8 = 5.6E8;//定义double类型变量score8,其值为$5.6*10^8$

注意:

① 浮点类型的数据,如果没有特别指明,默认是double型的。

② 定义float型的数据时,一定要指明是float型的,即必须在数值后面添加"F"或者"f"来表示。如:

float score2＝5.6f

③ 定义double型的数据时,可以不用指明,因为默认就是double型的;也可以通过在数字后面添加"D"或者"d"来表示。

(3) 字符类型。

字符类型(char类型)的数据用来表示单个字符。计算机处理字符类型数据时,是把这些字符看成不同的整数来处理的,因此,我们需要为每个字符分配一个与之对应的整数值,故字符类型数据也可看作整数类型数据的一种。Java语言使用的是Unicode字符编码集,Unicode编码又叫统一码、万国码或单一码,是一种在计算机上使用的字符编码,它为每种语言中的每个字符设定了统一并且唯一的二进制编码,以满足跨语言、跨平台进行文本转换、处理的要求。单个字符的unicode值表示如下:

'\u???? '

???? 为4个十六进制数。如小写字母a对应的unicode数值为'\u0061',即十进制数97。

字符类型数据的值除了可以用unicode值(或该unicode值对应的十进制、八进制等进制数值)表示外,还可以用单引号将该字符包含起来表示。如小写字母a的值也可以用'a'表示。

注意:

单引号只能包含单个字符。如'AB'是错误的。

字符类型(char类型)的变量可以用来保存单个英文字母或单个标点符号等任何单个字符。

【例2.4】字符类型变量定义。

char zimu1 = '\u0061';//定义字符变量zimu1,其值是字符小写字母a
char zimu2 = 97;//定义字符变量zimu2,其值是字符小写字母a
char zimu3 = 'a';//定义字符变量zimu3,其值是字符小写字母a

Java 中的字符类型数据除了可以使用上述的 Unicode 字符编码值和单引号包含起来表示外，一些特殊字符（如空格、回车等无法用单引号括起来的字符）还可以用转义字符表示。转义字符是指用一些普通字符的组合来代替一些特殊字符，由于其组合改变了原来字符表示的含义，因此称为"转义"。常见的转义字符如表 2-3 所示：

表 2-3

序号	转义字符	描述	Unicode 值	序号	转义字符	描述	Unicode 值
1	\f	换页	\u000c	5	\t	水平制表符	\u0009
2	\n	换行	\u000a	6	\'	单引号	\u0027
3	\r	回车	\u000d	7	\"	双引号	\u0022
4	\b	空格	\u0008	8	\\	反斜杠	\u005c

【例 2.5】字符类型变量定义。

char zifu1 = '\n';//定义字符类型变量 zifu1，其值为换行
char zifu2 = '\r';//定义字符类型变量 zifu2，其值为回车

注意：

① 字符类型数据只能表示一个字符，更多时候我们要使用一串字符（即多个字符作为一组使用），此时可以使用 Java 语言中的 String 类来表示多个字符，其表示方式是用双引号把要表示的字符串引起来，字符串里面的字符数量可以是任意多个。如：

String sname="FanLingYun";//定义字符串变量 sname，其值为 FanLingYun

② String 不是基本的数据类型，而是一个类（class），我们在后续任务中会具体讲解。

③ String 类型变量可以包含任意多个字符，而字符类型只能是一个。如"a"表示的是字符串，而'a'表示的是单个字符，它们具有不同的功能。

（4）逻辑类型（布尔类型）。

逻辑类型（boolean）数据的值只有 true（真）和 false（假）两种。即当将一个变量定义为逻辑类型（boolean）时，它的值只能是 true 或 false。

【例 2.6】逻辑类型变量定义。

boolean buer1 = true;//定义逻辑类型（布尔类型）变量 buer1，其值为 true
boolean buer2 = false;//定义逻辑类型（布尔类型）变量 buer2，其值为 false

注意：

有些语言（特别是 C 和 C++）允许将数字值转换成逻辑值（所谓"非零即真"），这在 Java 编程语言中是不允许的。在 Java 语言中，在整数类型和 boolean 类型之间无转换计算，boolean 类型数据只允许使用 boolean 值（true 或 false）。

3. 数据类型转换

Java 变量的数据类型在定义时就已经确定了，不能随意转换为其他数据类型，但 Java 允许用户有限度地作类型转换处理。数据类型的转换可分为自动类型转换和强制类型转换。在 Java 语言中，各数据类型（boolean 类型除外）按容量大小（取值范围大小）由小到大排列为：byte、short、char→int→long→float→double。进行数据类型转换时应遵循的原则如下：

① 容量小的类型自动转换为容量大的类型；

② 容量大的类型转换为容量小的类型时,要加强制转换符;
③ byte、short、char 之间不会互相转换,并且三者在计算时首先转换为 int 类型;
④ 实数常量默认为 double 类型,整数常量默认为 int 类型。
(1) 自动类型转换。

在 Java 程序中已经定义了数据类型的变量,在同时满足下列两个条件时,会自动作数据类型的转换:
① 转换前的数据类型与转换后的数据类型兼容;
② 转换后的数据类型的表示范围比转换前的数据类型范围大。

【例 2.7】自动类型转换示例。
char zf = 'a';//定义字符类型变量 zf,其值为小写字母 a
int zs = zf;
/*定义 int 类型变量 zs,将变量 zf 的值赋给 zs。先将变量 zf 的值的数据类型自动转换为 int 类型,然后将转换后得到的 int 类型数值赋给 zs,其值为 97*/
float f = zs + 6;
/*定义 float 类型变量 f,将变量 zf 与 6 相加的和赋给 f。第一步,将变量 zs 的值与 6 相加(因为 zs 是 int 型,而整数常量默认为 int 类型,二者数据类型一致,都是 int 类型,所以二者不需要数据类型转换);第二步,先将第一步相加得到的结果自动转换为 float 类型数据,然后将转换后得到的 float 类型的数值赋给变量 f,即 f 的值为 103.0*/
double sum = zf + zs + f;
/*定义 double 类型变量 sum,将变量 zf、zs、f 相加的和赋给 sum。第一步,先将变量 zf 的值转换为 int 型数据,然后与变量 zs 相加;第二步,先将第一步相加得到的 int 型中间结果转换为 float 类型,然后与变量 f 相加;第三步,先将第二步相加得到的 float 类型的结果转换为 double 类型数据,然后将转换后得到的 double 类型数值赋给变量 sum,即 sum 的值为 297.0*/

(2) 强制类型转换。

Java 语言是强类型语言,因此,当两个整数类型数据进行运算时,其运算的结果也将是整数类型数据。例如:两个整数相除,5/3 其结果为整数 1,而不是实际的 1.6666…。因此,在该例中,如果要得到浮点数的计算结果,就必须将数据类型作强制性的类型转换。强制类型转换的语法如下:
(欲转换的数据类型关键字)表达式;

【例 2.8】强制类型转换示例 1。

```
public class QiangZhi1{
    public static void main(String args[]){
        System.out.println((float)10/3);//执行强制转换,并输出结果
        System.out.println(10/3);//不执行强制转换,并输出结果
    }
}
```

程序运行结果：

```
3.3333333
3
```

【例 2.9】强制类型转换示例 2。

```java
public class QiangZhi2{
    public static void main(String args[]){
        double d1 = 333.333;//定义 double 类型变量 d1,其值为 333.333
        int    x1 = (int)d1;/* 执行强制转换,将变量 d1 的值强制转换为 int
型数据,然后赋值给变量 x1 */
        System.out.println(d1);//输出变量 d1 的值
        System.out.println(x1);//输出变量 x1 的值
    }
}
```

程序运行结果：

```
333.333
333
```

注意：

只要在变量前面加上要强制转换的数据类型,程序运行时就会自动将此行语句中的变量的值作类型转换处理,影响的范围只限于该行代码,并不影响变量原来定义的数据类型和其原来的变量值。

2.1.3 常量和变量

1. 常量

常量是程序运行过程中值不再发生变动的一种量。例如,计算圆的周长或面积时所用到的圆周率 π＝3.1415…,在程序中需要设置成常量。常量有两种主要的分类标准:根据数据类型的分类标准、根据数据表达式的分类标准。

根据数据的类型,常量可分为以下 5 种:

(1) 逻辑型常量又称布尔型常量,其取值只能为 true 或 false,代表一个逻辑量的两种不同状态值,其中 true 代表真,false 代表假。

(2) 整型常量表示一个不带有小数位的整数,数值可正可负。整型常量有十进制、八进制和十六进制 3 种数制表示方式。如 12(十进制整型常量)、012(八进制整型常量)、0x12(十六进制整型常量)。

(3) 浮点型常量又称为实型常量,用来表示有小数部分的十进制实数。如 12.3(十进制实数常量)。

(4) 字符型常量是指用单引号(")括起来的单个字符。如'a'(字符常量)。

(5) 字符串常量是指用双引号("")括起来的一个字符序列,字符串常量包含的字符个数称为它的有效长度。如"FanLingYun"(字符串常量)。

根据数据的表达方式,Java 的常量可分为值常量与符号常量两种:

(1) 值常量就是直接以特定值表达的量。如 true、3、'A'、3.2、"FanLingYun"。

(2) 符号常量是一种以标识符形式表示的常量,这类常量引用时以符号名称代替,但参与运算的是它的内容,即常量的值,这类常量必须先定义后才可使用。符号常量定义的格式为:

final〈类型〉〈符号常量标识符〉=〈常量值〉;

例如定义圆周率的语句为:

```
final float PI = 3.1415926;/* 在程序中可以用 PI 代替 3.1415926,程序运行时看到 PI,会自动调用值 3.1415926 */
```

注意:

建议符号常量的标识符全部大写并用下划线将词分隔。如 MAXIMUM_SIZE。

2. 变量

变量是指程序的运行过程中发生变化的量,通常用来存储中间结果,或者保存、输出结果。在 Java 语言中,所有的变量必须先定义后使用。变量有 3 个基本要素:变量名、变量的数据类型及变量值,其定义格式为:

〈数据类型〉 〈变量标识符〉[=〈初值〉][,〈变量标识符〉[=〈初值〉]…];

【例 2.10】变量定义示例。

```
int a,b;//定义整数类型变量 a 和 b
int a = 3,b;//定义整数类型变量 a 和 b,a 的初始值为 3
int a = 3,b = 4;//定义整数类型变量 a 和 b,a 的初始值为 3,b 的初始值为 4
```

注意:

① 变量名必须遵从标识符的所有规则。

② 变量名可以大小写混用,但建议首字符小写,词由大写字母分隔,并且在变量名中尽量不要使用下划线和 $ 符号。如 currentCustomer。

③ 变量可以先声明再赋值,如

```
int i;
i = 3;
```

也可以声明的同时进行赋值,如

```
int i = 3;
```

④ 全部大写并用下划线将词分隔。

Java 语言中变量起作用的范围称为作用域,变量按作用域可分为以下 4 种:

（1）局部变量：在一个方法或一对{}代码块内定义的变量称为局部变量。局部变量的作用域是整个方法或某个代码块。

（2）类变量：在类中声明且不在任何方法体中的变量称为类变量。类变量的作用域是整个类，类变量要加 static 关键字修饰。

（3）方法参数：方法参数定义了方法调用时传递的参数，其作用域就是所在的方法。

（4）异常处理器参数：异常处理器参数是 catch 语句块的入口参数。这种参数的作用域是 catch 语句后由{}表示的语句块，这种变量我们将在异常处理中讲解。

【例 2.11】变量作用域示例。

```
public class ZuoYongYu{
    static int x = 5;//定义变量x为类变量,其作用范围是整个类
    /*为自定义方法Test(),定义方法参数a,变量a只在方法Test()中起作用*/
    static void Test(int a)
    {
        System.out.println(a);
    }
    public static void main(String args[]){
        /*在主方法main()中,定义局部变量y,变量y只在主方法main()中起作用*/
        int y = 23;
        System.out.println(y);
        Test(6);//调用方法Test()
    }
}
```

x 为类变量，其作用域在 ZuoYongYu 类的类体内，即 ZuoYongYu 对应的那一对{}中；y 为局部变量，其作用域在 main()方法体内，即 main 对应的那一对{}中；a 为方法参数，其作用域在 Test()方法体内，即 Test 对应的那一对{}中。

任务 实施

2.1.4 数据表示的实现

学生成绩管理系统中要保存的数据如下：

学号，姓名，性别，课程名称，平时成绩，过程性考核成绩，终结性考核成绩，综合成绩，综合成绩等级等。

要实现上述数据的表示，我们首先要确定上述数据的数据类型；其次，确定上述各种类型的数据是变量还是常量；最后，根据确定的上述数据的数据类型及其是常量还是变量，分别按照符号常量和变量的定义格式，结合标识符的命名规则，完成数据表示。

1. 数据类型选择

学生成绩管理系统中要保存的数据的数据类型选择过程如下：

(1) 学号。

第一步，分析学生的学号数据特点，如本校学生的最大学号是 17 数媒 0031 班学生张晓慧：1761003001。由此，可知学生的学号是整数类型。

第二步，根据最大学号的整数数值(如 1761003001)确定应该选 int 型。由于 byte 的取值范围是 $-128\sim127$，short 的取值范围是 $-32768\sim32767$，int 的取值范围是 $-2^{31}\sim2^{31}-1$，long 的取值范围是 $-2^{63}\sim2^{63}-1$，根据已知的最大学号的数值 1761003001 可以判断学号的取值超过了 byte、short 类型的取值范围，但在 int 型的取值范围内，因此学号的数据类型应为 int 型。

(2) 姓名。

第一步，分析学生的姓名数据特点。如学生姓名：张晓慧，由此可知姓名是字符型数据。

第二步，进一步确认姓名数据的数据类型是 String。因为张晓慧是 3 个字符(一串字符)，而字符类型数据只能表示单个字符，因此姓名的数据类型选择字符串 String。

(3) 性别。

第一步，分析性别数据特点。如张晓慧的性别：女，由此可确定是字符型数据。

第二步，确定性别的数据类型为 char 型。由于性别数据的取值，只能是"男""女"二选一，因此性别的取值只有一个字符，因此可以确定性别是 char 类型数据。

(4) 依此类推，可知课程名称应是 String 类型，平时成绩是 byte 类型，过程性考核成绩是 byte 类型，终结性考核成绩是 byte 类型，综合成绩是 float 类型，综合成绩等级是 String 类型。

2. 变量、常量的确定

不同学生的学号不是一致的，即会有变化，因此学号数据是变量；不同学生的姓名也是变化的，因此姓名数据是变量；同样，不同学生的性别是变化的，因此性别数据是变量。依此类推，课程名称、平时成绩、过程性考核成绩、终结性考核成绩、综合成绩、综合成绩等级都是变量。

由此，确定学生基本信息要保存数据的数据类型如表 2-4 所示。

表 2-4

数据	数据类型	变量或常量
学号	int	变量
姓名	String	变量
性别	char	变量
课程名称	String	变量
平时成绩	byte	变量
过程性考核成绩	byte	变量
终结性考核成绩	byte	变量
综合成绩	double	变量
综合成绩等级	String	变量

3. 定义变量

学生基本信息要保存数据的变量定义如下：
int sno;//定义变量 sno,用于表示学生学号
String sname;//定义变量 sname,用于表示学生姓名
char sex;//定义变量 sex,用于表示学生性别
String Cname;//定义变量 Cname,用于表示课程名称
byte pingShiScore;//定义变量 pingShiScore,用于表示学生平时成绩
byte guoChengScore;//定义变量 guoChengScore,用于表示学生过程考核成绩
byte zhongJieScore;//定义变量 zhongJieScore,用于表示学生终结考核成绩
double zongHeScore;//定义变量 zongHeScore,用于表示学生综合成绩
String zongHeScoreJieBie;//定义变量 zongHeScoreJieBie,用于表示学生综合成绩的等级

任务 拓展

使用 Java 语言表示下图所示的银行账户信息。

用户名	账号	密码	身份证号码	余额
张三	179708061209	123456	647939208485940308	1000.0

2.2 任务2 学生课程的综合成绩计算

任务描述 及分析

在学生成绩管理系统中，小明需要根据学生平时成绩(百分制)、学生过程性考核成绩(百分制)、学生终结性考核成绩(百分制)计算出学生的综合成绩,学生平时成绩、学生过程性考核成绩、学生终结性考核成绩分别占综合成绩的 10%、40%、50%。要完成这个工作任务,第一步,我们先要掌握 Java 语言的运算符相关知识,然后根据任务的操作需要,选取相应的运算符;第二步,要掌握表达式的相关知识,然后根据任务需要,利用第一步选取的运算符,将 2.1 中定义的相关变量组合起来构成求解学生综合成绩的表达式;第三步,了解语句的知识,将第二步完成的表达式,变为 Java 程序的语句(Java 程序由一条一条的 Java 语句构成),完成 Java 程序的编写。

相关 知识

Java 程序是由许多语句组成的,语句的基本单位是表达式和运算符,而运算符又是表达式的重要组成部分,因此,我们首先讲解运算符的用法。

2.2.1 运算符

Java语言提供了多种运算符,根据其使用的类型,运算符可分为赋值运算符、算术运算符、关系运算符、逻辑运算符、位运算符等。

1. 赋值运算符

(1) 简单赋值运算符。

简单赋值运算符"=",将"="右边的值赋给左边的变量。右边的值可以是任何常数、变量或者表达式,只要能产生一个值就行,但左边的值必须是一个明确的、已命名的变量。例如:

```
int x;
x=3;//正确赋值方式
3=x;//错误赋值方式
```

(2) 复合赋值运算符。

Java语言允许使用复合赋值运算符,即在简单赋值符前加上其他运算符。复合赋值运算符是表达式的一种缩写。例如:a+=5等价于a=a+5。复合赋值运算符有11种,如表2-5所示。

表2-5

运算符	使用方法	等价表达式
+=	op1+=op2	op1=op1+op2
-=	op1-=op2	op1=op1-op2
*=	op1*=op2	op1=op1*op2
/=	op1/=op2	op1=op1/op2
%=	op1%=op2	op1=op1%op2
&=	op1&=op2	op1=op1&op2
|=	op1|=op2	op1=op1|op2
^=	op1^=op2	op1=op1^op2
>>=	op1>>=op2	op1=op1>>op2
<<=	op1<<=op2	op1=op<<op2
>>>=	op1>>>=op2	op1=op1>>>op2

注意:

① 复合赋值运算符右边的表达式是一个整体。例如:
x*=y+z;等价于x=x*(y+z);而不是x=x*y+z;

② 在使用赋值运算符时,应尽量使运算符右侧的表达式与左侧变量的类型一致。若不一致,要先将表达式的值转换成变量的数据类型,再进行赋值。

2. 算术运算符

Java 语言中的算术运算符分为一元运算符（只有一个操作数）和二元运算符（有两个操作数），算术运算符的操作数可以是整型或浮点型。

（1）一元运算符。

Java 语言中的一元运算符如表 2-6 所示：

表 2-6

类型	运算符	功能	用法举例
一元运算符	＋	正值	＋5
	－	负值	－5
	＋＋	自增	a＋＋,＋＋a
	－－	自减	a－－,－－a

注意：

① ＋（正值）和－（负值）运算符使操作数取正、负值。虽然它们应用的并不多，但有提升操作数数据类型的作用。例如：

```
byte a = 5,b;//定义 byte 类型变量 a 和 b,其中 a 的初始值为 5
b = -a;//取变量 a 的负值,然后赋给变量 b
```

上述语句将产生编译错误，原因是 a 经过－运算后已经由 byte 类型提升为 int 类型数据，而变量 b 仍是 byte 类型，二者数据类型不匹配。又由于－a 的 int 数据类型的范围大于变量 b 的 byte 类型，导致进行赋值运算时无法进行自动数据类型转换。而直接将 int 类型数据赋值给字节类型变量 b 是不允许的，因此导致编译错误。

② ＋＋（自增）、－－（自减）运算符的操作数必须是变量，不能是常量和表达式。单个变量单独使用时，它们可以放在变量前，也可以放在变量后，功能都是对变量增（减）1。例如：

```
int x = 3,y = 3;
x++ ;//x 自增 1
++y;//y 自增 1
```

变量 x、y 分别单独使用＋＋，＋＋在变量的前或后，二者功能一致，都是将变量值加 1，所以 x、y 的变量值都变为 4。

但若在表达式中使用＋＋（自增）、－－（自减），则前置、后置的意义是不同的。在表达式中，＋＋（自增）、－－（自减）前置时变量的值先增 1 或减 1，然后再用变量的新值参加表达式的计算；反之，＋＋（自增）、－－（自减）在表达式中后置时，变量的值先参加表达式的计算，然后变量再增 1 或减 1。例如：

```
int x = 3,y;//定义 int 类型变量 x、y,x 的初始值为 3
y = x + + ;/*由于在表达式中 + + 在变量 x 的后面(后置),因此应先使用变量 x 的
值参加运算,即先将变量 x 的值 3 赋给变量 y,因此 y 的值为 3,然后再将变量 x 加 1,因此
x 的值变为 4 */
```

执行后 y 的值为 3,x 的值为 4。

而

```
int x = 3,y;//定义 int 类型变量 x、y,x 的初始值为 3
y = + +x;/*由于在表达式中 + + 在变量 x 的前面(前置),因此应先将变量 x 的值
加 1,即变量 x 的值变为 4,然后再使用变量 x 的新值 4 参加运算,即将变量 x 的新值 4 赋
给变量 y,因此 y 的值为 4 */
```

执行后 y 的值为 4,x 的值为 4。

(2) 二元运算符。

二元运算符并不改变操作数的值而是返回一个表达式计算出的结果值。Java 语言中的二元运算符如表 2-7 所示:

表 2-7

类型	运算符	功能	用法举例
二元运算符	+	加	a+b
	-	减	a-b
	*	乘	a*b
	/	除	a/b
	%	求余	a%b

注意:

① "+"(加)运算符,不仅可以进行加法运算,还可以用来连接字符串。例如:

```
String xing = "Fan";//定义字符串变量 xing,其值为 Fan
String ming = "LingYun";//定义字符串变量 ming,其值为 LingYun
String name = xing + ming;/*定义字符串变量 name,其值为字符串 xing 和 ming 的
连接结果,即 name 的值为 FanLingYun */
```

则 name 值为"FanLingYun"。

② "/"(除)运算符。两个整数相除的结果是整数,但如果有一个操作数为小数则结果为小数。例如 5/2=2,5.0/2=2.5。

③ "%"取模运算符是求两个数相除的余数,如 17%3 的结果是 2。还可以对实数求余数,此时,余数的符号与被除数相同。例如 -3.5%3= -0.5,3.5% -3=0.5。

3. 关系运算符

关系运算符用于将两个值比较运算,关系运算的结果值为 true 或 false。Java 语言提供了 6 种关系运算符,它们都是双目运算符,如表 2-8 所示:

表 2-8

运算符	意义	运算符	意义
>	大于	>=	大于等于
<	小于	<=	小于等于
==	等于	!=	不等于

注意:

(1) 关系运算的结果是一个布尔类型值:true 或 false。如果关系成立,则表达式的值为 true;关系不成立,则表达式的值为 false。例如:3>2 的值为 true,4!=4 的值为 false。

(2) "=="等于运算符不要与"="赋值运算符混淆,"=="用于判断两个值是否相等,而"="用于将一个值赋给变量。例如:

int x=3,y=4,z;//定义 int 型变量 x、y、z,其中 x 的值为 3,y 的值为 4
if(x==y)//判断 x 是否等于 y,其结果是 false,因为二者不相等
z=x;//将变量 x 的值赋给 z,即 z 的值变为 3

(3) 关系运算符常用于 if 语句、循环语句的条件中。

4. 逻辑运算符

逻辑运算符可以对布尔类型的数据(布尔常量、布尔变量、关系表达式和布尔表达式等)进行运算,结果也为布尔类型。Java 语言中的逻辑运算符有 6 种,详见表 2-9,逻辑运算符的运算规则详见表 2-10。

表 2-9

运算符	名称	例子
!	逻辑非	!a
&&	简洁与	a&&b
||	简洁或	a||b
^	异或	a^b
&	非简洁与	a&b
|	非简洁或	a|b

表 2-10

op1	op2	!op1	op1&(&&)op2	op1|(||)op2	op1^op2
false	false	true	false	false	false

续　表

op1	op2	!op1	op1&(&&)op2	op1|(||)op2	op1^op2
false	true	true	false	true	true
true	false	false	false	true	true
true	true	false	true	true	false

注意：

Java 语言提供了两种与和或的运算符：&、| 和 &&、||。这两种运算符的区别是：

(1) &&、||：逻辑表达式求值过程中,对于或运算,先求左边表达式的值,如果为 true,则整个逻辑表达式的结果就是 true,从而不再对右边的表达式进行运算；同样对于与运算,如果左边表达式的值为 false,则整个逻辑表达式的结果就是 false,右边的表达式就不需要再进行运算了。

(2) &、|：利用 &、| 作运算时,运算符两边的表达式都要被判断,即不管第一个表达式的结果能否推测出整个逻辑表达式的结果,都要对第二个表达式进行运算。

5. 位运算符

位运算符用于对整型数据的二进制位进行测试、置位或移位处理。位运算符按功能可分为位逻辑运算符和位移位运算符。

(1) 位逻辑运算符。

位逻辑运算符是对操作数的每个二进制位进行相应的逻辑运算。位逻辑运算符共有 4 种,除 ~（非）为单目运算符外,其余 3 种 &（与）、|（或）、^（异或）为双目运算符,其逻辑运算规则同逻辑运算符。表 2-11 列出了 Java 语言的位逻辑运算符,表 2-12 列出了位逻辑运算符的运算规则。

表 2-11

运算符	位运算表达式	功能
~	~op1	按位取反
&	op1&op2	按位与
|	op1|op2	按位或
^	op1^op2	按位异或

表 2-12

op1	op2	~op1	op1&op2	op1|op2	op1^op2
0	0	1	0	0	0
0	1	1	0	1	1
1	0	0	0	1	1
1	1	0	1	1	0

上述每一种位逻辑运算符根据其运算规则都有一定的用途。

① 逻辑与的用途是清零：如果要将一个存储单元清零，即将其全部二进制位变为0，我们可以找一个新数，若原来的数中的某一个二进制位为1，则新数中对应的二进制位必须为0；若原来的数中的某一个二进制位为0，则新数中对应的二进制位既可以为0，也可以为1，然后再对两者进行 & 运算，就可以将其清零。例如：

```
原数：   11011010
新数： & 00100001
         00000000
```

② 逻辑或的用途是置1：如果要将一个存储单元的所有二进制位置1，我们可以找一个新数，若原来的数中的某一个二进制位为0，则新数中对应的二进制位必须为1；若原来的数中的某一个二进制位为1，则新数中对应的二进制位既可以为0，也可以为1，然后再对两者进行 | 运算，就可以将其所有二进制位置1。例如：

```
原数：   11011010
新数： | 00100101
        11111111
```

③ 逻辑异或的用途有两种：一种是使特定位翻转，即将欲翻转的位与1进行异或运算；另一种是使特定位保留原值，即将欲翻转的位与0进行异或运算。例如：

```
原数：   11011010
新数： ^ 00001111
        11010101
```

(2) 位移位运算符。

位移位运算是针对十进制整数对应的二进制数进行的，这里的二进制并不是该数本身的二进制码，而是要进行编码。计算机中常用的编码有：原码、反码、补码，Java语言使用补码表示二进制数。

① 原码：最高位为符号位，最高位为0表示正数，最高位为1表示负数，其余各位用该数的二进制码表示。例如：

+8 的原码为 00001000
−8 的原码为 10001000

② 反码：正数的反码与原码相同，负数的反码将其原码按位取反（不包括符号位）。例如：

+8 的反码为 00001000
−8 的反码为 11110111

③ 补码：正数的补码与原码相同，负数的补码将其反码加1。例如：

+8 的补码为 00001000
−8 的补码为 11111000

移位运算符是把操作数对应的二进制数向左或向右移动一定的位数。Java中的位运算符如表2-13所示。

表 2-13

运算符	位运算表达式	功能
<<	op1<<n	op1 左移 n 位
>>	op1>>n	op1 右移 n 位
>>>	op1>>>n	op1 无符号右移 n 位

对于位的移位运算符要进行如下说明：

① 左移运算中左移一位相当于乘以 2；反之，右移一位相当于除以 2。用移位运算实现乘除比算术运算符中的乘除运算符快。例如：

64<<2 结果是 $64*2^2=256$

−256>>4 结果是 $-256/2^4=-16$

② >>称为带符号位右移，进行右移运算时，最高位为 0，则左边补 0；最高位为 1，则左边补 1，即符号位不变。假设操作数按字节类型存储。例如：

−64>>2 等价于 11000000>>2 即 11110000

64>>2 等价于 01000000>>2 即 00010000

③ >>>称为无符号位右移，进行无符号右移运算时，左端出现的空位用 0 补。假设操作数按字节存储。例如：

−64>>>2 等价于 11000000>>>2 即 00110000

64>>>2 等价于 11000000>>>2 即 00110000

6. 条件运算符

条件运算符是三目运算符，其格式为：

表达式?语句 1:语句 2;

其中表达式的值是布尔类型，当表达式的值为 true 时执行语句 1，否则执行语句 2。要求语句 1 和语句 2 返回的数据类型必须相同，并且不能无返回值。例如：

int a=4,b=5,c=6,d=7;/* 定义 int 型变量 a、b、c、d,其初始值分别为 4、5、6、7 */
(a>b)?(c+2):(d+2);/* 由于 a>b 不成立，因此表达式 a>b 的值为 false,所以执行语句 2,即 d+2,因此结果为 9 */

2.2.2 表达式和语句

Java 的语句有很多形式，表达式就是其中一种。表达式是由操作数和运算符按一定的语法形式组成的符号序列，每个表达式运算后都会产生一个确定的值，称为表达式的值。表达式的值是有数据类型的，该数据类型称为表达式类型，由运算符和参与运算的数据的类型决定。一个常量或一个变量是最简单的表达式，其值即该常量或变量的值。而用运算符连接几个表达式构成的式子仍是表达式。

例如：

−47//表达式由一元运算符"−"与常量 47 构成

a+2//表达式由变量 a、算术运算符"+"和常量 2 构成

```
a+b*2/(3+d)-c//表达式由变量、运算符、常量构成
```

1. 数据类型转换(自动类型提升)

在一个表达式中可能有不同类型的数据进行混合运算,这是允许的,但在运算时,Java会将不同类型的数据转换成相同类型再进行运算。数据类型的转换分为自动转换和强制转换,我们在 2.1 的基本数据类型中已经讲解过,但有时在表达式的求值过程中,运算中间值的精度会超出操作数的取值范围。例如:

```
byte x = 30, y = 50, z = 100;
int a = x * y/z;
```

在运算 x * y 项时中间结果 1500 已经超出了操作数 byte 类型的范围。为解决这类问题,Java 语言在对表达式求值时,自动提升 byte 或 short 类型的数据为 int 类型。Java 语言对表达式求值的自动类型提升规则为:

(1) 所有 byte 和 short 类型提升为 int 类型;
(2) 若一个操作数是 long 类型,则整个表达式提升为 long 类型;
(3) 若一个操作数是 float 类型,则整个表达式提升为 float 类型;
(4) 若有 double 类型,则表达式值为 double 类型。

注意:

自动类型提升对数据的运算带来了方便,但也容易引起编译错误。例如:

```
byte    x = 30, a;
a = -x;//编译错误,不能向 byte 变量赋 int 值
```

因为使用一元运算符"-"对 x 进行取负值操作时,其中间结果自动提升为 int 型数据,然后通过简单赋值运算符"="将 int 型的中间结果赋给 byte 型变量 a 时,两者数据类型不一致;此外,由于 int 型数据的取值范围大于 byte 型,无法进行自动类型转换,因而出现编译错误。

2. 优先级

在一个表达式中可能有多种运算符,Java 语言规定了表达式中出现多种运算符时各种运算符进行运算的运算顺序,称为运算符的优先级。在表达式中优先级高的运算符先进行运算。

例如:

```
a = b + c * d/(c~d)
```

Java 处理该表达式时,将按照表 2-14 所示,从最高优先级到最低优先级进行计算。在上例中,因为括号优先级最高,所以先计算 c~d,接着是 c*d,然后除以 c~d,最后把上述结果与 b 的和存储到变量 a 中。

不论什么时候,若无法确定某种计算的执行次序时,可以使用加括号的方法为编译器指定运算顺序,这是提高程序可读性的一个重要方法。

例如：

```
a|4+c>>b&7||b>a%3//表达式1
(a|(((4+c)>>b)&7))||(b>(a%3))//表达式2
```

表达式1运算次序的可读性不如表达式2清晰，因为表达式2中用括号()显式地表明了运算次序。

表 2-14

优先级	运算符名称	运算符	结合性
1	限制符	()、[]、…	
2	自增、自减	++、--	自右向左
3	按位取反、逻辑非、负号	~、!、-	自右向左
4	强制转换、内存分配	(类型)表达式、new	自右向左
5	算术乘、除、取模	*、/、%	自左向右
6	算术加、减	+、-	自左向右
7	移位运算符	<<、>>、>>>	自左向右
8	关系运算	<、<=、>、>=	自左向右
9	相等性判断运算	!=、==	自左向右
10	按位与、非简洁与	&	自左向右
11	按位异或、逻辑异或	^	自左向右
12	按位或、非简洁或	|	自左向右
13	逻辑与(简洁与)	&&	自左向右
14	逻辑或(简洁或)	||	自左向右
15	条件运算	?:	自左向右
16	赋值运算	=、+=、-=*=、…	自左向右

提示：

初学者不必强行记忆各种运算符的优先级和结合性，可在后续的练习中多应用，以便熟悉。此外，运算优先级可采用小括号()显式指明运算顺序。

3. 结合性

当在表达式中出现多个相同优先级的运算符时，就需要考虑结合性。结合性用于确定同级运算符的运算顺序，运算符有左结合性和右结合性两种。

左结合性指的是从左向右使用运算符。例如：

```
a+b-c
```

该表达式运算时,"+"和"-"是同级运算符,由于二元算术运算符具有左结合性(自左向右结合),所以,计算时 b 先与左边的+结合,计算 a+b 后,其和再与 c 相减。

而右结合性是从右向左使用运算符。例如:

```
a = b = c
```

该表达式运算时,两个赋值运算符"="同级,由于"="具有右结合性(自右向左结合),所以,计算时,先执行 b=c,再执行 a=b。

4. 语句

Java 语言中的语句可分为空语句、标识语句、声明语句、表达式语句、分支语句、循环语句、跳转语句、同步语句、异常语句等,其中表达式语句和声明语句是最基本的语句。

(1) 表达式语句。

表达式语句是由表达式加分号构成的语句。例如:

```
m++;
System.out.println("Hello World!");
```

Java 语言中,语句用分号";"终止,但并不是所有的表达式都可以构成语句。例如:

```
a<=b;
```

该表达式加分号构成的语句无意义。满足下列条件之一的表达式可以通过添加分号,构成有意义的表达式语句:

① 赋值表达式包含赋值运算符=或复合赋值运算符之一;
② ++或--的前后缀形式;
③ 方法调用:无论它是否有返回值;
④ 对象创建表达式:用 new 来创建一个对象的表达式。

(2) 声明语句。

声明语句声明一个变量,并可为其赋初值。声明语句可以出现在任意块内,定义在方法内或块内的局部变量在使用前必须赋初值,即要么在声明变量时进行初始化,要么在声明变量后进行赋值操作。

5. 块

用一对花括号{}将零个或多个语句括起来,就构成一个块(也称复合语句)。在 Java 语言的块中允许包含另一个块(块嵌套),并且允许一个块出现在任何单一语句可以出现的地方。回顾前面的内容,可以知道类体和方法体都是块。块体现了 Java 面向对象程序设计的封装概念,在一个块中声明的局部变量的作用域是从该变量的声明位置到最小的包含其声明的块结束。

任务 实施

2.2.3 综合成绩计算

1. 编写程序框架

```
public class ZongHeScore{
    public static void main(String args[]){

    }
}
```

2. 编写实现综合成绩计算的指令代码

```
public class ZongHeScore{
    public static void main(String args[]){
    byte pingShiScore;//用于表示学生平时成绩
    byte guoChengScore;//用于表示学生过程考核成绩
    byte zhongJieScore;//用于表示学生终结考核成绩
    double zongHeScore;//用于表示学生综合成绩
    pingShiScore = 85;//某学生的平时成绩为 85
    guoChengScore = 82;//某学生的过程考核成绩为 82
    zhongJieScore = 75;//某学生的终结考核成绩为 75
    //计算综合成绩
    zongHeScore = pingShiScore * 0.1 + guoChengScore * 0.4 + zhongJieScore * 0.5;
    //输出综合成绩
    System.out.println(zongHeScore);
    }
}
```

如图 2-2 所示。

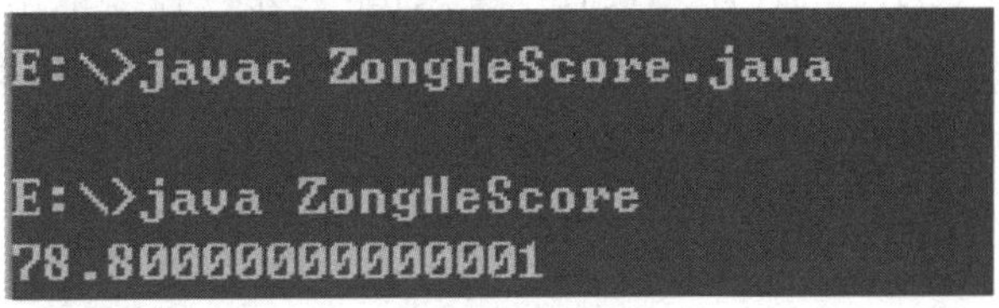

```
E:\>javac ZongHeScore.java

E:\>java ZongHeScore
78.80000000000001
```

图 2-2

任务 拓展

1. 实现自助银行服务系统初始化 3 个用户信息的操作,且将用户信息显示出来,效果如图所示:

用户名	账号	密码	身份证号码	余额
王丽丽	179708064356	1234	210050619890808185	1000.0
张颖颖	179708064359	4321	210050619891231127	2000.0
刘华	179708064368	4567	410207198904051271	3000.0

2. 假设银行的存款利率如下:
年利率:1.25%
6 个月:0.85%
3 个月:0.45%

上例中王丽丽、张颖颖、刘华 3 人的存款时间分别为 3 个月、6 个月、1 年,请编写程序求解这 3 个人的存款到期后的本息金额。

任务 3 学生课程的综合成绩等级判断

任务描述 及分析

小明需要编写程序,实现综合成绩等级的判断:优(90~100 分)、良(80~89 分)、中(70~79 分)、及格(60~69 分)、不及格(0~59 分)。要完成这个工作任务,我们需要将综合成绩按照给出的不同等级的条件进行判断,如果满足相应条件就输出相应的等级。因此,我们首先要学习 Java 语言条件分支语句(if、switch)的相关知识,然后根据需要选取相应的条件分支结构实现综合成绩等级的判断。

相关 知识

分支语句在 Java 程序中的作用是使程序更灵活,它允许程序根据不同的情况、不同的条件等,采取不同的动作、进行不同的操作,实现选择结构。在 Java 语言中使用的分支语句有 if-else 语句和 switch 语句。

2.3.1 if-else 语句

1. 用单个 if 语句实现单选、二选一结构

用一个 if-else 语句,可根据一个关系或逻辑表达式的值是 true 还是 false,进行不同操作,其语法结构如下:

```
if(expression)
statement1
[else
statement2]
```

执行上述语句,首先要计算表达式 expression,若其值为真(true),则执行语句 statement1;反之,若其值为假(false),则执行语句 statement2,实现二选一结构。

注意:

(1) statement1 和 statement2 都可以是复合语句;

(2) if-else 语句的 else statement2 部分可省略。省略时,若表达式 expression 的值为假(false),则不执行任何语句,实现单选结构。

【例2.12】判断 double 类型变量 x 的值是否大于变量 y 的值,如果大于,就输出"x 大于 y"。

```java
public class If1{
public static void main(String args[]){
double x = 5.5;//定义 double 类型变量 x
double y = 3.4;//定义 double 类型变量 y
if(x>y){
System.out.println("= = = = =判断结果如下= = = = =");
System.out.println("        x 大于 y        ");
System.out.println("= = = = = = = = = = = = = = = = = = = = =");
}
}
}
```

如图 2-3 所示。

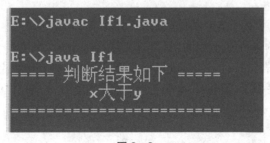

图 2-3

【例2.13】判断 double 类型变量 x 的值和变量 y 的大小。如果大于,就输出"x 大于 y";如果小于,就输出"x 小于 y"。

```java
public class If2{
public static void main(String args[]){
```

```
double x = 2.5;//定义 double 类型变量 x
double y = 3.4;//定义 double 类型变量 y
if(x>y){
System.out.println(" = = = = 判断结果如下 = = = = ");
System.out.println("         x 大于 y         ");
System.out.println(" = = = = = = = = = = = = = = = = = = = = = = ");
}
else{
System.out.println(" = = = = 判断结果如下 = = = = ");
System.out.println("         x 小于 y         ");
System.out.println(" = = = = = = = = = = = = = = = = = = = = = = ");
}
}
}
```

如图 2-4 所示。

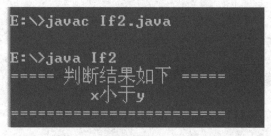

图 2-4

2. 多个 if 语句实现多选结构

(1) if 语句并列。

在多个 if 语句中,可书写多个条件(每一个 if 语句都可表示一种条件)。若这些 if 语句并列且这些 if 语句中的条件互不相同,就可以实现多选结构。

【例 2.14】使用 if 语句并列,判断 double 类型变量 x 的值和变量 y 的大小。如果大于,就输出"x 大于 y";如果小于,就输出"x 小于 y";如果 x 等于 y,输出"x 等于 y"。

```
public class If3{
public static void main(String args[]){
double x = 3.2;//定义 double 类型变量 x
double y = 2.3;//定义 double 类型变量 y
System.out.println(" = = = = 比较结果如下 = = = = = ");
if(x>y){
System.out.println("x 大于 y!");
```

```
}
if(x<y){
System.out.println("x 小于 y!");
}
if(x = = y){
System.out.println("x 等于 y!");
}
System.out.println(" = = = = =比较完成 = = = = =");
}
}
```

如图 2-5 所示。

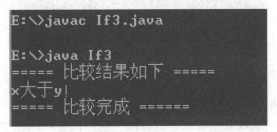

图 2-5

(2) if 语句嵌套。

在 if-else 语句中,若 statement1 或 statement2 又是 if-else 语句时,就构成了 if 语句嵌套。采用 if 语句嵌套的程序设计方法,也可实现多选操作。

【例 2.15】使用 if 语句嵌套,判断 double 类型变量 x 的值和变量 y 的大小。如果大于,就输出"x 大于 y";如果小于,就输出"x 小于 y";如果 x 等于 y,输出"x 等于 y"。

```
public class If4{
public static void main(String args[]){
double x = 2.3;//定义 double 类型变量 x
double y = 2.3;//定义 double 类型变量 y
System.out.println(" = = = = =比较结果如下 = = = = =");
if(x>y){
System.out.println("x 大于 y!");
}else if(x<y){
System.out.println("x 小于 y!");
}
else{
System.out.println("x 等于 y!");
```

```
    }
    System.out.println("======比较完成=======");
  }
}
```

如图 2-6 所示。

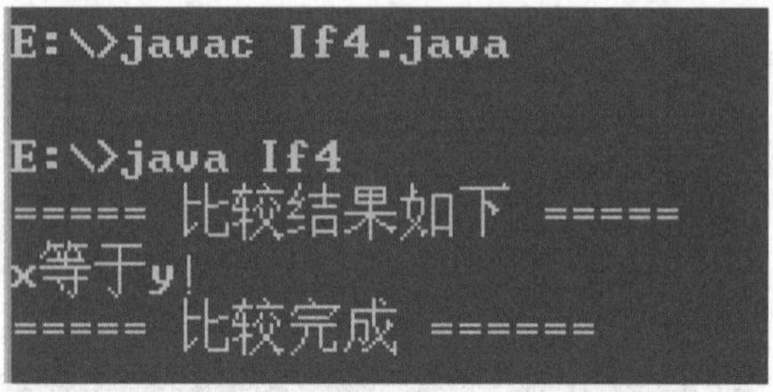

```
E:\>javac If4.java

E:\>java If4
===== 比较结果如下 =====
x等于y!
===== 比较完成 =====
```

图 2-6

2.3.2 switch 语句

switch 多分支语句的结构本质上也是一种 if-else 结构,但 switch 语句的判断条件易于表示,特别是条件有较多选项且比较简单的时候。switch 语句的语法结构如下:

```
switch(expression){
case value1:statement1;break;
case value2:statement2;break;
...
case valueN:statementN;break;
default:defaultstatement;
}
```

执行 switch 语句时,首先计算表达式 expression 的值,其类型必须是整型或字符型,并与各个 case 之后的常量值 value 的类型相同;然后,将表达式 expression 的值,依次与每个 case 对应的常量值 value 进行是否相等判断,若相等,则程序流程转入常量值 value 后对应的语句 statement,对应的语句 statement 执行完毕后,通过 break 语句退出 switch 语句。反之,若表达式 expression 的值与任何一个 case 后的常量值 value 都不相等,就执行 default 后的语句 defaultstatement;若没有 defaultstatement 子句,则什么都不执行。

注意:

(1) 各个 case 后对应的常量值 value 应各不相同。

(2) 通常在每一种 case 情况后都应使用 break 语句,即每一种 case 情况后对应的语句 statement 执行完毕后,都必须使用 break 语句退出 switch,否则会继续向下执行,直到遇到 break 语句或执行完整个 switch 语句。但在 switch 语句最后的 default 对应的语句 statement 的后面可以不加 break。例如:

```
switch(x){
case 1:   System.out.println("1");break;
case 2:   System.out.println("2");break;
default:  System.out.println("不是1或2");
}
```

（3）各个分支的 statement 可以是一条或多条语句，不必使用复合语句，即不必用"{ }"括起来。例如：

```
switch(x){
case 1:   System.out.println("1");System.out.println("这是2种情况中的第一种");break;
case 2:   System.out.println("2");System.out.println("这是2种情况中的第二种");break;
default:  System.out.println("不是1或2");System.out.println("不符合要求的情况");
}
```

（4）不同 case 后的语句 statement 相同时，可以合并多个 case 子句，即只在最后一个 statement 相同的 case 子句中加入 statement。例如：

```
switch(x){
case 1:
case 2:
case 3:   System.out.println("1或2或3");break;
case 4:
case 5:   System.out.println("4或5");break;
default:  System.out.println("不是1、2、3、4、5");
}
```

（5）switch 语句在用表达式的值比较每一个 case 后的常量值 value 时，是从前往后顺序进行的，若各个常量值 value 互不相同，则 case 子句的顺序可任意，但通常将 default 子句放在 switch 结构的最后，即 switch 语句最后执行 default 子句。

【例 2.16】根据整型变量 x 的值判断今天是星期几，判断要求如下：

（1）x＝1，输出"今天是星期一"；
（2）x＝2，输出"今天是星期二"；
（3）x＝3，输出"今天是星期三"；
（4）x＝4，输出"今天是星期四"；
（5）x＝5，输出"今天是星期五"；
（6）x＝6，输出"今天是星期六"；

(7) x=7,输出"今天是星期天";
(8) x 为其他值,输出"不合理的输入"。

```java
public class switch1{
public static void main(String args[]){
int x = 4;
switch(x){
case 1:{
System.out.println("今天是星期一");
break;
}
case 2:{
System.out.println("今天是星期二");
break;
}
case 3:{
System.out.println("今天是星期三");
break;
}
case 4:{
System.out.println("今天是星期四");
break;
}
case 5:{
System.out.println("今天是星期五");
break;
}
case 6:{
System.out.println("今天是星期六");
break;
}
case 7:{
System.out.println("今天是星期天");
break;
}
default:{
System.out.println("不合理的输入");
break;
```

            }
          }
        }
      }

如图 2-7 所示。

图 2-7

【例 2.17】根据 double 类型变量 x 的值，输出 x 对应的等级，等级判断规则如下：
(1) 0~59 分的等级是不及格；
(2) 60~69 分的等级是及格；
(3) 70~79 分的等级是中；
(4) 80~89 分的等级是良；
(5) 90~100 分的等级是优；
(6) 其他值输出"不合理的成绩!"

```java
public class switch2{
public static void main(String args[]){
double x = 83.5;
int y = (int)x/10;//求解成绩 x 的百位和十位对应的整数,作为判断条件
switch(y){
case 0:
case 1:
case 2:
case 3:
case 4:
case 5:{
System.out.println("您的成绩等级是不及格");
break;
}
case 6:{
System.out.println("您的成绩等级是及格");
break;
```

```
}
case 7:{
System.out.println("您的成绩等级是中");
break;
}
case 8:{
System.out.println("您的成绩等级是良");
break;
}
case 9:
case 10:{
System.out.println("您的成绩等级是优");
break;
}
default:{
System.out.println("不合理的输入");
break;
}
}
}
}
```

如图 2-8 所示。

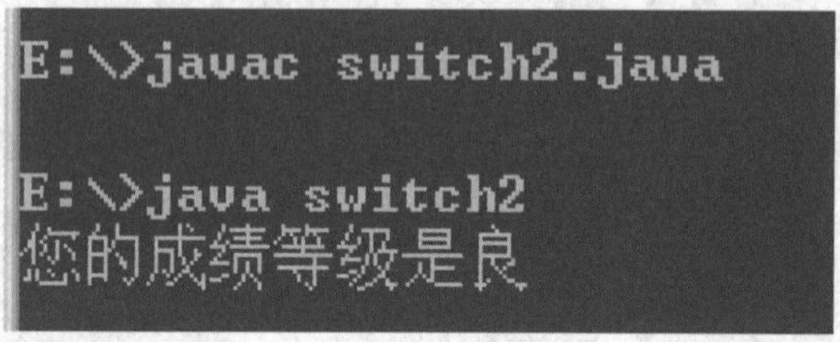

```
E:\>javac switch2.java

E:\>java switch2
您的成绩等级是良
```

图 2-8

任务 实施

2.3.3 使用 if 语句实现综合成绩等级判断

```
public class ZongHeScoreJiBie1{
public static void main(String args[]){
```

```java
double x = 83.5;
if(x<60){
System.out.println("您的成绩等级是不及格");
}else if(x<70){
System.out.println("您的成绩等级是及格");
}else if(x<80){
System.out.println("您的成绩等级是中");
}
else if(x<90){
System.out.println("您的成绩等级是良");
}
else if(x<=100){
System.out.println("您的成绩等级是优");
}
}
}
```

如图 2-9 所示。

```
E:\>javac ZongHeScoreJiBie1.java

E:\>java ZongHeScoreJiBie1
您的成绩等级是良
```

图 2-9

2.3.4 使用 switch 语句实现综合成绩等级判断

```java
public class ZongHeScoreJiBie2{
public static void main(String args[]){
double x = 83.5;
int y = (int)x/10;//求解成绩 x 的百位和十位对应的整数,作为判断条件
switch(y){
case 0:
case 1:
case 2:
case 3:
case 4:
case 5:{
```

```
System.out.println("您的成绩等级是不及格");
break;
}
case 6:{
System.out.println("您的成绩等级是及格");
break;
}
case 7:{
System.out.println("您的成绩等级是中");
break;
}
case 8:{
System.out.println("您的成绩等级是良");
break;
}
case 9:
case 10:{
System.out.println("您的成绩等级是优");
break;
}
default:{
System.out.println("不合理的输入");
break;
}
}
}
}
```

如图 2-10 所示。

```
E:\>javac ZongHeScoreJiBie2.java

E:\>java ZongHeScoreJiBie2
您的成绩等级是良
```

图 2-10

任务 拓展

已知王丽丽、张颖颖、刘华 3 人的银行账户信息如下：

用户名	账号	密码	身份证号码	余额
王丽丽	179708064356	1234	210050619890808185	1000.0
张颖颖	179708064359	4321	210050619891231127	2000.0
刘华	179708064368	4567	410207198904051271	3000.0

现在，王丽丽想提取 3000 元，张颖颖想提取 400 元，刘华想提取 2500 元，请先分别判断其取款要求是否合理，如合理请计算取款后的账户余额。

2.4 任务 4 学生课程的综合成绩等级分布统计

任务描述 及分析

小明需要完成一门课程的综合成绩的等级分布统计，以便用户了解学生的课程知识掌握情况。要完成这个工作任务，我们首先要学习文本扫描类 Scanner 的基础应用知识，以便用其实现从键盘输入每个学生的综合成绩。其次，我们需要掌握循环程序结构（for、while、do-while），然后根据需要选取合适的循环结构，用循环结构程序依次判断每个学生的综合成绩的等级，每判断一次，就将相应等级的人数加 1，以便统计每个等级的人数。

相关 知识

2.4.1 Scanner 类

java.util.Scanner 是 Java 5 的新特性，我们可以通过 Scanner 类获取用户的输入。Scanner 类属于引用数据类型，引用数据类型的使用与定义基本数据类型变量不同，它的变量定义及赋值有自己相对固定的格式：

```
数据类型  变量名  =  new 数据类型();
```

此外，每种引用数据类型都有其不同的功能，我们可以通过该类型的实例（变量）调用相应的功能：

```
变量名.方法名();
```

Scanner 类是引用数据类型的一种，我们可以使用该类来完成从键盘录入数据。Scanner 类的使用步骤如下：

(1) 导入头文件"java.util.Scanner"：

```
import java.util.Scanner;
```

(2) 创建对象实例：

```
Scanner  对象实例名 = new Scanner(System.in);
```

例如：

```
Scanner sc = new Scanner(System.in);
```

(3) 通过对象实例，调用相应的方法：
对象实例名.方法名();
例如：

```
sc.nextInt();//用来接收从键盘录入的整数
int i = sc.nextInt();/*用来接收从键盘录入的整数,并将接收到的整数赋给int型变量i*/
```

注意：
(1) Scanner类有多种不同的方法，用于读取用户在命令行输入的各种类型数据：

```
nextByte()//接收用户输入的byte型整数
nextShort()//接收用户输入的short型整数
nextInt()//接收用户输入的int型整数
nextLong()//接收用户输入的long型整数
nextFloat()//接收用户输入的float型浮点数
nextDouble()//接收用户输入的double型浮点数
nextLine()//接收用户输入的一串字符
next()//接收用户输入的数据
```

(2) 上述方法执行时，都会造成堵塞，即等待用户在命令行输入数据，并按回车键确认。

(3) 数据从键盘录入后，对于每种数据类型都有对应的方法（如 hasNextInt()、hanNextFloat()等），确认输入的数据是哪一种数据类型。例如：用户从键盘输入"12.34"，按回车键后，hasNextFloat()的值是true(输入的是浮点数)，而hasNextInt()的值是false。

【例2.18】 使用nextInt()接收整数数据，并验证接收到的数据是否是整型，如果不是输出提示信息。

```
import java.util.Scanner;
public class Scanner1{
public static void main(String args[]){
Scanner scan = new Scanner(System.in);//从键盘接收数据
```

```
int x = 0;
System.out.print("请输入整数数据:");
if(scan.hasNextInt()){//判断输入的是否是整数
x = scan.nextInt();//接收整数
System.out.println("接收到整数数据:" + x);
}else{
System.out.println("输入的不是整数!");
}
}
}
```

如图 2-11 所示。

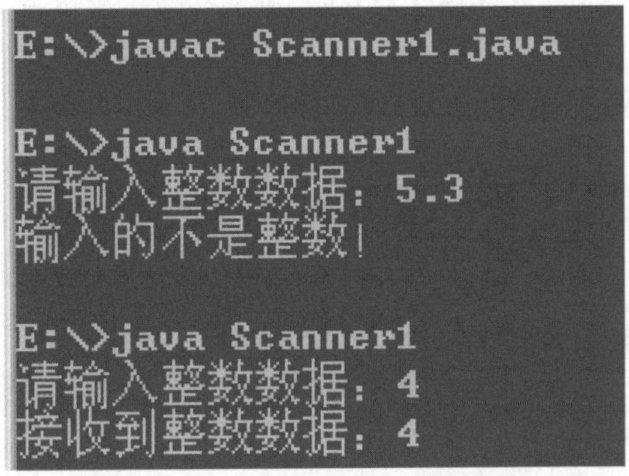

图 2-11

2.4.2 循环语句

循环语句的作用是使某一段程序根据需要重复执行多次。循环语句由循环体和循环条件两部分构成,循环体是要重复执行的语句,循环条件决定循环的开始重复执行以及结束循环。循环语句实现的循环(或称重复)结构是一种封闭结构,当循环条件被满足时,重复执行循环结构内的操作;当循环条件不被满足时,退出循环结构。

循环一般包括 4 个部分:

(1) 初始化部分,用来设置循环的一些初始条件,如累加器清零等。

(2) 循环体部分,重复执行的一段程序,可以是一条语句,也可以是一块语句。

(3) 迭代部分,在当前循环结束、下一次循环开始前执行的语句,其常用的形式为一个计数器值在增减。

(4) 终止部分,一般为布尔表达式,每一次循环都要对该表达式求值,以检查是否满足循环终止条件。

Java 语言提供了 3 种形式的循环语句:while 循环语句、do-while 循环语句和 for 循环语句。

1. while 语句

while 语句的一般格式为：

```
while(expression)
statement
```

Java 执行 while 循环语句时，先检查 expression 表达式（循环条件）的值是否为 true。若为 true，则执行给定语句 statement（即循环体），然后再检查 expression 表达式的值；反复执行上述操作，直到 expression 表达式的值为 false，就退出循环结构。

while 语句的执行是先判断条件，根据条件再决定是否继续执行循环体（简称先判断后执行）。每执行一次循环体后，循环条件均应发生相应的变化，使得执行若干次循环后，循环条件会从 true 变为 false，以便能够结束循环。若执行循环时，循环条件总是 true，则不能终止循环，这种死循环在程序设计中是要注意避免的。

若首次执行 while 语句时循环条件为 false，则循环体一次也未执行，即 while 语句循环体最少执行次数为 0 次。

【例 2.19】使用 while 循环实现 $1 \times 2 \times 3 \times \cdots \times 9$。

```java
public class while1{
public static void main(String args[]){
int i = 1;
int x = 1;//保存累乘的结果
while(i<=9){
x = x * i;//执行累乘操作
i++;
}
System.out.println("1—9 累乘的结果为:" + x);
}
}
```

如图 2-12 所示。

图 2-12

2. do-while 语句

while 语句在执行循环体前，先检查 expression 表达式（循环条件）。但在有些情况下，不管条件表达式的值是 true 还是 false,都希望把循环体至少执行一次,此时,就应使用 do-while 循环。

do-while 循环语句的一般格式为:

```
do
statement
while(expression);
```

do 循环语句首先执行给定的语句 statement(循环体),然后再计算 expression 表达式(判断循环条件)。若表达式值为 false,则结束循环,否则重复执行循环体。

do 语句的循环体至少被执行一次,这是 do 循环与 while 循环最大的区别。一般称 while 循环为"当型循环"(先判断后执行),do 循环为"直到型循环"(先执行后判断)。

【例 2.20】使用 do-while 循环求解 1+2+3+…+50。

```
public class dowhile1{
public static void main(String args[]){
int i = 1;
int sum = 0;//保存累加的结果
do{
sum = sum + i;//执行累加操作
i++;
}while(i<= 50);
System.out.println("1—50 累加的结果为:" + sum);
}
}
```

如图 2-13 所示。

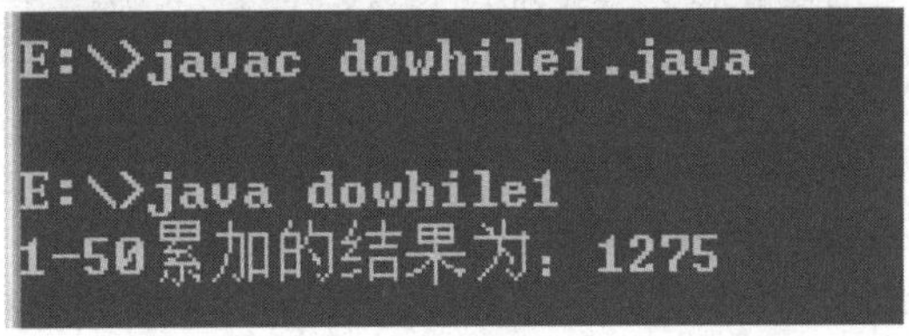

```
E:\>javac dowhile1.java

E:\>java dowhile1
1-50 累加的结果为: 1275
```

图 2-13

3. for 语句

for 循环语句的格式与用法在几种循环语句中最灵活,它的一般格式为:

```
for([expression1];[expression2];[expression3])statement
```

其中,表达式 expression1 指出 for 循环的循环初值;表达式 expression2 是一个关系或逻辑表达式,其值为 false 时循环结束;表达式 expression3 指出每次循环时所进行的计算和更新。3个表达式在使用中可根据需要,部分或全部不写均可。

执行 for 语句时,先计算表达式 expression1(只计算一次,默认时表示无初始内容)。接着检查表达式 expression2 的值是 true 还是 false。若为 false,则不执行语句 statement(循环体),退出循环;若为 true,就执行给定的语句 statement,再计算表达式 expression3。然后,又检查 expression2 表达式的值,再根据值为 true 或为 false 决定是否执行循环体。

可以在 for 循环的表达式 expression1 中说明仅在循环中使用的变量。例如:

```
for(int i=10;i>=0;i--)
sum=sum+i;
```

【例 2.21】使用 for 循环求解 1+2+3+…+50。

```java
public class for1{
public static void main(String args[]){
int sum=0;//保存累加的结果
for(int i=1;i<=50;i++){
sum=sum+i;//执行累加操作
}
System.out.println("1—50 累加的结果为:"+sum);
}
}
```

如图 2-14 所示。

```
E:\>javac for1.java

E:\>java for1
1-50累加的结果为: 1275
```

图 2-14

(1) 省略表达式 expression1。

若 for 语句的初值部分在 for 循环外已全部设置,则 for 语句的相关部分可省略。

【例 2.22】使用 for 循环求解 1+2+3+…+50。

```java
public class for2{
public static void main(String args[]){
int i=1;//初值 i 在循环外已经定义
int sum=0;//保存累加的结果
for( ;i<=50;i++){//省略 expression1
sum=sum+i;//执行累加操作
```

```
    }
    System.out.println("1—50 累加的结果为:" + sum);
    }
}
```

如图 2-15 所示。

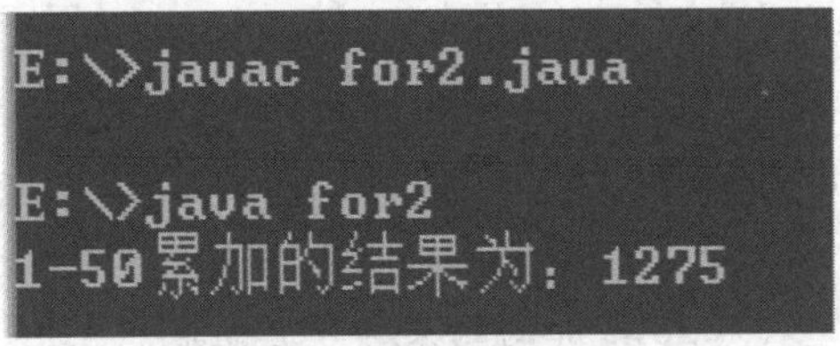

```
E:\>javac for2.java

E:\>java for2
1-50累加的结果为：1275
```

图 2-15

(2) 在表达式 expression1 中包含变量定义等更多的内容。

在 for 语句的初值部分可包含变量定义(仅在 for 循环中使用)、赋值等内容,各项之间用逗号分隔。

【例 2.23】使用 for 循环求解 1+2+3+…+5。

```
public class for3{
public static void main(String args[]){
for(int i=1,sum=0;i<=5;i++){//expression1 中不仅定义了初始值 i
//还包括存放累加结果变量 sum 的定义,二者使用逗号间隔开
sum=sum+i;//执行累加操作
System.out.println("第"+i+"次累加的结果为:"+sum);
//注意 sum 的输出位置发生变化
//因为在 expression1 中定义的 sum,其有效作用范围在 for 循环里
    }
   }
}
```

如图 2-16 所示。

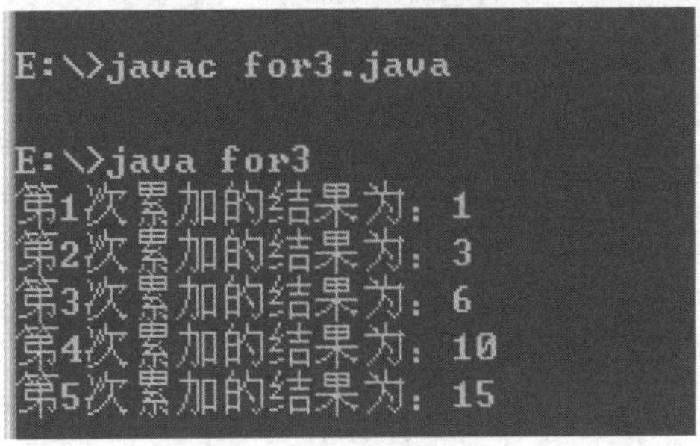

```
E:\>javac for3.java

E:\>java for3
第1次累加的结果为：1
第2次累加的结果为：3
第3次累加的结果为：6
第4次累加的结果为：10
第5次累加的结果为：15
```

图 2-16

（3）省略表达式 expression3。

可将表达式 expression3 写到循环体中，即省略了 expression3 部分。

【例 2.24】使用 for 循环求解 1+2+3+…+50。

```
public class for4{
public static void main(String args[]){
int sum = 0;//保存累加的结果
for(int i = 1;i< = 50;){//省略 expression3
sum = sum + i;//执行累加操作
i + + ;//expression3 写入表达式
}
System.out.println("1—50 累加的结果为:" + sum);
}
}
```

如图 2-17 所示。

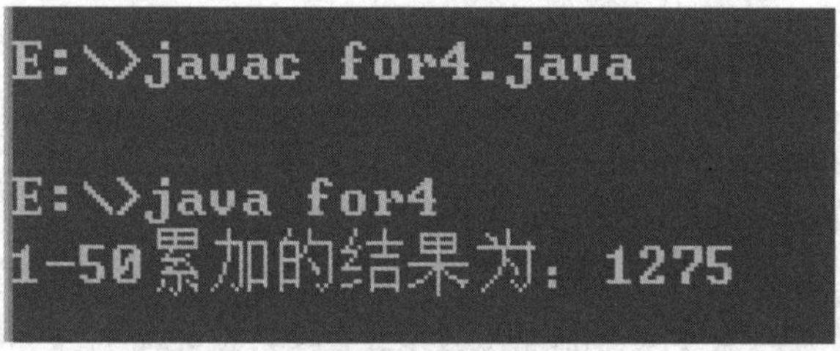

图 2-17

4. 循环嵌套

循环嵌套是指在某个循环语句的循环体中又包含另一个循环语句,也称多重循环。外面的循环语句称为"外层循环",外层循环的循环体中包含的循环称为"内层循环"。

设计循环嵌套结构时,要注意内层循环语句必须完整地包含在外层循环的循环体中,不得出现内外层循环体交叉的情况。Java 语言中的 3 种循环语句都可以组成多重循环。

【例 2.25】打印九九乘法表。

```
public class qiantao{
public static void main(String args[]){
for(int i = 1;i< = 9;i + + ){//外层循环,控制行
for(int j = 1;j< = i;j + + ){//内层循环,控制列
System.out.print(i + " * " + j + " = " + (i * j) + "    ");
}
System.out.println();
```

```
        }
    }
}
```

如图 2-18 所示。

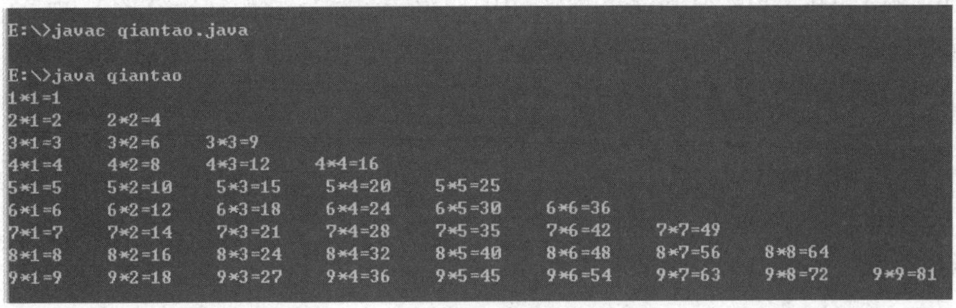

```
E:\>javac qiantao.java

E:\>java qiantao
1*1=1
2*1=2    2*2=4
3*1=3    3*2=6    3*3=9
4*1=4    4*2=8    4*3=12   4*4=16
5*1=5    5*2=10   5*3=15   5*4=20   5*5=25
6*1=6    6*2=12   6*3=18   6*4=24   6*5=30   6*6=36
7*1=7    7*2=14   7*3=21   7*4=28   7*5=35   7*6=42   7*7=49
8*1=8    8*2=16   8*3=24   8*4=32   8*5=40   8*6=48   8*7=56   8*8=64
9*1=9    9*2=18   9*3=27   9*4=36   9*5=45   9*6=54   9*7=63   9*8=72   9*9=81
```

图 2-18

5. break 语句和 continue 语句

(1) break 语句。

break 语句可以强迫程序中断循环。当程序执行到 break 语句时,就会离开循环,继续执行循环外的下一个语句。如果 break 语句出现在嵌套循环中的内层循环,则 break 语句只会跳出当前层的循环。

【例 2.26】输出 1~10 之间的整数,但是当遇到能被 3 整除的整数时,停止输出,退出循环。

```
public class break1{
public static void main(String args[]){
for(int i=1;i<=10;i++){
if(i%3==0){
break;
}else{
System.out.print(i+"    ");
}
}
}
}
```

如图 2-19 所示。

```
E:\>javac break1.java

E:\>java break1
1    2
```

图 2-19

（2）continue 语句。

continue 语句只能用在循环中，它的功能是使得程序跳过循环体中 continue 语句后剩下的部分（即短路），用于终止当前这一轮循环的执行。continue 语句的格式如下：

```
continue;
```

continue 语句在 while 或 do-while 语句中使流程直接跳到循环条件的判断上；在 for 语句中则直接计算表达式 3 的值，再根据表达式 2 的值是 true 或 false，决定是否继续循环。

【例 2.27】 跳过整数 5，循环输出 1~9 的所有整数。

```
public class continue1{
public static void main(String args[]){
for(int i=1;i<10;i++){
if(i==5){
continue;
}
System.out.println("i=" + i);
}
}
}
```

如图 2-20 所示。

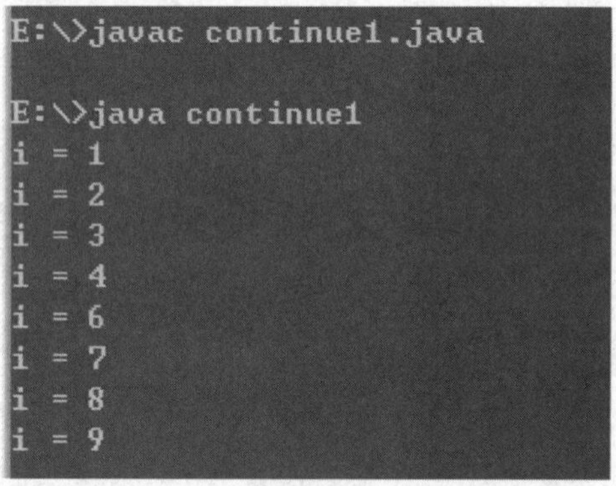

```
E:\>javac continue1.java

E:\>java continue1
i = 1
i = 2
i = 3
i = 4
i = 6
i = 7
i = 8
i = 9
```

图 2-20

任务 实施

2.4.3 综合成绩等级分布统计实现

```
import java.util.Scanner;
```

```java
public class ZongHeScoreJiBieCount{
public static void main(String args[]){
//变量a、b、c、d、e用于存放优、良、中、及格、不及格的人数
//变量f用于存放总人数
double a = 0,b = 0,c = 0,d = 0,e = 0,f = 0;
//变量x用于依次接收每个学生的综合成绩,当x = -1时停止输入成绩
double x = 0;
System.out.println("请输入第" + (f + 1) + "个学生的成绩:");
while(x! = -1){
Scanner scan = new Scanner(System.in);//从键盘接收数据
x = scan.nextDouble();
f = f + 1;
if(x<60){
e = e + 1;
}else if(x<70){
d = d + 1;
}else if(x<80){
c = c + 1;
}
else if(x<90){
b = b + 1;
}
else if(x< = 100){
a = a + 1;
}
System.out.println("请输入第" + (f + 1) + "个学生的成绩:");
}
System.out.println("学生总人数为" + f + "人");
System.out.println("成绩为不及格的人数为" + e + "人,比率为:" + (e/f * 100.0) + "%");
System.out.println("成绩为及格的人数为" + d + "人,比率为:" + (d/f * 100.0) + "%");
System.out.println("成绩为中的人数为" + c + "人,比率为:" + (c/f * 100.0) + "%");
System.out.println("成绩为良的人数为" + b + "人,比率为:" + (b/f * 100.0) + "%");
System.out.println("成绩为优的人数为" + a + "人,比率为:" + (a/f * 100.0) + "%");
```

```
    }
}
```

如图 2-21 所示。

```
E:\>java ZongHeScoreJiBieCount
请输入第1.0个学生的成绩:
45
请输入第2.0个学生的成绩:
76
请输入第3.0个学生的成绩:
87
请输入第4.0个学生的成绩:
62
请输入第5.0个学生的成绩:
97
请输入第6.0个学生的成绩:
54
请输入第7.0个学生的成绩:
77
请输入第8.0个学生的成绩:
-1
请输入第9.0个学生的成绩:
学生总人数为8.0人
成绩为不及格的人数为3.0人，比率为：37.5%
成绩为及格的人数为1.0人，比率为：12.5%
成绩为中的人数为2.0人，比率为：25.0%
成绩为良的人数为1.0人，比率为：12.5%
成绩为优的人数为1.0人，比率为：12.5%
```

图 2-21

任务 拓展

1. 实现银行自助服务系统菜单的切换功能，当输入 0，结束循环操作，效果如图所示：

```
===========================欢迎使用自动银行服务===========================
1. 开户  2. 存款  3. 取款  4. 转账  5. 查询余额    6. 修改密码    0. 退出
========================================================================
请输入选择：2
====进入银行存款操作====
===========================欢迎使用自动银行服务===========================
1. 开户  2. 存款  3. 取款  4. 转账  5. 查询余额    6. 修改密码    0. 退出
========================================================================
请输入选择：3
====进入银行取款操作====
===========================欢迎使用自动银行服务===========================
1. 开户  2. 存款  3. 取款  4. 转账  5. 查询余额    6. 修改密码    0. 退出
========================================================================
请输入选择：
```

2. 实现银行自助服务系统添加多个账号信息的方法，选择"y"继续添加账户信息。实现效果如图所示：

```
====进入银行开户操作====
请输入用户的姓名：
zhangsan
请输入用户的密码：
1234
请输入用户的身份证号码：
213123
请输入用户的开户金额：
233
开户成功！！！你的账户信息如下：
用户名        账号              密码      身份证号码    余额
zhangsan    179708065426    1234     213123      233.0
是否继续开户？y/n
```

2.5 任务5 学生课程的综合成绩的保存、排序（数组）

任务描述 及分析

小明需要实现一门课程的综合成绩的保存并按升序排序后输出。根据之前的学习内容，如果要完成这个工作任务，我们可以定义多个 double 类型变量（变量可以保存数据），用于存放综合成绩，然后使用循环结构依次比较判断，按升序确定顺序，再依次输出显示。但这种方法非常烦琐，若有 100 个学生，就要定义 100 个 double 类型的变量等。我们可以通过定义 float 类型的数组来解决这个问题。此外，Java 语言还提供了一些数组操作的常用方法，如用于数组的排序 sort() 方法，可轻松实现数组数据的排序。因此，我们首先需要学习数组的定义和引用，其次需要掌握数组操作的常用方法，最后我们根据任务要求，定义数组并利用循环程序依次输入综合成绩，通过 sort() 方法实现升序排序，然后使用循环依次输出排序后的数据。

相关 知识

数组是一种最简单的复合数据类型，是一组有相同数据类型的有序数据的集合。数组中的一个数据成员称为数组元素，数组元素可以用一个统一的数组名和下标序号来唯一确定。根据数组下标是一个还是多个，数组可分为一维数组和多维数组。

2.5.1 数组

1. 一维数组

一维数组中的各个元素排成一行，通过数组名和一个下标就能访问一维数组中的元素。

（1）一维数组的定义。

数组的定义包括数组声明和为数组分配空间等内容。

① 一维数组的声明。

声明一个一维数组的一般形式为：

```
type arrayName[];
```

或

```
type[]arrayName;
```

其中，类型 type 可以是 Java 中任意的基本数据类型或引用类型，数组名 arrayName 是一个合法的标识符，[]指明该变量是一个数组变量。

例如：

```
int intArray[];(或 int[]intArray;)//声明一个整型数组
double decArray[];(或 double[]decArray;)//声明一个双精度实型数组
String strArray[];(或 String[]strArray;)//声明一个字符串数组
Button btn[];(或 Button[]btn);//声明一个按钮数组
```

一个数组声明语句可同时声明多个数组变量。此时，后一种声明格式写起来简单些。例如：

```
int[]a,b,c;
```

相当于

```
int a[],b[],c[];
```

与其他高级语言不同，Java 在数组声明时并不为数组分配存储空间，因此在声明数组的[]中，不能指出数组中元素的个数（数组长度），而且上例中声明的数组是不能访问它的任何元素的，必须经过初始化、分配存储空间创建数组后，才能访问数组的元素。当仅有数组声明而未分配存储空间时，数组变量中只是一个值为 null 的空引用（指针）。

② 一维数组的空间分配。

为数组分配空间有两种方法：数组初始化和使用 new 运算符。为数组分配空间后，数组变量中存储的是数组存储空间的引用地址。

- 数组初始化。

数组初始化是指在声明数组的同时指定数组元素的初始值，一维数组初始化的形式如下：

```
type arrayName[] = {element1[,element2…]}
```

其中 element1、element2、…为 type 类型数组 arrayName 的初始值，基本类型和字符串类型等可以用这种方式创建数组空间。

例如：

```
int intArray[ ] = {1,2,3,4,5};
double decArray[ ] = {1.1,2.2,3.3};
String strArray[ ] = {"Java","BASIC","FORTRAN"};
```

从上述例子可以看出，一维数组的初始化即在前面数组声明的基础上在大括号中给出数组元素的初值，系统将自动按照所给初值的个数计算出数组的长度并分配相应的存储空间。

- 使用 new 运算符。

通过使用 new 运算符可为数组分配存储空间和指定初值。若数组已经声明，为已声明数组分配空间的一般形式如下：

```
arrayName = new type[arraySize];
```

其中 arrayName 是已声明的数组变量，type 是数组元素的类型，arraySize 是数组的长度，可以为整型常量或变量。即通过数组运算符 new 为数组 arrayName 分配 arraySize 个 type 类型大小的空间。

若数组未声明，则可在数组声明的同时用 new 运算符为数组分配空间：

```
type arrayName[ ] = new type[arraySize];
```

例如：

```
int a[ ];
a = new int[10];//给数组 a 分配 10 个整型数据空间
double b[ ] = new double[5];//给数组 b 分配 5 个双精度实型数据空间
String s[ ] = new String[2];//给数组 s 分配 2 个元素的引用空间
```

一旦数组初始化或用 new 分配空间以后，数组的长度即固定下来，不能变化，除非用 new 运算符重新分配空间。

在 Java 语言中，用 new 运算符为数组分配空间是动态的，即可根据需要随时用 new 为已分配空间的数组再重新分配空间。但须注意，对一个数组再次动态分配空间时，若该数组的存储空间的引用没有另外的存储，则该数组的数据将会丢失。例如：

```
int a[ ] = {1,2,3};
a = new int[5];//为 a 数组重新分配空间，原 a 数组的值 1,2,3 将丢失
```

用 new 进行数组的动态空间分配时，若未指定初值，则使用各类数据的默认初值，即对数值类型其默认初值是相应类型的 0，对字符型默认初值为'\u0000'，对布尔型默认初值为 false，对复合数据类型默认初值为 null。

经过上述操作，就完成了基本类型的数组和已经初始化的字符串数组的定义，接着就可以访问数组或存取数组元素了。但对复合类型的数组还要对数组元素分配空间、初始化。

③ 复合类型数组元素的动态空间分配和初始化。

一般情况下,复合类型的数组需要进一步对数组元素用 new 运算符进行空间分配并初始化操作。假设已声明一个复合类型的数组:

```
type arrayName[];//type 是一个复合数据类型
```

对数组 arrayName 的动态空间分配步骤如下:
a. 为数组分配每个元素的引用空间。

```
arrayName = new type[arraySize];
```

b. 为每个数组元素分配空间。

```
arrayName[0] = new type(paramList);
… …
arrayName[arraySize-1] = new type(paramList);
```

其中,paramList 参数表用于数组元素初值的指定。

例如,下面是一个图形界面应用程序中所用按钮数组的定义:

```
Button btn[];//声明一个 Button 按钮类型的数组 btn
btn = new Button[2];//给数组 btn 分配 2 个元素的引用空间
btn[0] = new Button("确定");//为 btn[0]分配空间并赋显示文本"确定"
btn[1] = new Button("退出");//为 btn[1]分配空间并赋显示文本"退出"
```

当然,在比较简单的情况下,上述操作可简化为:

```
Button btn[] = {new Button("确定"),new Button("退出")};
```

(2) 一维数组的引用。

一维数组的引用分为数组元素的引用和数组的引用,大部分时候都是数组元素的引用。一维数组元素的引用方式为:

```
arrayName[index]
```

其中,index 为数组下标,是 int 类型的量,也可以是 byte、short、char 等类型,但不允许为 long 类型;下标的取值从 0 开始,直到数组的长度减 1;一维数组元素的引用与同类型的变量相同,每一个数组元素都可以用在同类型变量被使用的地方。对前面建立的数组变量 intArray,有 5 个数组元素,通过使用不同的下标来引用不同的数组元素 intArray[0]、intArray[1]、…、intArray[4]。

Java 对数组元素要进行越界检查以保证安全性。若数组元素下标小于 0、大于或等于数组长度将产生 ArrayIndexOutOfBoundsException 异常。

Java 语言对于每个数组都有一个指明数组长度的属性 length,它与数组的类型无关。

例如 a.length 指明数组 a 的长度。

对一维数组元素的逐个处理，一般用循环结构的程序。

【例 2.28】将整型数组 a 的 5 个整型数据加 10 后，依次输出显示。

```java
public class Array1{
public static void main(String args[]){
int a[] = {87,65,74,66,82};//声明数组
System.out.print("加 10 之前的数据为:");
for(int i = 0;i<a.length;i++){
System.out.print("a["+i+"] = "+a[i]+"  ");
}
System.out.print("\n");
System.out.print("加 10 之后的数据为:");
for(int i = 0;i<5;i++){
a[i] = a[i]+10;
}

for(int i = 0;i<5;i++){
System.out.print("a["+i+"] = "+a[i]+"  ");
}
}
}
```

如图 2-22 所示。

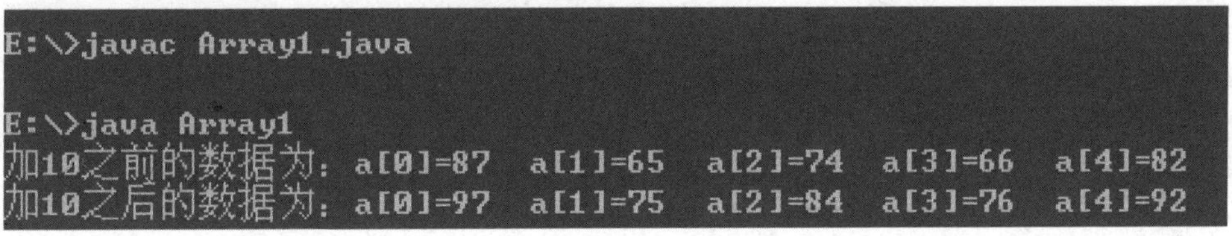

```
E:\>javac Array1.java

E:\>java Array1
加10之前的数据为: a[0]=87  a[1]=65  a[2]=74  a[3]=66  a[4]=82
加10之后的数据为: a[0]=97  a[1]=75  a[2]=84  a[3]=76  a[4]=92
```

图 2-22

2. 多维数组

Java 也支持多维数组。在 Java 语言中多维数组被看作数组的数组，例如二维数组可看作一个特殊的一维数组，其中每个元素又是一个一维数组。下面的讨论主要针对二维数组，更高维数的数组情况类似于二维数组。使用二维数组可方便地处理表格形式的数据。

(1) 二维数组的定义。

二维数组的定义与一维数组类似，包括数组声明、为数组和数组元素分配空间、初始化等内容。

① 二维数组的声明。

声明二维数组的一般形式为：

type arrayName[ ][ ];

或

type[ ][ ]arrayName;

或

type[ ]arrayName[ ];

其中，type 是数组的类型，可以是简单类型，也可以是复合类型。
例如：

char c[ ][ ];//声明一个二维 char 类型的数组 c
float f[ ][ ];//声明一个二维 float 类型的数组 f

与一维数组时的情况一样，对数组的声明不分配数组的存储空间。
② 二维数组的空间分配。
• 二维数组的初始化。
二维数组的初始化也是在声明数组的同时就为数组元素指定初值。例如：

int intArray[ ][ ] = {{1,2},{3,4},{5,6,7}};

Java 系统将根据初始化时给出的初始值的个数自动计算出数组每一维的大小。在这个例子中，二维数组 intArray 由 3 个一维数组组成，这 3 个一维数组的元素个数分别为 2、2、3。在 Java 语言中，由于把二维数组看作数组的数组，数组空间不一定连续分配，所以不要求二维数组每一维的大小相同。

【例 2.29】使用二维数组存放如下所示的"＊"字符，并显示到屏幕。

＊
＊ ＊
＊ ＊ ＊
＊ ＊ ＊ ＊

```
public class Array2{
public static void main(String args[ ]){
char score[ ][ ] = {
{'*'},{'*','*'},{'*','*','*'},{'*','*','*','*'}
};//静态初始化完成，每行的数组元素个数不一样
for(int i = 0;i<score.length;i++){
for(int j = 0;j<score[i].length;j++){
System.out.print(score[i][j]);
```

```
            }
            System.out.print("\n");
        }
    }
}
```

如图 2-23 所示。

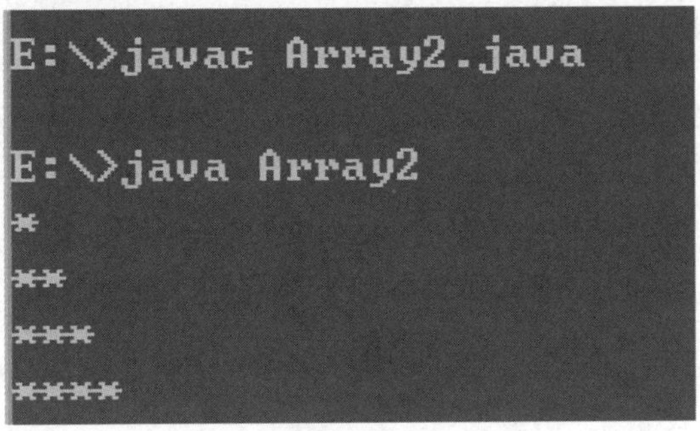

```
E:\>javac Array2.java

E:\>java Array2
*
**
***
****
```

图 2-23

- 使用 new 运算符。

对二维数组,用 new 运算符分配空间有两种方法。一种方法是直接为二维数组的每一维分配空间。若数组已经声明,为已声明数组分配空间的一般形式如下:

```
arrayName = new type[arraySize1][arraySize2];
```

其中,arrayName 是已声明的数组名,type 是数组元素的类型,arraySize1 和 arraySize2 分别是数组第一维和第二维的长度,可以为整型常量或变量。通过数组运算符 new 为数组 arrayName 分配 arraySize1×arraySize2 个 type 类型大小的空间。

若数组未声明,则可在数组声明的同时用 new 运算符为数组分配空间:

```
type arrayName[ ][ ] = new type[arraySize1][arraySize2];
```

例如:

```
int a[ ][ ];
a = new int[3][4];//给数组 a 分配 12 个整型数据空间
double b[ ][ ] = new double[2][5];//给数组 b 分配 10 个双精度实型数据空间
String s[ ][ ] = new String[2][2];//给数组 s 分配 4 个 String 元素的引用空间
```

另一种方法是从最高维开始,分别为每一维分配空间,格式为:

```
arrayName = new type[arrayLength1][ ];
arrayName[0] = new type[arrayLength20];
```

```
arrayName[1] = new type[arrayLength21];
… …
arrayName[arrayLength1-1] = new type[arrayLength2n];
```

例如:

```
int a[][] = new int[2][];//定义二维数组 a 由两个一维数组构成
a[0] = new int[3];//二维数组 a 的第 1 个一维数组有 3 个元素
a[1] = new int[5];//二维数组 a 的第 2 个一维数组有 5 个元素
```

在 Java 语言中,必须首先为最高维分配引用空间,然后再顺次为低维分配空间。

注意:

在用 new 进行二维数组动态空间分配时可以先只确定第一维的大小,其余维的大小可以在以后分配。在 Java 语言中,对二维数组不允许有如下的形式:

```
int a[][] = new int[][2];
```

- 复合类型数组元素的动态空间分配和初始化。

与一维数组相同,对于复合类型的数组,要为每个数组元素单独分配空间。

例如:

```
String s[][] = new String[2][];
s[0] = new String[2];
s[1] = new String[2];
s[0][0] = new String("Java");
s[0][1] = new String("Program");
s[1][0] = new String("Applet");
s[1][1] = new String("Application");
```

在上述二维数组的定义中,数组的初始化方式和 new 方式是分别使用的,实际上,还经常见到这两种方式的混合使用方式。例如:

```
int a[][] = {new int[2], new int[3], new int[4]};
```

即二维数组 a 由 3 个一维数组组成,初始化时 new 的个数即一维数组个数,而每个一维数组的数据个数与存储空间用 new 运算符动态分配。

(2) 二维数组的引用。

大多数情况是引用二维数组的元素。对二维数组中的每个元素,引用方式为:

```
arrayName[index1][index2]
```

其中,index1 和 index2 为下标,可用类型同一维数组,如 c[2][3]等。同样,每一维的下标都从 0 开始。对二维数组元素的逐个处理，一般用嵌套循环结构的程序。

【例 2.30】 从键盘依次给 3 行 2 列的二维整型数组赋值，然后依次显示数组内容。

```java
import java.util.Scanner;
public class Array3{
public static void main(String args[]){
Scanner scan = new Scanner(System.in);
int score[][] = new int[3][2];//声明并实例化二维数组
System.out.print("开始输入数据:\n");
for(int i = 0;i<3;i++){
for(int j = 0;j<2;j++){
score[i][j] = scan.nextInt();
}
}
System.out.print("输入二维数组的数据为:\n");
for(int i = 0;i<3;i++){
for(int j = 0;j<2;j++){
System.out.print("score[" + i + "]" + "[" + j + "] = " + score[i][j] + "   ");
}
System.out.print("\n");
}
}
}
```

如图 2-24 所示。

图 2-24

2.5.2 数组的常用方法

在 Java 语言中提供了一些对数组进行操作的类和方法，掌握它们的用法，可方便数组程序的设计。

1. 类 System 的静态方法 arraycopy( )

系统类 System 的静态方法 arraycopy( )可用来进行数组复制,其格式和功能如下:

```
public static void arraycopy(Object src,int src_position,Object dst,int dst_position,int length)
```

从源数组 src 的 src_position 处,复制到目标数组 dst 的 dst_position 处,复制长度为 length。

【例 2.31】{11,22,33,44,55,66}是整形数组 array1 存放的内容,使用 arraycopy( )方法将 array1 从 22 处开始的 3 个整数复制到整型数组 array2,并输出显示整型数组 array2 的内容。

```
public class arraycopy1{
public static void main(String args[]){
int array1[] = {11,22,33,44,55,66};
int array2[] = {0,0,0};
System.arraycopy(array1,1,array2,0,3);
System.out.println("复制完成后数组 array2 的内容为:");
for(int i = 0;i<array2.length;i + + )
System.out.print(array2[i] + "");
}
}
```

如图 2 - 25 所示。

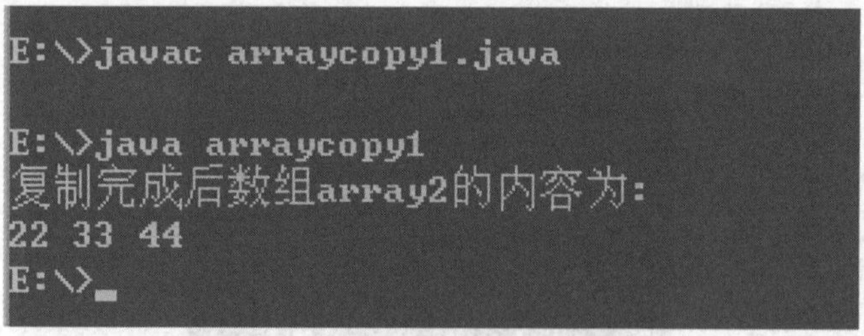

```
E:\>javac arraycopy1.java

E:\>java arraycopy1
复制完成后数组array2的内容为:
22 33 44
E:\>
```

图 2 - 25

2. 类 Arrays 中的方法

java.util.Arrays 类中提供了对数组的排序 sort( )等静态方法,sort( )方法有重载,以适应对不同类型数组 a 的递增排序,其格式如下:

```
void sort(Object[]a)
```

【例 2.32】 使用 sort( )方法对存放成绩的 double 型数组 score 进行排序,score 数组的内容为{54.2,37.8,87.6,76.5,99.4}。

```java
import java.util.Arrays;
public class sort1{
    public static void main(String args[]){
        double score[] = {54.2,37.8,87.6,76.5,99.4};
        Arrays.sort(score);
        System.out.println("sort排序后的score数组内容如下:");
        for(int i = 0;i<score.length;i++)
            System.out.print(score[i]+"  ");
    }
}
```

如图 2-26 所示。

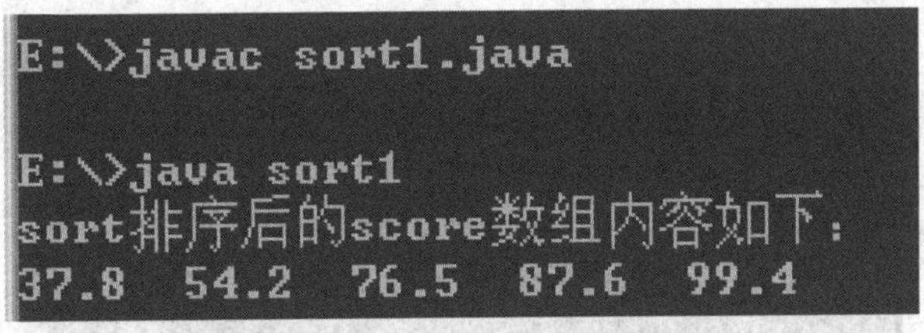

```
E:\>javac sort1.java

E:\>java sort1
sort排序后的score数组内容如下:
37.8  54.2  76.5  87.6  99.4
```

图 2-26

sort( )方法的另一类重载为数组的部分元素的递增排序,其格式为:

```
void sort(Object[ ]a,int fromIndex,int toIndex)
```

排序范围为 fromIndex 至 toIndex-1。例如:

```
int a[ ] = {8,6,7,3,5,4};
Arrays.sort(a,2,5);//a数组元素的顺序为 8 6 3 5 7 4
```

任务 实施

2.5.3 综合成绩的保存、排序实现

```java
import java.util.Scanner;
import java.util.Arrays;
public class ZongHeScoreSave{
```

```java
public static void main(String args[]){
//定义 double 类型数组存放学生成绩
//假设总共 5 个学生
double[ ]scores = new double[5];
Scanner scan = new Scanner(System.in);//从键盘接收数据
for(int i = 0;i<scores.length;i + + ){//依次输入学生的综合成绩并保存
    System.out.println("请输入第" + (i + 1) + "个学生的成绩:");
    scores[i] = scan.nextDouble();
}
Arrays.sort(scores);//对数组进行升序排序
System.out.println("学生综合成绩升序排序后的结果如下:");
for(int i = 0;i<scores.length;i + + ){//依次输出排序后的学生成绩
    System.out.println("升序排序后,第" + (i + 1) + "个学生的成绩为:" + scores[i]
 + "   ");
}
}
}
```

如图 2-27 所示。

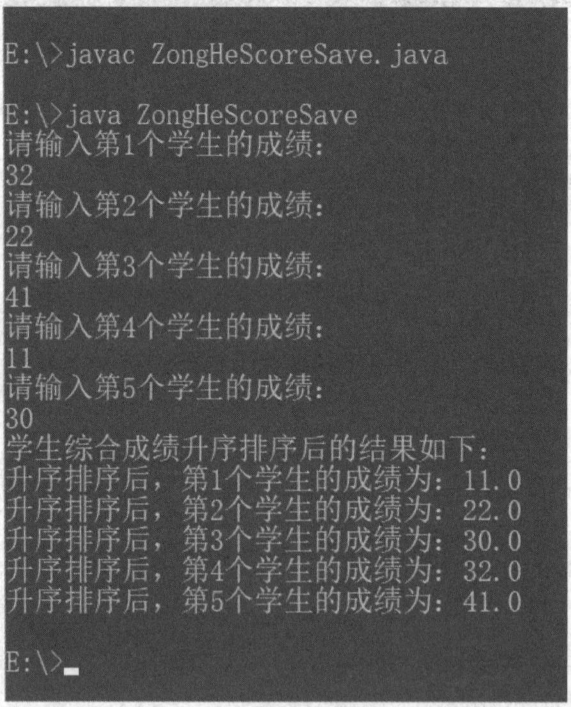

图 2-27

任务 拓展

1. 实现自助银行服务系统初始化 3 个用户信息的操作,且将用户信息显示出来,效果如图所示:

```
用户名    账号              密码      身份证号码              余额
王丽丽    179708064356     1234     210050619890808185     1000.0
张颖颖    179708064359     4321     210050619891231127     2000.0
刘华      179708064368     4567     410207198904051271     3000.0
```

2. 实现自助银行服务系统验证取款用户是否存在的操作,实现效果如图所示:

```
==========================欢迎使用自动银行服务==========================
1. 开户  2. 存款  3. 取款  4. 转账  5. 查询余额     6. 修改密码      0. 退出
======================================================================
请输入选择：3
====进入银行取款操作====
请输入取款的账号：
21321414
请输入账号的密码：
1234
取款的账户或密码不正确！！请检查

==========================欢迎使用自动银行服务==========================
1. 开户  2. 存款  3. 取款  4. 转账  5. 查询余额     6. 修改密码      0. 退出
======================================================================
请输入选择：3
====进入银行取款操作====
请输入取款的账号：
179708064368
请输入账号的密码：
4567
请输入你的取款金额：
300
取款成功！！你的账户余额为：2700.0
```

2.6 习 题

一、选择题

1. x 和 y 为 int 类型,值分别是 17 和 5,分别计算 x/y 和 x%y 的值为(　　)和(　　)。(选择两项)
   A. 3　　　　　　　B. 2　　　　　　　C. 1　　　　　　　D. 2

2. 以下哪些变量名是合法的?(　　)(选择两项)
   A. 3D　　　　　　B. sum　　　　　　C. name　　　　　D. int

3. 表达式(11＋3＊8)/4%3 的值是(　　)。

A. 31　　　　　　B. 0　　　　　　C. 1　　　　　　D. 2

4. 下列代码输出结果正确的是(　　)。

```
double b = 75.6;
b++;
int c = b/2;
```

A. 42　　　　　　　　　　　　　　　B. 编译错误,更改为 int c＝(int)b/2;
C. 43　　　　　　　　　　　　　　　D. 编译错误,更改为 int c＝int(b)/2;

5. Java 中关于 if 选择结构描述错误的是(　　)。

A. if 选择结构是根据条件判断之后再作处理的一种语法结构
B. 关键字 if 后小括号里必须是一个条件表达式,表达式的值必须是布尔类型
C. if 后小括号里表达式的值为 false 时,程序需要执行大括号里的语句
D. if 语句可以和 else 一起使用

6. 有 else if 的选择结构是(　　)。

A. 基本 if 选择结构　　　　　　　　　B. if-else 选择结构
C. 多重 if 选择结构　　　　　　　　　D. switch 选择结构

7. 下面代码的运行结果是(　　)。

```
public class Weather{
    public static void main(String[]args){
        int shiDu = 45;//湿度
        if(shiDu >= 80){
            System.out.println("要下雨");
        }else if(shiDu >= 50){
            System.out.println("天很阴");
        }else if(shiDu >= 30){
            System.out.println("很舒适");
        }else if(shiDu >= 0){
            System.out.println("很干燥");
        }
    }
}
```

A. 要下雨　　　　　B. 天很阴　　　　　C. 很舒适　　　　　D. 很干燥

8. 以下说法正确的是(　　)。

A. 如果 while 循环结构的循环条件语句始终为 true,则一定会出现死循环
B. 程序执行时加入断点会改变程序的执行流程
C. do-while 循环结构的循环体至少无条件执行一次
D. while 循环结构的循环体有可能一次都不执行

9. 对以下代码,下面说法正确的是(　　)。

```
int k = 10;
while(k = = 0){
    k = k - 1
}
```

  A. 循环将执行 10 次   B. 死循环,将一直执行下去
  C. 循环将执行一次   D. 循环一次也不执行

10. 定义一个数组 String[]cities={"北京","天津","上海","重庆","武汉","广州","香港"},数组中的 cities[6]指的是( )。

  A. 北京   B. 广州   C. 香港   D. 数组越界

11. 下列数组初始化正确的是( )。(选择两项)

  A. int score={90,12,34,77,56}
  B. int[]score=new int[5]
  C. int[]score=new int[5]{90,12,34,77,56}
  D. int score[]=new int[]{90,12,34,77,56}

12. 以下代码输出结果是( )。

```
public class Test{
    public static void main(String[]args){
        double[]price = new double[5];
        price[0] = 98.10;
        price[1] = 32.18;
        price[0] = 77.74;
        for(int i = 0;i<5;i + + ){
            System.out.print((int)price[i] + "");
        }
    }
}
```

  A. 98 32 77 0 0   B. 98 32 78 0 0
  C. 77 32 0 0 0   D. 编译出错

13. 阅读下面代码,它完成的功能是( )。

```
String[]a = {"我们","你好","小河边","我们","读书","自习"};
for(int i = 0;i<a.length;i + + ){
    if(a[i].equals("我们")){
        a[i] = "他们";
    }
}
```

  A. 查找   B. 查找并替换   C. 增加   D. 删除

14. 下面代码的运行结果是(　　)。

```java
public class Example{
    public static void main(String[]args){
        int[]a = new int[3];
        int[]b = new int[]{1,2,3,4,5};
        a = b;
        for(int i = 0;i<a.length;i++){
            System.out.print(a[i]+"  ");
        }
    }
}
```

A. 程序报错　　　B. 1 2 3　　　C. 1 2 3 4 5　　　D. 0 0 0

二、简答题

1. 简述 Java 中变量的命名规则。

2. 小强左右两手分别拿一个带有编号的桌球,分别是黑 8、15 号,现在交换手中的球,但是手不够大。用程序模拟这一过程:两个整数分别保存在两个变量中,将这两变量的值互换,并输出互换后的结果,程序运行如下:

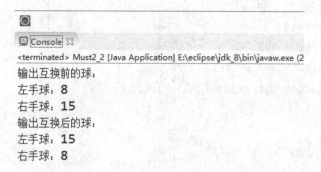

提示:互换两个变量的值需要借助第三个变量,前两个变量用来存储两个整数,第三个变量作为中间变量,借助这一中间变量,将两个变量的值进行互换。

3. 从键盘上输入圆的半径,输出圆的面积和周长。运行效果如下图所示:

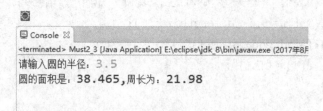

4. 银行提供了整存整取定期储蓄项目,存期为 1 年、2 年、3 年、5 年,到期支取本息,年

利率如下表：

存期	年利率(%)	存期	年利率(%)
1年	2.25	3年	3.24
2年	2.7	5年	3.6

编写一程序，输入存入金额，计算存期1年、2年、3年或5年取款时银行支付的本息是多少。运行效果如下图所示：

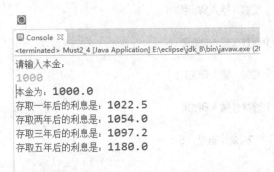

提示：利息＝本金×年利率×存期，本息＝本金＋利息。

5. 说明什么情况下可以使用switch选择结构代替多重if选择结构。

6. 编程实现从键盘上输入一个整数，判断是否能被3或5整除。如果能则输出"该数是3或5的倍数"，否则输出"该数不是3或5的倍数"。

7. 使用switch结构实现：制定学习计划"星期一、星期三、星期五学习编程，星期二、星期四、星期六学习英语，星期日休息"。

8. 某公司为纪念英雄联盟游戏7周年，进行英雄皮肤打折活动，具体如下：

- 星期一：大魔王 提莫 9折
- 星期二：冬季仙境 卡尔玛 8折
- 星期三：虎痴之拳蔚 8折
- 星期四：冰雪节 格雷福斯 7.5折
- 星期五：圣诞老人 布隆 5折
- 星期六：张辽 文远 1折
- 星期日：司马懿 仲达 2折

编写一个程序，输入今天是星期几(1~7)，输出今天的特价皮肤是什么。运行效果如下图所示：

9. 从键盘上接收一批整数，比较并输出其中的最大值和最小值，输入 0 时结束循环。运行效果如下图所示：

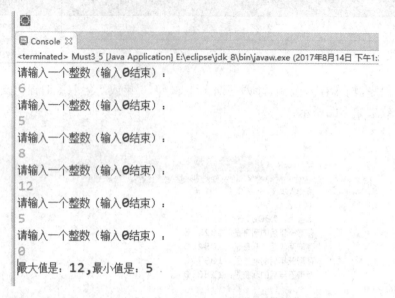

提示：声明两变量 max 和 min，分别记录最大值和最小值。将用户输入的数字 num 和上面的两个变量比较，使得 max 始终保持当前输入的最大值，min 始终保持当前的最小值。

10. 依次输入 5 句话，然后将它们逆序输出，运行结果如下图所示：

提示：创建一个字符串数组，每句话作为字符串数组的一个元素，然后从该数组的末尾开始循环输出。

11. 某百货商场当日消费积分最高的 8 名顾客，他们的积分分别是 18,25,7,13,2,89,12,66，编写程序找出最低的积分及他在数组中的原始位置（下标）。

提示：创建数组，存储 8 名顾客的积分，定义变量 min 保存数组最低积分，定义变量 index 存储最低积分的下标。假设第一个元素为最低积分，下标为 0，遍历数组，将数组元素和 min 的值进行比较。

12. 从键盘上输入 10 个整数，合法值为 1，2，3，不是这 3 个数的则为非法数字。试编程统计每个整数和非法数字的个数。程序运行结果如下图所示：

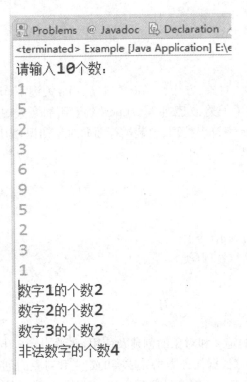

提示：创建数组，长度为 10，存储用户输入的数字；创建数组，长度为 4，存储 3 个合法数字及非法数字的个数。循环输入数字，利用 switch 判断数字的值，根据不同的值对数组中的不同元素进行累加。

面向对象的学生成绩信息处理

项目 3 首先介绍了类、对象、方法、static 关键字、this 关键字、final 关键字等面向对象基础知识,其次进一步讲解了子类、方法重写、super 关键字、抽象类、接口、包、成员访问控制权限、常用的 Java API、Java 字符串处理、包装类等面向对象知识,为后续各种应用类型的 Java 面向对象程序设计奠定基础。

工作任务

(1) 学生成绩信息的表示(类)。
(2) 学生成绩的运算与处理(类)。

学习目标

(1) 掌握 Java 中类的定义和对象的创建及使用,熟悉 static、this 等关键字的用法,理解面向对象的基本概念,能够根据任务需求完成类的定义和对象的创建。
(2) 掌握类的继承和多态的概念及实现方法,了解抽象类和接口的简单应用,了解常用 Java API 和包装类,掌握 Java 字符串处理知识,能够用面向对象思想解决实际问题。

3.1 任务1 学生信息、课程信息和学生成绩的表示(类)

 任务描述及分析

小明在利用数组存储学生成绩信息时发现,若每个学生有各种类型的信息需要保存(学号、姓名、性别、课程名称、平时成绩、过程考核成绩、综合成绩、综合成绩等级等),单纯用一维或二维数组很难表示出来,因此,他准备使用对象数组存储学生成绩信息。要完成这个工作任务,第一步,我们要掌握类的定义的相关知识,根据任务需要完成相关的类的定义;第二步,掌握对象创建和使用的相关知识,根据定义的类,创建相应的对象,实现学生成绩信息的存储。

相关知识

之前我们曾介绍过 Java 语言中系统定义的特定类的使用(如 String 类),很多时候用户程序还需要针对特定问题的特定逻辑来定义自己的类。接下来我们将介绍 Java 语言中重要的自定义的复合数据类型类的相关知识。

3.1.1 类

类是对具有共同特征的一些对象或一系列对象的描述。如果不同的对象可以根据其类似的属性特征和行为操作划分成不同类型的话,那么决定它们属性特征和行为操作的东西就是类。例如:对于园丁所种植的香蕉、橘子、葡萄等不同的水果对象,它们有共同的名称属性,以及种植和收获的操作,因此可将它们划归为果园系统中的水果类。

当用户创建一个面向对象程序时,是如何建立对象的呢? 当然是通过类。类是用来创建对象的模板,类抽象出具体对象的相似性,定义它们的共同特征,包括数据和操作。我们可以通过类声明来定义类,然后使用类来创建用户需要的对象。类声明是用来创建对象的抽象规格说明。当用户编写自己的 Java 程序时,主要工作就是编写类。当程序运行时,已声明的类用来创建新对象。由类创建对象的过程称为实例化(instantiation),每个对象都是类的一个新实例(instance)。

类是一种抽象数据类型,在 Java 中类也被当作一个数据类型来定义,它的语法是:

```
[修饰符]class 类名    [extends 基类]  {
//类的方法成员
构造方法 1;
构造方法 2;
……
方法 1;
方法 2;
……
//类的数据成员
字段 1;
字段 2;
……
}
```

类的语法结构包括关键字 class、跟在后面的类名称。如果其继承自某个基类,还需要使用 extends 关键字加基类名称。类成员位于类体中,并分成了两大类:数据成员、方法成员,并用{}包括。

【例 3.1】定义一个 Person 类。

```
class Person{//Person 类
```

```
//数据成员包括 name、age、sex(即人有姓名、年龄、性别3个属性)
String name;//表示姓名
int age;//表示年龄
String sex;//表示性别
//Person 类共有3个方法 Print_Name()、Print_Age()、Print_Sex()
//方法 Print_Name()用于输出人的姓名
public void Print_Name(){
System.out.println("姓名:"+name);
}
//方法 Print_Age()用于输出人的年龄
public void Print_Age(){
System.out.println("年龄:"+name);
}
//方法 Print_Sex()用于输出人的性别
public void Print_Sex(){
System.out.println("性别:"+sex);
}
}
```

在一个程序中,有时需要定义多个类,多个类的定义形式有两种:并列和嵌套。常见的多个类定义形式为并列定义,即一个类接着一个类进行定义,它们之间是并列的关系;另一种形式为嵌套定义,即在一个类中定义另外的类,它们之间是包含和被包含的关系,可分别称为包含类和内部类(或嵌套类)。采用何种形式定义多个类,由类之间的访问关系确定。

类定义了一个类型(type),与 Java 语言提供的几种基本类型一样,可用来声明、定义该类型的变量。例如下面的语句:

```
Person p1;
```

声明变量 p1 的类型为类 Person。类型为类的变量与基本类型变量有所不同,类是一种引用类型,实际上 p1 是一个对类型为类 Person 的对象的引用,p1 不是对象本身,可理解为一个指针,上述声明仅生成一个空引用。

需要说明的是,类通常不需要从头生成。相反,它们可以从其他的类派生而来,继承父类的可用类成员,包括字段、方法等。即使是从头创建的新类也必须是继承自 Object 类,只不过我们可以合法省略 extends Object 而已。在 Java 语言中,Object 类是所有类的根。Object 类定义在 java.lang 包中,它是所有 Java 类的基类,Java 的任何类都是 Object 类的派生类。即 java.lang 包可由编译器自动加入,无须手工导入(import)。

3.1.2 类的对象

一旦定义了所需的类,就可以创建该类的变量。创建类的变量称为类的实例化,类的变量也称为类对象、类的实例等。

类的对象是在程序运行中创建生成的,其所占的空间在程序运行中动态分配。当一个类的对象完成了它的使命,为节省资源,Java 的垃圾收集程序就会自动回收这个对象所占的空间,即类对象有自己的生命周期。

1. 创建对象

创建类的对象须用 new 运算符,一般形式为:

```
objectName = new className()
```

new 运算符用指定的类在内存中分配空间,并将存储空间的引用存入语句中的对象变量 objectName。例如:

```
p1 = new Person();
```

new 运算符也可以与类声明一起使用,来创建类的对象。例如:

```
Person p1 = new Person();
```

2. 引用对象

在创建了类的对象后,就可以对对象的各个成员进行访问,进行各种处理。访问对象成员的一般形式为:

```
objectName.fieldName
objectName.methodName()//方法名带圆括号
```

运算符"."在这里称为成员运算符,在对象名 objectName 和成员名 fieldName、methodName() 之间起到连接的作用,指明是哪个对象的哪个成员。

例如,设已经定义了例 3.1 中的 Person 类,可以用如下的方法来引用对象的成员:

```
Person p = new Person();//定义类的对象
p.name,p.age//引用对象的数据成员
p.name = "zhangsan";//向数据成员赋值
p.age = 20;
p.Print_Name()//引用对象的成员方法
```

【例 3.2】创建、引用多个对象示例。

```
class Person{
String name;//声明姓名属性
int age;//声明年龄属性
public void print(){//取得信息
System.out.println("姓名:" + name + ",年龄:" + age);
}
```

```java
};
public class Class_Test1{
public static void main(String args[]){
Person p1 = null;//声明 per1 对象
Person p2 = null;//声明 per2 对象
p1 = new Person();//实例化 per1 对象
p2 = new Person();//实例化 per2 对象
p1.name = "张三";//设置 per1 中的 name 属性内容
p1.age = 30;//设置 per1 中的 age 属性内容
p2.name = "李四";//设置 per2 中的 name 属性内容
p2.age = 33;//设置 per2 中的 age 属性内容
System.out.print("p1 对象中的内容 - ->");
p1.print();//调用类中的方法
System.out.print("p2 对象中的内容 - ->");
p2.print();//调用类中的方法
}
}
```

如图 3-1 所示。

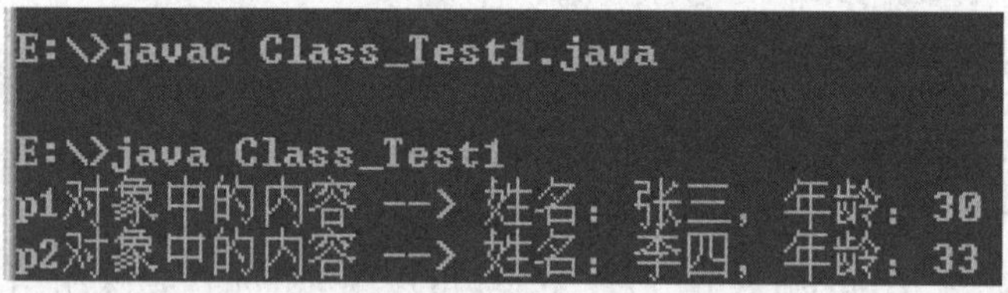

```
E:\>javac Class_Test1.java

E:\>java Class_Test1
p1对象中的内容 --> 姓名：张三，年龄：30
p2对象中的内容 --> 姓名：李四，年龄：33
```

图 3-1

3.1.3 方法成员

类成员包括数据成员和方法成员。数据成员是面向对象的术语，用于表示类中的数据变量，即 Java 中的字段(field)。方法成员也是面向对象的术语，用于表示类中的操作。Java 的函数成员包括方法和构造方法：

- 方法是一些封装在类中的过程和函数，用于执行类的操作，完成类的任务。
- 构造方法是一种特殊的方法，用于对象的创建和初始化。

1. 方法

在 Java 语言中，一个类中可以根据需要设计多个方法，在程序设计时可将一个程序中完成特定功能的程序段定义为方法，在需要使用这些功能时，可调用相应的方法，特别是在某些功能多次被使用时，采用方法可大大提高程序代码的可复用性。使用方法要掌握方法定义、方法调用、方法参数传送等方面的内容。

(1) 方法的定义。

方法的定义是描述实现某个特定功能所需的数据及进行的运算和操作。定义形式如下：

```
[modifier]returnType methodName([Parameter list]){
//methodBody 方法体
}
```

其中，用方括号括住的项目是可选的。方法的类型 returnType 指的是方法的返回值类型，若方法完成的功能是计算值，则计算结果值或计算值的表达式一般要书写到方法体里的 return 语句中，而且类型一般应与 returnType 指明的类型一致，返回值类型可以是基本类型数组类等。若方法完成的功能不返回值，returnType 处应为 void，而且方法体中的 return 语句不能带表达式或不用 return 语句。方法名 methodName 是一个标识符，是对方法的命名。可选的参数表必须用圆括号括起来，参数表 Parameter list 由 0 个或多个用逗号分隔的参数构成，每个参数由类型和参数名组成，参数可以是基本数据，也可以是数组或类实例，参数类型可以是基本数据类型或类。方法用参数来与外界发生联系（数据传送）。

方法定义中{}括起来的部分称为方法体，在其中书写方法的实现语句，包括数据定义和执行语句。方法定义前面的修饰符 modifier 用关键字表示，修饰符是可选的，用来说明方法的某些特性，可用的修饰符有 public、static、private 等。

Java 语言允许一个类中定义多个方法，方法定义形式为并列形式，其先后顺序无关紧要。

【例3.3】 定义计算圆的周长的方法。

```
double ZhouChang(double r){//参数 r 为圆的半径
double x = 2 * Math.PI * r;//方法参数 r 在方法体中可直接引用，x 为圆的周长
return(x);//返回圆的周长
}
```

(2) 方法的调用。

在程序中需要某个方法的功能时，就要调用该方法，此时，要用实际参数替换方法定义中的参数表中的形式参数，实际参数的个数、类型、顺序都必须与形式参数一致。

【例3.4】 调用上例定义的 ZhouChang 方法计算半径为变量 a 的圆的周长。

```
public class Class2{
public static void main(String args[]){
double r = 2.0,circle;
circle = ZhouChang(r);//用实参 r 调用方法 ZhouChang,方法的返回值存入 circle 变量
System.out.println(circle);
}
static double ZhouChang(double r){//参数 r 为圆的半径
```

```
        double x = 2 * Math.PI * r;//方法参数 r 在方法体中可直接引用,x 为圆的周长
        return(x);//返回圆的周长
    }

}
```

如图 3-2 所示。

```
E:\>javac Class2.java

E:\>java Class2
12.566370614359172
```

图 3-2

在上例中,方法 ZhouChang 的返回值类型为基本类型 double,在调用方法 main 中,用一个与方法返回值相同类型的变量 circle 来接收返回值。也可以对方法调用的返回值直接输出,如上例中输出周长可改为 System.out.println(ZhouChang(r))。

注意:

在方法 ZhouChang 的声明中用了修饰符 static,它说明该方法是一个类方法。与类方法 main()相同,类方法可以直接调用而不需要创建实例对象。若方法未用 static 修饰符修饰,这个方法就是实例方法,实例方法不能像类方法那样被直接调用。

(3) 方法调用中的数据传送。

调用方法与被调方法之间往往需要进行数据传送。例如,调用方法传送数据给被调方法,被调方法得到数据后进行计算,计算结果再传送给调用方法。一般说来,方法间传送数据有如下的几种方式:值传送方式、引用传送方式、返回值方式、实例变量和类变量传送方式。方法的参数可以是基本类型的变量、数组和类对象等。通过实参与形参的对应数据传送给方法体使用。

① 值传送方式。

值传送方式是将调用方法的实参的值计算出来赋予被调方法对应形参的一种数据传送方式。在这种数据传送方式下,被调方法对形参的计算、加工与对应的实参已完全脱离关系。当被调方法执行结束时,形参中的值可能发生变化,但是返回后,这些形参中的值将不会被带到对应的实参中。因此,值传送方式的特点是数据的单向传送。

使用值传送方式时,形式参数一般是基本类型的变量,实参可以是常量、变量,也可以是表达式。

② 引用传送方式。

使用引用传送方式时,方法的参数类型一般为复合类型(引用类型),复合类型变量中存储的是对象的引用,所以在参数传送中是传送引用,方法接收参数的引用,因此任何对形参

的改变都会影响对应的实参。因此,引用传送方式的特点是"数据的双向传送"。

③ 返回值方式。

返回值方式不是在形参和实参之间传送数据,而是被调方法通过方法调用后直接返回值到调用方法中。使用返回值方式时,方法的返回值类型不能为 void,且方法体中必须有带表达式的 return 语句,其中表达式的值就是方法的返回值。

④ 实例变量和类变量传送方式。

实例变量和类变量传送方式也不是在形参和实参之间传送数据,而是利用在类中定义的实例变量和类变量是类中诸方法共享的变量的特点来传送数据。类变量 static 变量可直接访问,使用较简单。下面是一个使用类变量的例子:

【例 3.5】根据半径 r 和高度 h 求解圆柱体的体积。

```
public class Class3{
static double r = 2.0,h = 5.0,mianji,tiji;//定义类变量
public static void main(String args[]){
Compute_Area();//调用 Compute_Area()方法
Compute_Vol();//调用 Compute_Vol()方法
System.out.println("圆柱体的底面积 = " + mianji);
System.out.println("圆柱体的体积 = " + tiji);
}
static void Compute_Area(){
mianji = Math.PI * r * r;//计算半径为 r 的圆柱体的底面积 mianji
}
static void Compute_Vol(){
tiji = mianji * h;//计算高度为 h 的圆柱体的体积 tiji
}
}
```

如图 3-3 所示。

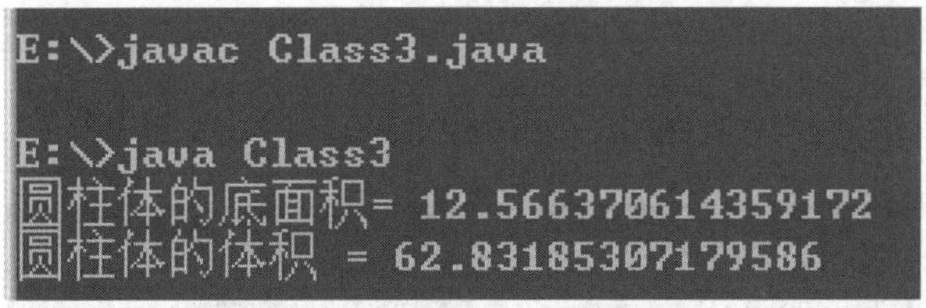

```
E:\>javac Class3.java

E:\>java Class3
圆柱体的底面积= 12.566370614359172
圆柱体的体积 = 62.83185307179586
```

图 3-3

2. 构造方法

在 Java 中,任何变量在被使用前都必须先设置初值。Java 提供了为类的成员变量赋初值的专门功能:构造方法。构造方法是一种特殊的成员方法,它的特殊性反映在如下 4 个方面:

(1) 构造方法名与类名相同；
(2) 构造方法不返回任何值且没有返回类型；
(3) 每一个类可以有零个或多个构造方法；
(4) 构造方法在创建对象时自动执行，一般不能显式地直接调用。

【例 3.6】利用半径 r、高度 h 求解圆柱体的体积，要求使用构造方法初始化圆柱体的半径 r 和高度 h。

```java
class vol{
double r;
double h;
vol(double x,double y){//构造方法设置圆的半径r和高度h
r = x;
h = y;
}
double Compute_Vol(){
return Math.PI * r * r * h;
}
}
class Class4{
public static void main(String args[]){
double tiji;//存放圆柱体的体积
vol v1 = new vol(2,5);//初始化实例对象v1的半径r和高度h
vol v2 = new vol(4,3);//初始化实例对象v2的半径r和高度h
tiji = v1.Compute_Vol();//调用Compute_Vol()方法计算第一个圆柱体的体积
System.out.println("第一个圆的体积是:" + tiji);
tiji = v2.Compute_Vol();//调用Compute_Vol()方法计算第二个圆柱体的体积
System.out.println("第二个圆的体积是:" + tiji);
}
}
```

如图 3-4 所示。

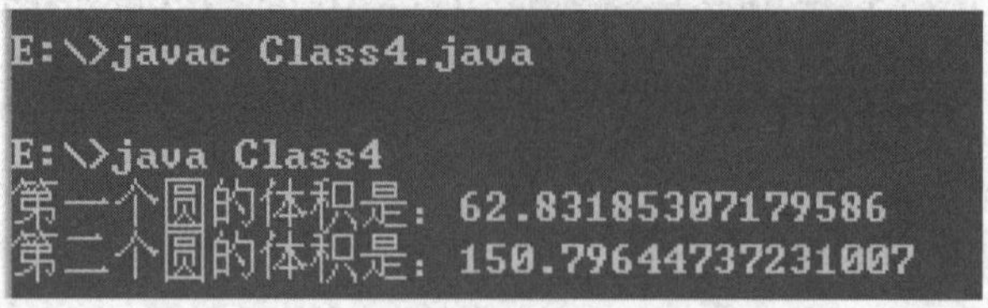

```
E:\>javac Class4.java

E:\>java Class4
第一个圆的体积是：62.83185307179586
第二个圆的体积是：150.79644737231007
```

图 3-4

需要注意的是，当方法虚拟参数名与成员变量名相同时，使用时会产生混淆，在 Java 语言中，可用 this 关键字表示本对象。例如：

```
Rectangle(double length,double width){//使用 this 避免命名空间冲突
this.length = length;//明确参数向成员变量赋值
this.width = width;
}
```

在有多个构造方法时，一个构造方法可以调用另一个构造方法，调用的方法是：

```
this(实际参数表)
```

这个语句调用是与 this 的形参列表中的参数匹配的构造方法。

没有参数的构造方法叫作无参数构造方法。一个类若没有任何用户定义的构造方法，Java 会自动提供一个空无参数构造方法，在创建对象时，使用这个无参的构造方法为类对象的成员变量赋数据类型的默认值。一旦用户定义了自己的构造方法，无参构造方法就不能再被使用。

3. 方法的重载

Java 语言允许在一个类中定义几个同名的方法，但要求这些方法具有不同的参数集合，即方法参数的个数、类型和顺序要不同，这种做法称为方法的重载。当调用一个重载的方法时，Java 编译器可根据方法参数的个数、类型和顺序的不同，来调用相应的方法。

注意：

(1) Java 根据方法名及参数集合的不同来区分不同的方法，若调用的两个同名方法中参数个数、类型及顺序均一样，仅仅返回值类型不同，则编译时会产生错误；

(2) 重载方法可以有不同类型的返回值；

(3) 在 Java 语言中也允许构造方法重载，即允许定义多个构造方法。

【例 3.7】方法重载示例(利用方法重载求解长方形、正方形的面积)。

```
class Compute_Area{
static double area(double x){
return x * x;
}
static double area(double x,double y){
return x * y;
}
public static void main(String args[]){
double mianji1 = area(3.0);
System.out.println("正方形的面积 = " + mianji1);
double mianji2 = area(3.0,4.0);
System.out.println("矩形的面积 = " + mianji2);
}
}
```

如图 3-5 所示。

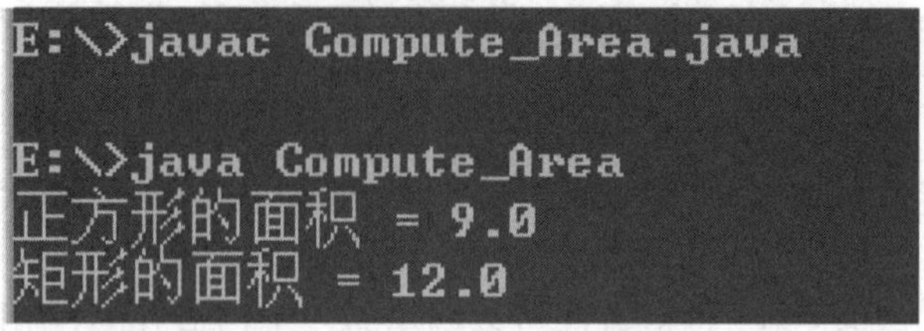

```
E:\>javac Compute_Area.java

E:\>java Compute_Area
正方形的面积 = 9.0
矩形的面积 = 12.0
```

图 3-5

使用构造方法并非只是为了给对象置初值方便,更重要的是,它是确保一个对象有正确起始状态的必要手段。另外,通过使用非 public 的构造方法,可以防止程序被其他人错误地使用与扩展。

3.1.4 类和类成员的修饰符

在类和类的成员定义时可以使用一些修饰符,来对类和成员的使用作某些限定。一般将修饰符分为两类:访问控制符和非访问控制符。访问控制符有 public、protected、private 等,它们的作用是给予对象一定的访问权限,实现类和类中成员的信息隐藏,详见 3.2。非访问控制符作用各不相同,有 static、final、abstract、synchronized 等。

1. static 修饰符

之前已经使用 static 修饰符来修饰类的成员变量和方法成员,使它们成为静态成员,也称为类成员。静态成员存储于类的存储区,属于整个类而不属于一个具体的类对象。因为静态成员属于整个类,所以它被所有该类对象共享。在不同的类对象中访问静态成员,访问的是同一个。

对静态成员的使用要注意以下两点:

(1) 静态方法不能访问属于某个对象的成员变量,而只能处理属于整个类的成员变量,即静态方法只能处理静态变量;

(2) 可以用两种方式调用静态成员,它们的作用相同。

变量:类名.变量、类对象.变量

方法:类名.方法名()、类对象.方法名()

【例 3.8】静态成员的应用示例。

```
class Static_Test{
static int a = 32;//静态变量
static int b = 67;
static void print(){//静态方法
System.out.println("a = " + a);
}
}
class Static1{
```

```
public static void main(String args[]){
Static_Test.print();//无须创建对象,通过类名直接调用静态方法
System.out.println("b = " + Static_Test.b);//通过类名直接调用静态变量
  }
}
```

如图 3-6 所示。

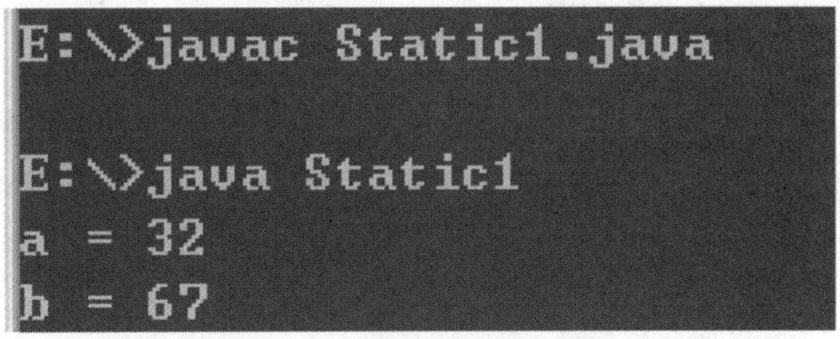

```
E:\>javac Static1.java

E:\>java Static1
a = 32
b = 67
```

图 3-6

2. final 修饰符

final 修饰符可应用于类、方法和变量。final 类不能被继承,即 final 类无子类。final 方法不能被覆盖,即子类的方法名不能与父类的 final 方法名相同。final 变量实际上是 Java 语言的符号常量,可在定义时赋初值或在定义后的其他地方赋初值,但不能再次赋值,习惯上使用大写的标识符表示 final 变量。例如:

```
final double PI = 3.1416;
final double G = 9.18;
```

因为 final 变量不能改变,所以没有必要在每个对象中进行存储,可以将 final 变量声明为静态的,以节省存储空间。例如:

```
static final double PI = 3.1416;
```

3. abstract 修饰符

abstract 修饰符可应用于类和方法,称为抽象类和抽象方法。抽象类需要继承、抽象方法需要在子类中实现才有意义,进一步的讨论见 3.2 节。

3.1.5 类的使用

1. 私有成员的访问

为了降低类间的偶合性,可以为类成员指定 private 修饰符,表示该成员只能在该类内部访问。若需要在其他类中访问私有成员,只能通过取数和送数的方法来访问,这样的方法常命名为 getXxx()和 setXxx()等。

【例 3.9】私有成员访问示例。

```java
class Person{
private String name;//声明姓名属性
private int age;//声明年龄属性
public void setName(String x){//设置姓名
name = x;
}
public void setAge(int y){//设置年龄
age = y;
}
public String getName(){//取得姓名
return name;
}
public int getAge(){//取得年龄
return age;
}
public void print(){
System.out.println("姓名:" + name + ";年龄:" + age);
}
}
public class fengzhuang{
public static void main(String args[]){
Person p1 = new Person();//声明并实例化对象
p1.setName("赵明星");//调用 setName 设置姓名
p1.setAge(20);//调用 setAge 设置年龄
String name1 = p1.getName();//调用 getName 获取 p1 的姓名
int age1 = p1.getAge();//调用 getAge 获取 p1 的年龄
System.out.println("姓名:" + name1 + ";年龄:" + age1);
p1.print();//输出姓名、年龄信息
}
}
```

如图 3-7 所示。

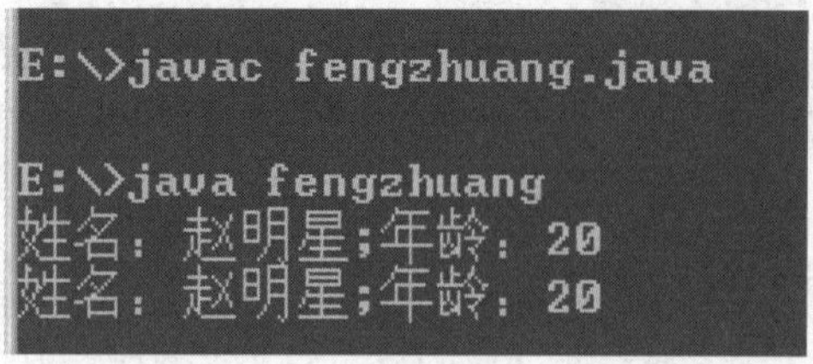

```
E:\>javac fengzhuang.java

E:\>java fengzhuang
姓名:赵明星;年龄:20
姓名:赵明星;年龄:20
```

图 3-7

2. 方法参数是类的对象

在 Java 语言中，方法的参数类型除了可以是基本类型外，还可以是引用类型——类。因为在类的对象中实际存储为对象的引用，所以在调用类参数时方法间传送的是引用。尽管 Java 采用值传送，引用从调用方法单向传送到被调方法，但由于调用方法与被调方法对应类参数的引用相同，它们引用同一对象，所以，若在被调方法中修改了引用类型形式参数的取值，则调用方法对应的实际参数也将发生相应的变化，即调用方法与被调方法之间是"引用单向传送，数据双向传送"。应用引用类型的方法参数，可在方法间传送数据。

【例3.10】引用类型的参数示例。

```java
class Compute_Area{
    double width,length,area;
    Compute_Area(double x,double y){
        width = x;
        length = y;
    }
    void Compute(Compute_Area c){
        c.area = c.width * c.length;
    }
}
class Area{
    public static void main(String args[]){
        Compute_Area c1 = new Compute_Area(2,3);
        c1.Compute(c1);
        System.out.println("矩形面积为:" + c1.area);
    }
}
```

如图 3-8 所示。

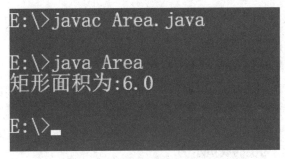

图 3-8

3. 方法返回值为类的对象

在 Java 语言中，方法的返回值类型也可以为引用类型，如类。

【例 3.11】 方法的返回值类型为引用类型示例。

```java
class Area{//圆面积
double r,area;//定义变量半径 r 和圆面积 area
Area(double r1){//构造函数,获取半径 r 的值
r = r1;
}
Area calArea(Area a){//声明方法的返回值类型为引用类型,其方法参数也是类
a.area = Math.PI * a.r * a.r;
return a;//返回值为引用类型的对象
}
}
class Class5{
public static void main(String args[]){
Area aa = new Area(2);
aa = aa.calArea(aa);
System.out.println("圆的面积为:" + aa.area);
}
}
```

如图 3-9 所示。

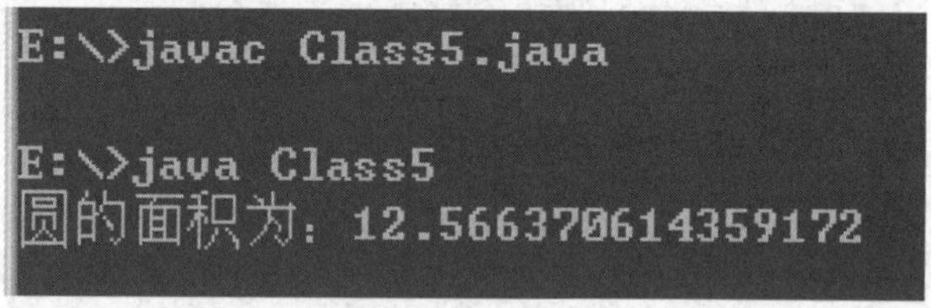

```
E:\>javac Class5.java

E:\>java Class5
圆的面积为:12.566370614359172
```

图 3-9

注意:

当方法返回值类型声明为引用类型时,方法中 return 语句的表达式类型也应为该引用类型,return 的对象应是该类的对象。

4. 类对象作为类的成员

类的数据成员也可以是引用类型的数据,如数组、字符串和类等。若一个类的对象是一个类的成员时,要用 new 运算符为这个对象分配存储空间。在包含类数据成员的类及类的实例中可以访问类数据成员的成员。

【例 3.12】 类对象作为类的成员示例。

```java
class Area{
double r;
```

```
    Area(double r1){
    r = r1;
    }
}
class Area1{//具有两个成员的类
    Area a = new Area(2);//类对象a作为类Area1的成员
    double area;//基本类型成员
}
class Class6{
    public static void main(String args[]){
        Area1 aa = new Area1();
        aa.area = Math.PI * aa.a.r * aa.a.r;
        System.out.println("圆的面积为:" + aa.area);
    }
}
```

如图3-10所示。

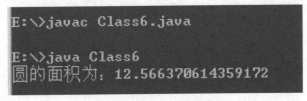

图3-10

任务 实施

3.1.6 学生成绩信息的表示实现(类)

```
class StudentScore{
    int sno;//定义变量sno,用于表示学生学号
    String sname;//定义变量sname,用于表示学生姓名
    char sex;//定义变量sex,用于表示学生性别
    String cname;//定义变量Cname,用于表示课程名称
    byte pingShiScore;//定义变量pingShiScore,用于表示学生平时成绩
    byte guoChengScore;//定义变量guoChengScore,用于表示学生过程考核成绩
    byte zhongJieScore;//定义变量zhongJieScore,用于表示学生终结考核成绩
    double zongHeScore;//定义变量zongHeScore,用于表示学生综合成绩
```

```java
        //定义变量 zongHeScoreJiBie,用于表示学生综合成绩的等级
        String zongHeScoreJiBie;

public StudentScore(){//空构造函数
    }

    public StudentScore(int sno, String sname, char sex, String cname, byte pingShiScore, byte guoChengScore, byte zhongJieScore, double zongHeScore, String zongHeScoreJiBie){//带参数的构造函数
        this.sno = sno;
        this.sname = sname;
        this.sex = sex;
        this.cname = cname;
        this.pingShiScore = pingShiScore;
        this.guoChengScore = guoChengScore;
        this.zhongJieScore = zhongJieScore;
        this.zongHeScore = zongHeScore;
        this.zongHeScoreJiBie = zongHeScoreJiBie;
    }

    public int getSno(){
        return sno;
    }

    public void setSno(int sno){
        this.sno = sno;
    }
    public String getSname(){
        return sname;
    }
    public void setSname(String sname){
        this.sname = sname;
    }
    public char getSex(){
        return sex;
    }
    public void setSex(char sex){
        this.sex = sex;
```

```java
    }
    public String getCname(){
        return cname;
    }
    public void setCname(String cname){
        this.cname = cname;
    }
    public byte getPingShiScore(){
        return pingShiScore;
    }
    public void setPingShiScore(byte pingShiScore){
        this.pingShiScore = pingShiScore;
    }
    public byte getGuoChengScore(){
        return guoChengScore;
    }
    public void setGuoChengScore(byte guoChengScore){
        this.guoChengScore = guoChengScore;
    }
    public byte getZhongJieScore(){
        return zhongJieScore;
    }
    public void setZhongJieScore(byte zhongJieScore){
        this.zhongJieScore = zhongJieScore;
    }
    public double getZongHeScore(){
        return zongHeScore;
    }
    public void setZongHeScore(double zongHeScore){
        this.zongHeScore = zongHeScore;
    }
    public String getZongHeScoreJiBie(){
        return zongHeScoreJiBie;
    }
    public void setZongHeScoreJiBie(String zongHeScoreJiBie){
        this.zongHeScoreJiBie = zongHeScoreJiBie;
    }
}
```

任务 拓展

(1) 实现银行用户账户信息类的创建和封装。
(2) 将用户操作和银行操作的代码封装到对应的方法中。

任务 2　学生成绩查询(类)

任务描述 及分析

小明要利用 3.1 创建的类,实现从键盘输入 10 个学生的成绩信息,并可以根据学号查询学生的成绩。本次工作任务,我们将进一步深入了解子类、方法重写、super 关键字、抽象类、接口、包、成员访问控制权限、常用的 Java API、Java 字符串处理、包装类等面向对象编程知识,然后利用 3.1 创建的类,实现从键盘输入 10 个学生的成绩信息,并可以根据学号查询学生的成绩,巩固并加深对面向对象编程知识的理解,学会用面向对象思想解决实际问题。

相关 知识

3.2.1　子类

Java 中的继承是通过 extends 关键字来实现的,在定义新类时使用 extends 关键字指明新类的父类,就在两个类之间建立了继承关系,其语法格式为:

```
[修饰符]class 类名    extends 父类  {
//类体
}
```

一般地,在类声明中,加入 extends 子句来创建一个类的子类。extends 后即为父类名,若父类名又是某个类的子类,则定义的类也是该类的间接子类。若无 extends 子句,则该类为 java.lang.Object 的子类。

1. 类继承的传递性

类继承具有传递性,即子类继承父类的所有非私有成员,也继承父类的父类直至祖先所有的非私有成员。

2. 类的成员覆盖

在类的继承中,若子类新增的成员名称与父类成员相同,则称为成员覆盖。在子类中定义与父类同名成员的目的是修改父类的属性和行为。在子类中,通过名称仅能直接访问本身的成员;若要访问父类的同名成员,可用关键字 super。

(1) 成员变量的覆盖。

若子类声明了与父类同名的变量,则父类的变量被隐藏起来,直接使用的是子类的变量,但父类的变量仍占据空间,可通过 super 或父类名来访问。

【例 3.13】成员变量覆盖示例。

```
class Test{
int a = 10;
}
class Class7 extends Test{
int a = 15;//在子类中定义与父类同名变量 a
void print(){
System.out.println("子类变量 a 的值:" + a);//直接输出为子类变量
System.out.println("父类变量 a 的值:" + super.a);//父类变量用 super 访问
}
public static void main(String args[]){
(new Class7()).print();
}
}
```

如图 3-11 所示。

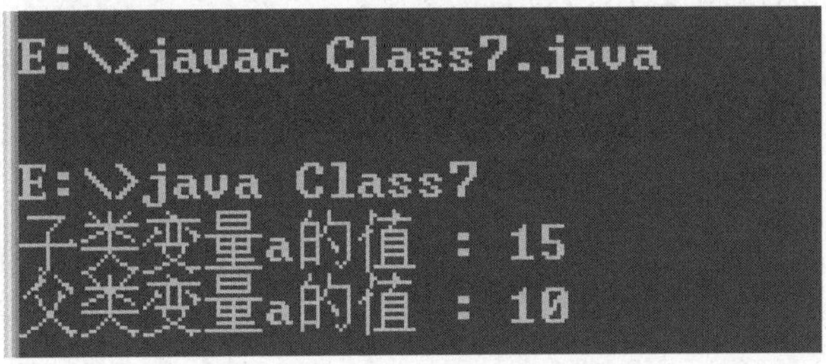

图 3-11

注意:

super 不能用于静态方法中,因为静态方法只能访问静态变量。在静态方法中,父类的静态变量可通过类名前缀来引用。

(2) 成员方法重写(覆盖)。

子类也可以定义与父类同名的方法,实现对父类方法的重写(覆盖)。方法成员的重写(覆盖)与成员变量的隐藏的不同之处在于:子类隐藏父类的成员变量,只是使得它不可见,父类的同名成员变量在子类对象中仍然占据自己的存储空间;而子类成员方法对父类同名方法的重写(覆盖),将清除父类方法占用的内存空间,从而使得父类的方法在子类对象中不复存在。

方法的重写(覆盖)中须注意的是,子类在重新定义父类已有的方法时,应保持与父类完全相同的方法头声明,即应与父类有完全相同的方法名、返回值和参数列表,否则就不是方法的重写（覆盖），而是子类定义自己的与父类无关的成员方法，父类的方法未被重写（覆盖），所以仍然存在。

重写（覆盖）方法要遵守下列规则：
① 覆盖方法的返回类型必须与它所覆盖的方法相同；
② 覆盖方法不能比它所覆盖的方法访问性差；
③ 覆盖方法不能比它所覆盖的方法抛出更多的异常。

3. 子类的初始化

在创建子类的对象时，使用子类的构造方法对其初始化，不但要对自身的成员变量赋初值，还要对父类的成员变量赋初值。因为成员变量赋初值通常在构造方法中完成，所以在Java语言中，允许子类继承父类的构造方法。构造方法的继承遵循如下的原则：

（1）若父类是无参数的构造方法，则子类无条件地继承该构造方法。

（2）若子类无自己的构造方法，则它将继承父类的无参构造方法作为自己的构造方法；若子类有自己的构造方法，则在创建子类对象时，它将先执行继承自父类的无参构造方法，然后再执行自己的构造方法。

（3）若父类是有参数的构造方法，子类可以通过在自己的构造方法中使用super关键字来调用它，但这个调用语句必须是子类构造方法的第一个可执行语句。

【例3.14】继承父类构造方法初始化成员变量示例。

```java
class Area{//父类
double r;
Area(double r1){//构造方法，初始化变量r
r = r1;
}
double Cal_Area(){//计算圆的面积
return Math.PI * r * r;
}
}




class Vol extends Area{//子类Vol继承父类Area
double height;
Vol(double r,double h){
super(r);//调用父类的构造方法是子类构造方法中第一个可执行语句
height = h;
}
void Cal_Vol(){
System.out.println("圆柱体的体积 = " + Cal_Area() * height);
}
}
```

```
public class Class8{
public static void main(String args[]){
Vol v = new Vol(2,2);
v.Cal_Vol();
}
}
```

如图 3-12 所示。

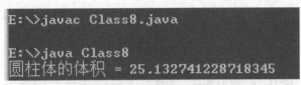

图 3-12

注意：

在本例子类的构造方法 Vol(double r,double h)中,用 super()调用父类的构造方法对父类的成员变量赋初值,注意该 super()语句必须是子类构造方法中的第一条语句。

3.2.2 抽象类和抽象方法

abstract 关键字修饰的类称为抽象类,抽象类需要子类继承,并在子类中实现抽象类中的抽象方法,抽象类只有在其子类中实现抽象方法后才有实际意义。抽象方法是只有返回值类型、方法名、方法参数而不定义方法体的一种方法,抽象方法的方法体在子类中才编写实现。

注意：

(1) 不能用 abstract 修饰构造方法、静态方法和私有 private 方法；

(2) 抽象类是一种未实现的类,抽象类不能用 new 实例化一个对象。

【例 3.15】抽象类定义示例。

```
abstract class Xing_Zhuang{//定义抽象类 Xing_Zhuang 和抽象方法 print()
abstract void print();
}
class Yuan extends Xing_Zhuang{
void print(){//实现抽象类的方法
System.out.println("圆形");
}
}
class Ju_Xing extends Xing_Zhuang{
void print(){//实现抽象类的方法
```

```
        System.out.println("矩形");
    }
}
class Class9{
    public static void main(String args[]){
        (new Yuan()).print();
        (new Ju_Xing()).print();
    }
}
```

如图 3-13 所示。

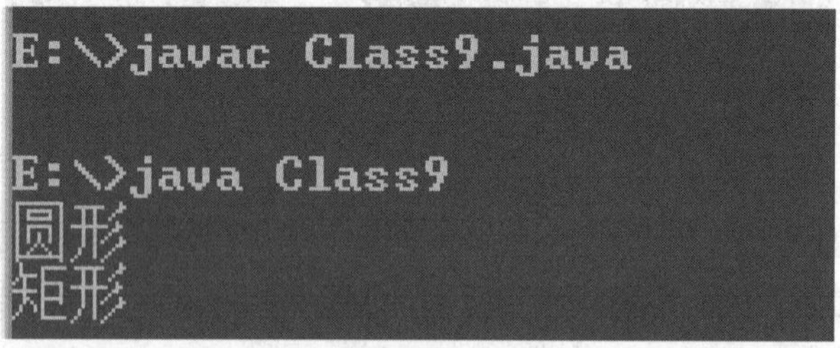

图 3-13

3.2.3 接口

接口是若干完成某一特定功能的没有方法体的方法(抽象方法)和常量的集合。接口仅提供了方法协议的封装,为了获取接口功能和真正实现接口功能,需要使用类来继承该接口。在继承接口的类中,通过定义接口中抽象方法的方法体来实现接口功能。

Java 语言使用接口来实现类间多重继承的功能,从而弥补了 Java 语言只支持类间单重继承、描述复杂实际问题处理不方便的不足。

1. 接口的定义和实现

(1) 接口的定义。

在 Java 语言中,用关键字 interface 来定义接口。接口有类似类的结构,其定义格式如下:

```
[modifier] interface interfaceName [extends superInterfaceNames]{
    //interfaceBody
}
```

从接口定义的格式可以看到,接口定义包括两个方面的内容:定义接口名和接口体。接口名 interfaceName 是一个合法的标识符,接口体 interfaceBody 同抽象类相似,为变量和抽象方法的集合,但没有构造方法和静态初始化代码。接口体中定义的变量均为终极的(final)、静态的(static)和公共的(public),接口体中定义的方法均为抽象的和公共的。由于接口所有成员均具有这些特性,相关的修饰符可以省略。

在 Java 系统中也定义了不少的接口,如用于数据输入输出的 DataInput 接口和 DataOutput 接口、用于事件处理的 ActionListener 接口等,这些都是本书后面要学习使用的接口。

(2) 接口的实现。

在某个继承接口的子类中为接口中的抽象方法书写语句并定义实在的方法体,称为实现这个接口。子类实现哪个或哪些接口用 implements 说明,不能用 extends 说明。

子类在实现接口时还要注意:若实现接口的类不是抽象类,则在该类的定义部分必须实现指定接口的所有抽象方法。方法体可以由 Java 语言书写,也可以由其他语言书写,因为是覆盖方式,所以方法头部分应该与接口中的定义完全一致,即有完全相同的参数表和返回值。

【例 3.16】接口示例:求解正方形的面积。

```
interface Area{//定义接口
double a = 4;
void Cal();
}
class Zheng_Fang implements Area{//定义实现接口的类
public void Cal(){
System.out.println("边长为 4 的正方形面积为"+a*a);
}
}
public class Interface_Test{//定义主类创建接口类对象
public static void main(String args[]){
Zheng_Fang z = new Zheng_Fang();
z.Cal();
}
}
```

如图 3-14 所示。

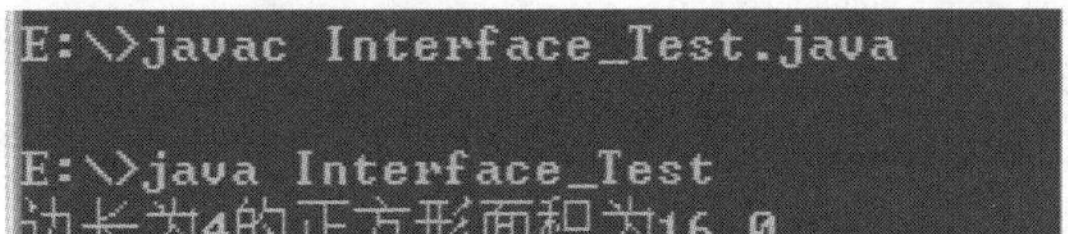

```
E:\>javac Interface_Test.java

E:\>java Interface_Test
边长为4的正方形面积为16.0
```

图 3-14

2. 接口的继承和组合

接口也可以通过关键字 extends 继承其他接口,子接口将继承父接口中所有的常量和抽象方法。此时,子接口的非抽象子类不仅需要实现子接口的抽象方法,而且需要实现继承来的抽象方法,且不允许存在未被实现的接口方法。

【例 3.17】接口的继承示例。

```
interface Test1{//定义接口 Test1
```

```java
String a = "接口 Test1";
void PrintT1();
}
interface Test2 extends Test1{//定义接口 Test2,它继承接口 Test1
String b = "接口 Test2";
void PrintT2();
}
interface Test3 extends Test2{//定义接口 Test3,它继承接口 Test2
String c = "接口 Test3";
void PrintT3();
}
class Interface_Test implements Test3{//定义实现接口 Test3 的类
public void PrintT1(){
System.out.println(a);
}//实现 public 方法
public void PrintT2(){
System.out.println(b);
}
public void PrintT3(){
System.out.println(c);
}
}
public class Interface_Test2{
public static void main(String args[]){
Interface_Test i = new Interface_Test();
i.PrintT1();
i.PrintT2();
i.PrintT3();
}
}
```

如图 3-15 所示。

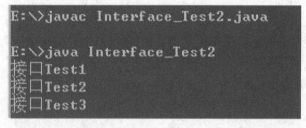

图 3-15

在本例中,实现接口 C 的派生类 InterfaceABC 中定义抽象方法的方法体时,一定要声明方法为 public 的,否则编译将显示如下的出错信息:

attempting to assign weaker access privileges;

它的中文含义为:企图缩小方法访问权限范围。

接口继承不允许循环继承或继承自己,接口与类有些方面不同。所有类的根类为类 Object,而接口没有所谓的共同根接口;接口可以同时继承多个接口,还可以通过 extends 将多个接口组合成一个接口。例如:

```
public interface Myall extends interface1,interface2{
void doSomethingElse();
}
```

3. 接口的多态

接口的使用使得方法的描述说明和方法功能的实现分开考虑,这有助于降低程序的复杂性,使程序设计灵活,便于扩充修改,这也是 Java 面向对象程序设计方法中多态特性的体现。

【例 3.18】多态示例:计算 $1×2×3×\cdots×n$ 和 $1+2+3+\cdots+n$。

```
interface Duo_Tai{
int cal(int n);
}
class Lei_Jia implements Duo_Tai{//继承接口
public int cal(int n){//实现接口中的 cal 方法
int m = 0,i;
for(i = 1;i<= n;i++)m+= i;
return m;
}
}
class Lei_Cheng implements Duo_Tai{//继承接口
public int cal(int n){//实现接口中的 cal 方法
int x = 1,i;
for(i = 1;i<= n;i++)x *= i;
return x;
}
}
public class Interface_Test3{
public static void main(String args[]){
int n = 5;
Lei_Jia L1 = new Lei_Jia();
```

```
Lei_Cheng L2 = new Lei_Cheng();
System.out.println("1 至 n 的和 = " + L1.cal(n));
System.out.println("1 至 n 的积 = " + L2.cal(n));
}
}
```

如图 3-16 所示。

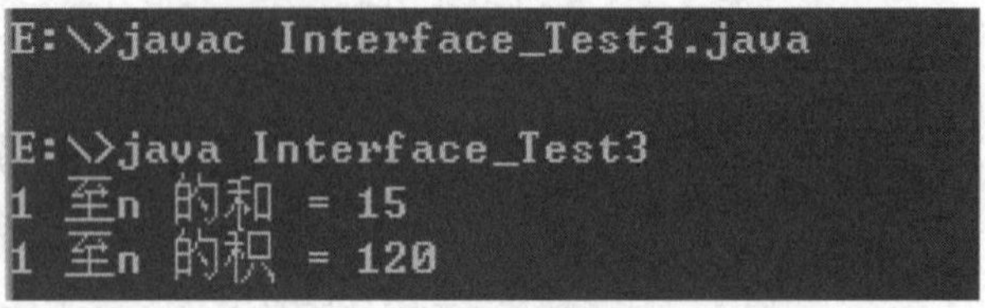

```
E:\>javac Interface_Test3.java

E:\>java Interface_Test3
1 至n 的和 = 15
1 至n 的积 = 120
```

图 3-16

4. 接口类型的使用

在 Java 语言中,接口可以作为一种类型来使用,它是一种引用类型。在 Java 语言中,任何实现接口的类的实例都可以存储在该接口类型的变量中。通过这些变量可以访问类所实现的接口中的方法,Java 运行时系统动态地确定应该使用哪个类中的方法。将接口作为一种数据类型可以不需要了解对象所对应的具体的类,而着重于它的交互界面。

【例 3.19】接口类型的使用示例。

```
interface Duo_Tai{//定义接口 Duo_Tai
int cal(int n);
}
class Lei_Jia implements Duo_Tai{//继承接口
public int cal(int n){//实现接口中的 cal 方法
int m = 0,i;
for(i = 1;i<=n;i++)m+=i;
return m;
}
}
class Lei_Cheng implements Duo_Tai{//继承接口
public int cal(int n){//实现接口中的 cal 方法
int x = 1,i;
for(i = 1;i<=n;i++)x*=i;
return x;
}
}
```

```
public class Interface_Test4{
public static void main(String args[]){
int n = 5;
Duo_Tai dt1;
Lei_Jia L1 = new Lei_Jia();
dt1 = L1;//在接口类型变量 dt1 中存储 Lei_Jia 类的实例 L1
System.out.println("1 至 n 的和为:" + dt1.cal(n));
Lei_Cheng L2 = new Lei_Cheng();
dt1 = L2;//在接口类型变量 dt1 中存储 Lei_Cheng 类的实例 L2
System.out.println("1 至 n 的乘积为:" + dt1.cal(n));
}
}
```

如图 3-17 所示。

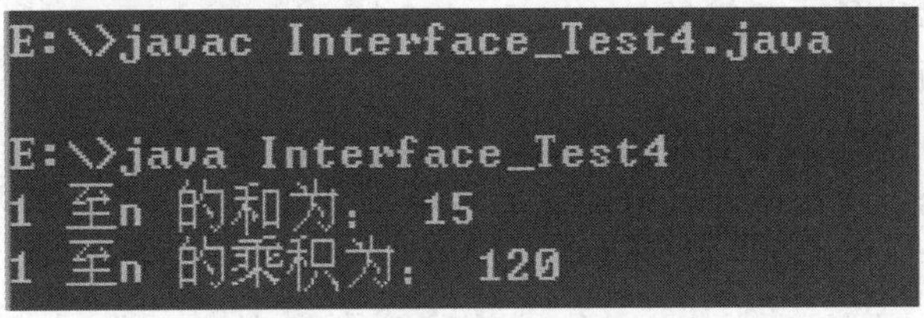

```
E:\>javac Interface_Test4.java

E:\>java Interface_Test4
1 至n 的和为: 15
1 至n 的乘积为: 120
```

图 3-17

3.2.4 包

一组相关的类和接口集合称为包。包体现了 Java 语言面向对象特性中的封装机制,它将 Java 语言的类和接口有机地组织成层次结构,这个层次结构与具体的文件系统的目录树结构层次一致。因此,Java 包就是具有一定相关性、在文件系统中可准确定位的 Java 文件的集合。

1. 创建包

包由包语句 package 创建,其语法格式如下:

```
package[package1[.package2[.[ … ]]]];
```

关键字 package 后的 package1 是包名,在 package1 下允许有次一级的子包 package2, package2 下可以有更次一级的子包 package3,等等,各级包名之间用"."号分隔。通常情况下,包名称的元素被整个地小写。

在 Java 程序中,package 语句必须是程序的第一条非空格、非注释语句。通过 package 语句,可将 Java 程序分层次地存放在不同的目录下,目录名称与包的名称相同。

【例3.20】将类 package1（package1 实现输出"Hello World！"）放入包 ch3 的次一级子包 Package。

```
package ch3.Package;
public class package1{
    public static void main(String args[]){
        System.out.println("Hello World!");
    }
}
```

2. 使用包

将类组织为包的目的是为了更好地利用包中的类。一般情况下，一个类只能引用与它在同一个包中的类。

在 Java 程序中，若要用到某些包中的类或接口，一种方法是在程序的开始部分写出相应的引入（import）语句，指出要引入哪些包的哪些类；另一种方法不用引入语句，直接在要引入的类和接口前给出其所在包名。

（1）使用 import 语句。

import 语句用于灵活地实现在编译单元中使用外部类和接口的引入机制，引入语句不必考虑类和接口的引入顺序以及是否被多次引入。

import 语句的格式与意义如下：

```
import PackageName//引入 PackageName 包
import PackageName.Identifier//引入 PackageName 包中的类和接口
import PackageName.*//引入 PackageName 包中的全部类和接口
```

（2）直接使用包。

这种方法一般用在程序中引用类和接口次数较少的时候，在要引入的类和接口前直接给出其所在包名。例如：

```
java.applet.Applet ap = new java.applet.Applet();
```

在一些 Java 程序中，还使用全局唯一包名的引用形式。全局是相对于 Internet 和 Intranet 而言的，全局唯一包名通常用一个 Internet 域名经过简单变换命名，如 sun.com 和 ibm.com 等，将域名前后颠倒，得到 com.sun 和 com.ibm 等，这些作为引用包名的前缀，再加上组织部门、项目、硬件系统名称等。例如：

```
com.sun.java.io.*;
```

3.2.5 类及类成员的访问权限

Java 程序将数据和对数据的处理代码封装为类，并以类为程序的基本单位，但类又被封装在包中。要访问类或封装在类中的数据和代码,必须清楚在什么情况下它们是可访问的。

一个类总可以访问和调用自己的变量和方法,但这个类之外的程序其他部分是否能访问这些变量和方法,则由该变量和方法以及它们所属类的访问控制符决定。

1. 类成员的访问权限

Java 将类的成员可见性(可访问性)划分为 5 种情况,按照可见性的范围大小从小到大列出如下:

(1) 仅在本类内可见;

(2) 在本类及其子类内可见;

(3) 在同一包内可见;

(4) 在同一包内及其子类不同包内可见;

(5) 在所有包内可见。

类成员的可访问性与定义时所用的修饰符 private(私有)、protected(保护)和 public(公共)有关。声明为 private 的类成员仅能在本类内被访问;声明为 protected 的类成员可以在本类、本包、本类的子类内被访问;声明为 public 的类成员可以在所有包内被访问;未用修饰符声明的类成员则隐含为在本包内可被访问。

2. 类的访问权限

类通常只用两种访问权限:默认和 public。类声明为 public 时,可以被任何包的代码访问;默认时,可被本包的代码访问。因为类封装了类成员,所以类成员的访问权限也与类的访问权限有关。例如,public 访问权限的类成员封装在默认修饰符的类中,则该类成员只能在本包内被访问。

为清楚起见,将类成员的可访问性总结在表 3-1 中,其中"√"表示允许使用相应的变量和方法。

表 3-1

	无修饰符	private	protected	public
同类	√	√	√	√
同包,子类	√		√	√
同包,非子类	√		√	√
不同包,子类			√	√
不同包,非子类				√

3.2.6 Java 的应用程序接口(API)

Java 的应用程序接口 API 是以包的形式提供的,每个包内包含大量相关的类、接口和异常。这些是 Java 程序设计时要充分利用的资源,编写 Java 程序不需要从头开始,只须针对所要解决的问题,用自己编写的类来继承系统提供的有关标准类,这样可以提高编程效率,降低代码出错的可能。

下面介绍几个 Java API 的主要包。

1. java.lang

java.lang 是 Java 语言的核心包,有 Java 程序所需要的最基本的类和接口,包括 Object 类、基本数据类型、包装类、数学类、异常处理类、线程类、字符串处理类、系统与运行类和类操作类等,这个包由编译器自动引入。

2. java.awt

java.awt 是 Java 抽象窗口工具箱包,包含许多字体和颜色设置、几何绘图、图像显示、图形用户接口操作的类和接口。

3. java.io

java.io 是 Java 语言的标准输入/输出类库,包含实现 Java 程序与操作系统、外部设备以及其他 Java 程序作数据交换所使用的类,如基本输入/输出流、文件输入/输出流、过滤输入/输出流、管道输入/输出流、随机输入/输出流等,还包含了目录和文件管理类等。

4. java.net

java.net 是 Java 网络包,实现网络功能。

5. java.util

java.util 包含了 Java 语言中的一些低级的实用工具,如处理时间的 Date 类,处理变长数组的 Vector 类,实现栈和杂凑表的 Stack 类和 HashTable 类等。

使用包中系统类的方法有 3 种:

(1) 继承系统类在用户程序中创建系统类的子类,例如 Java Applet 程序的主类作为 java.applet 包中 Applet 类的子类。

(2) 创建系统类的对象,例如创建包装类的对象。

(3) 直接使用系统类,例如程序中常用的 System.out.println()方法,就是系统类 System 的静态属性 out 的方法。

3.2.7 包装类

Java 语言有 byte、short、int、long、float、double、char 和 boolean 等基本数据类型,而有时候确实需要将这些基本类型作为类来处理。为此,Java 语言中专门提供了包装类,这些类将基本数据类型包装成类,基本类型与它们对应的包装类如表 3-2 所示。

表 3-2

基本类型	基本类型包装类	基本类型	基本类型包装类
byte	Byte	float	Float
short	Short	double	Double
int	Int	boolean	Boolean
long	Long	void	Void
char	Char		

使用包装类的方法与其他类一样,定义对象的引用,用 new 运算符创建对象,用方法来对对象进行操作。

例如：

```
Integer i = new Integer(10);//i 是 Integer 类的一个对象值为 10
Integer j = new Integer(20);//j 是 Integer 类的一个对象值为 20
```

包装类中各类的方法虽然不完全相同，但有一些是类似的。下面的例子以 Integer 类为例：

【例 3.21】输出 int 类型的最大值与最小值。

```
public class Class10{
public static void main(String args[]){
System.out.println(Integer.MAX_VALUE);//int 类型的最大值
System.out.println(Integer.MIN_VALUE);//int 类型的最小值
}
}
```

如图 3-18 所示。

```
E:\>javac Class10.java

E:\>java Class10
2147483647
-2147483648
```

图 3-18

3.2.8 常用的字符串类

字符串是字符的序列，它是组织字符的基本数据结构。Java 将字符串当作对象来处理，它提供了一系列的方法对整个字符串进行操作，使得字符串的处理更加容易和规范。Java 语言中的包 java.lang 中封装了 final 类 String 和 StringBuffer，其中类 String 对象是字符串常量，建立后不能改变，而 StringBuffer 对象类似于一个字符缓冲区，建立后可以修改。

1. String 类

（1）String 字符串定义。

① 由于每个字符串常量（用双引号括起来的一串字符）实际上都是 String 对象，如字符串"Hello World"在编译后即成为 String 对象，因此，可以用字符串常量直接初始化一个 String 对象。例如：

```
String s = "Hello World.";
```

【例 3.22】使用直接赋值实例化 String 对象。

```
public class String1{
public static void main(String args[]){
String name = "张三";//实例化 String 对象
System.out.println("姓名:" + name);
}
}
```

如图 3-19 所示。

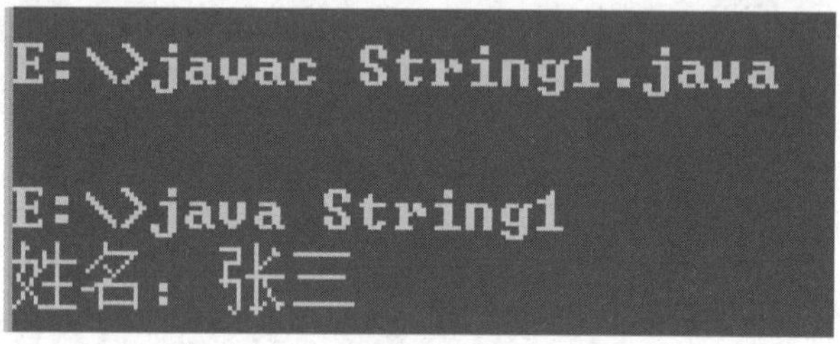

```
E:\>javac String1.java

E:\>java String1
姓名:张三
```

图 3-19

② 通过类 String 提供的构造方法,可以生成一个空字符串(不包含任何字符的字符串),也可以由字符数组或字节数组来生成一个字符串对象。默认的构造方法不需要任何参数,它生成一个空字符串。例如:

String s = new String();//建立一个空字符串对象

其他创建 String 对象的构造方法有:

• String(String value)用已知串 value 创建一个字符串对象。

• String(char chars[])用字符数组 chars 创建一个字符串对象。

• String(char chars[], int startIndex, int numChars)用字符数组 chars 的 startIndex 位置开始的 numChats 个字符,创建一个字符串对象。

• String(byte ascii[], int hiByte)用字节数组 ascii 创建一个字符串对象,Unicode 字符的高位字节为 hiByte,通常应该为 0。

• String(byte ascii[], int hiByte, int startIndex, int numChars)用字节数组 ascii 创建一个字符串对象,其参数的意义同上。

由于在互联网上通常使用的字符都为 8 位的 ASCII 码,Java 提供了从字节数组来初始化字符串的方法,并且用 hiByte 来指定每个字符的高位字节。对 ASCII 码来说,hiByte 应为 0;对于其他非拉丁字符集,hiByte 的值应该非 0。

【例 3.23】使用关键字 new 实例化 String 对象。

```
public class String2{
public static void main(String args[]){
String name = new String("张三");//实例化 String 对象
```

```
System.out.println("姓名:" + name);
    }
}
```

如图 3-20 所示。

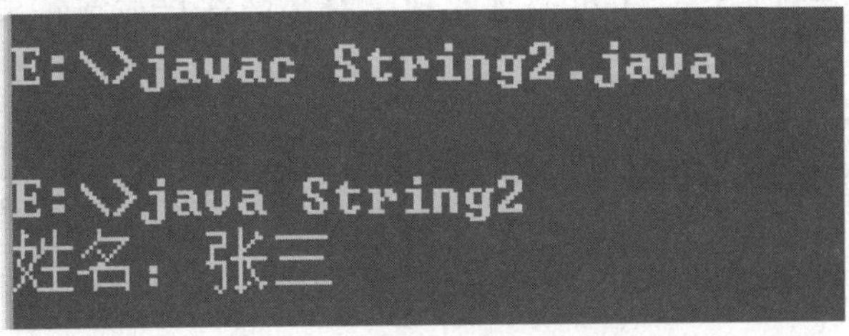

```
E:\>javac String2.java

E:\>java String2
姓名：张三
```

图 3-20

(2) String 类的常用方法。
① int length()。

方法 length() 的功能为返回类 String 字符串对象的长度。例如：

```
int len = "Hello World".length();
```

将返回字符串的长度 12,字符串的长度即字符串中字符的个数。

【例 3.24】使用 length() 方法返回字符串的长度示例。

```
public class String3{
public static void main(String args[]){
String str1 = "zhangsan";//定义字符串变量
String str2 = new String("lisi");//实例化 String 对象
System.out.println("字符串 str1 的长度为:" + str1.length());
System.out.println("字符串 str2 的长度为:" + str2.length());
    }
}
```

如图 3-21 所示。

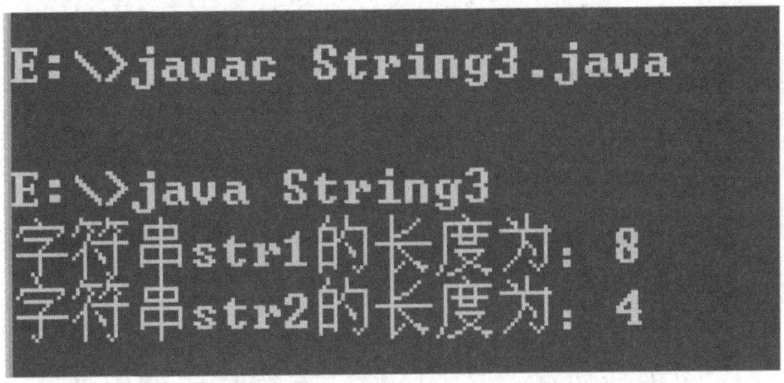

```
E:\>javac String3.java

E:\>java String3
字符串str1的长度为: 8
字符串str2的长度为: 4
```

图 3-21

② String concat(String str)。

方法 concat() 的功能是将 str 连接到调用串对象的后面,即进行字符串的连接操作。例如:

```
System.out.println("41177".concat("abc"));//输出为 41177abc
```

在 Java 语言中,运算符"+"也可用来实现字符串的连接或字符串与其他类对象的连接。例如:

```
String s = "Java" + "程序设计";//s = "Java 程序设计"
int age = 20;
String s1 = "他" + age + "岁";//s1 = "他 20 岁"
```

【例 3.25】字符串连接示例。

```
public class String4{
    public static void main(String args[]){
        String str1 = "连接后的字符串为:";//定义字符串变量
        String str2 = new String("我是新字符串");//实例化 String 对象
        System.out.println(str1.concat(str2));
        System.out.println(str1 + str2);
    }
}
```

如图 3-22 所示。

图 3-22

③ boolean equals(Object anObject) 和 equalsIgnoreCase(String anotherString)。

String 类中提供了方法 equals() 和 equalsIgnoreCase() 用来比较两个字符串的值是否相等,不同的是后者忽略字母的大小写。例如:

```
System.out.println("abc".equals("abc"));//输出为 true
System.out.println("abc".equalsIgnoreCase("ABC"));//输出为 true
```

注意:

它们与运算符"=="实现的比较是不同的,运算符"=="比较两个字符串对象是否引用同一个实例对象,而 equals( )和 equalsIgnoreCase( )则比较两个字符串中对应的每个字符是否相同。例如:

```
String s = "abc";
String s1 = "abc";
System.out.println(s = = s1);//输出为 false
System.out.println("abc" = = "abc");//输出为 true
```

【例 3.26】分别用 equals( )和 equalsIgnoreCase( )方法判断字符串"Hello World!"和"hello world!"对应的每个字符是否相同。

```
public class String5{
public static void main(String args[]){
String str1 = "Hello World!";//定义字符串
String str2 = "hello world!";//定义字符串
//区分大小写的比较
System.out.println("str1 equals str2:" + str1.equals(str2));
//不区分大小写的比较
System.out.println("str1 equalsIgnoreCase str2:" + str1.equalsIgnoreCase(str2));
}
}
```

如图 3-23 所示。

```
E:\>javac String5.java

E:\>java String5
str1 equals str2:false
str1 equalsIgnoreCase str2:true
```

图 3-23

④ char charAt(int index)。
该方法的功能是返回字符串 index 处的字符,index 的值从 0 到串长度减 1。例如:

```
System.out.println("China".charAt(1));//输出为 h
```

【例 3.27】使用 charAt(int index)方法返回"Hello World!"字符串的 index 为 3 的字符。

```
public class String5{
public static void main(String args[]){
String str1 = "Hello World!";//定义 String 对象
```

```
        System.out.println("字符串 str1 的 index 为 3 的字符是:" + str1.charAt(3));
    }
}
```

如图 3-25 所示。

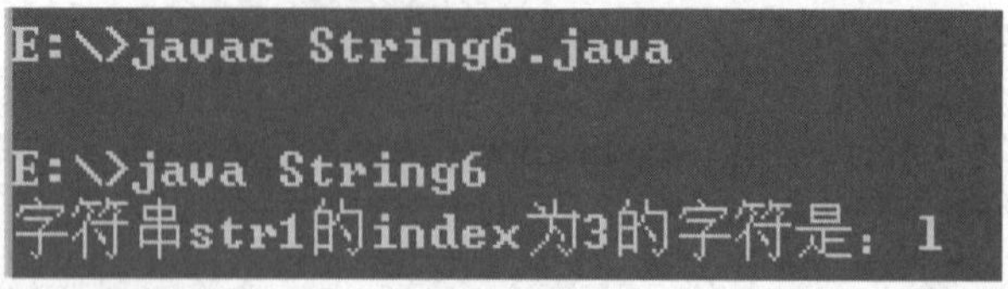

图 3-24

⑤ indexOf()。

方法 indexOf()有重载:
- int indexOf(int ch, int fromIndex)
- int indexOf(String str)
- int indexOf(String str, int fromIndex)

以上方法的功能是返回字符串对象中指定的字符和子串首次出现的位置,从串对象开始处或从 fromIndex 处开始查找,若未找到则返回-1。

【例 3.28】使用 indexOf()方法判断字符"W"在字符串"Hello World!"中的位置。

```
public class String7{
    public static void main(String args[]){
        String str1 = "Hello World!";//声明字符串
        System.out.println("字符 W 在字符串中的位置为:" + str1.indexOf("W"));//查到返回位置
    }
}
```

如图 3-25 所示。

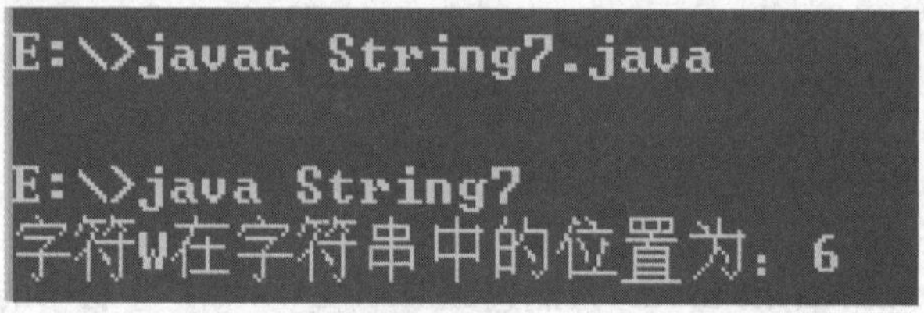

图 3-25

⑥ String substring(int beginindex)和 substring(int beginindex, int endindex)。

该方法的功能是返回子字符串。前者返回从位置 beginindex 开始处到串尾的子字符串；后者返回从 beginindex 开始到 endindex－1 为止的子字符串，子串长度为 endindex－beginindex。

【例 3.29】分别使用 substring( )方法和 substring(int beginIndex，int endIndex)方法从字符串"Hello World!"中截取字符。

```
public class String8{
    public static void main(String args[]){
        String str1 = "Hello World!";//定义字符串
        System.out.println(str1.substring(5));//从 index 为 5 的位置开始截取字符
        System.out.println(str1.substring(1,4));//截取 1～3 位置上的字符
    }
}
```

如图 3-26 所示。

```
E:\>javac String8.java

E:\>java String8
 World!
ell
```

图 3-26

⑦ String toLowerCase( )和 String toUpperCase( )。

方法 toLowerCase( )和 toUpperCase( )的功能是分别将字符串中的字母转换为小写和大写。例如：

```
System.out.println("abcde".toUpperCase());//输出为 ABCDE
System.out.println("ABCDE".toLowerCase());//输出为 abcde
```

【例 3.30】将字符串"str1"全部转为大写，将"STR2"全部转为小写。

```
public class String9{
    public static void main(String args[]){
        String s1 = "str1";
        String s2 = "STR2";
        System.out.println("将 s1 转成大写为:" + s1.toUpperCase());
        System.out.println("将 s2 转成小写为:" + s2.toLowerCase());
    }
}
```

如图 3-27 所示。

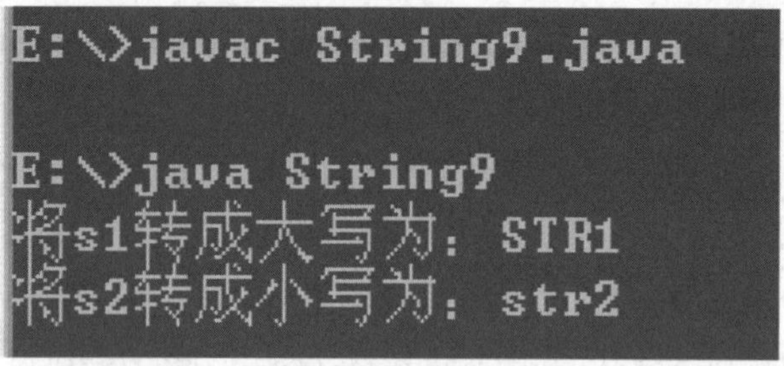

图 3-27

⑧ String trim()。

trim() 方法的功能是截去字符串两端的空白字符。例如：

```
String s = "Java";
String s1 = "他 10 岁";
System.out.println(s.trim() + s1);//输出为 Java 他 10 岁
```

【例 3.31】使用 trim 方法将字符串"　Hello World!　"两端的空格去掉。

```
public class String10{
public static void main(String args[]){
String str1 = "  Hello World  ";//定义字符串
System.out.println(str1.trim());//去掉字符串左右空格后输出
}
}
```

如图 3-28 所示。

图 3-28

⑨ static String valueOf(Object obj)。

该方法的功能是将其他类的参数转换为字符串对象返回。例如：

```
System.out.println(String.valueOf(Math.PI));//PI 转换为 String 对象
```

注意：
- 其他类也提供了方法 valueOf() 把一个字符串转换为相应的类对象。例如：

```
String piStr = "3.14159";
Float pi = Float.valueOf(piStr);//String 对象转换为 Float 对象
```

- java.lang.Object 中还提供了方法 toString()，该方法也可以把其他对象转换为字符串。例如：

```
String s = Integer.toString(123,8);//将十进制的 123 转换为八进制数据
System.out.println(s);//输出为 173
```

【例 3.32】将数字字符串"123"转换为整数数值。

```java
public class String11{
    public static void main(String args[]){
        String str = "123";
        Integer i = Integer.valueOf(str);
        int j = i.intValue();
        System.out.println(j);
    }
}
```

如图 3-29 所示。

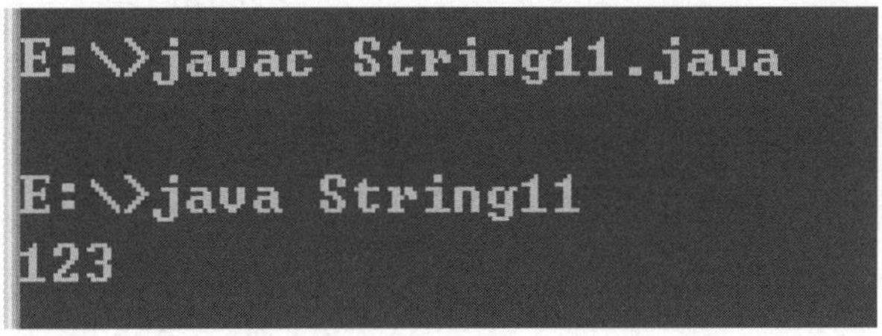

```
E:\>javac String11.java

E:\>java String11
123
```

图 3-29

2. StringBuffer 类

（1）StringBuffer 字符串的定义。

Stringbuffer 类也是操作字符串的，但是与 String 类不同。String 类的内容一旦声明之后则不可改变，改变的只是其内存地址的指向；而 StringBuffer 是使用缓冲区的，其内容是可以改变的。对于 StringBuffer 而言，其本身是一个具体的操作类，所以不能像 String 那样采用直接赋值的方式进行对象的实例化，必须通过构造方法完成对象的实例化。

StringBuffer 类的对象是一种可以改变（如增长）的字符串对象，使用起来比 String 类更加方便。在 Java 语言中支持字符串的加运算，实际上就是使用了 StringBuffer 类。

StringBuffer 类提供了 3 种构造方法对一个可变的字符串对象进行初始化：

① StringBuffer()建立空的字符串对象。
② StringBuffer(int len)建立长度为 len 的字符串对象。
③ StringBuffer(String s)建立一个初值为 String 类对象 s 的字符串对象。
例如：

```
StringBuffer sb1 = new StringBuffer();
StringBuffer sb2 = new StringBuffer(30);
StringBuffer sb3 = new StringBuffer("StringBuffer");
```

注意：
若不给任何参数，则系统为字符串分配 16 个字符大小的缓冲区，这是默认的构造方法，如上例的 sb1；参数 len 则指明字符串缓冲区的初始长度，如上例的 sb2；参数 s 给出字符串的初始值，系统按照初始值分配相应大小的空间后，还要再为该串分配 16 个字符大小的空间，如上例的 sb3，其初始值为"StringBuffer"（即长度为 12 个字符），系统在分配了 12 字符的空间后，还会再为其分配 16 个字符的空间，由此，sb3 的缓冲区大小为 28。

(2) StringBuffer 类的常用方法。
① StringBuffer append(Object obj)。
方法 append()用于在串缓冲区的字符串末尾添加各种类型的数据。

【例 3.33】使用 append()方法向 StringBuffer 中添加"Hello"，然后再次使用 append()方法追加"World!"。

```
public class String_Buffer1{
public static void main(String args[]){
StringBuffer s_buf = new StringBuffer();//声明 StringBuffer 对象
s_buf.append("Hello");//向 StringBuffer 中添加"Hello"
s_buf.append("World!");//向 StringBuffer 中追加"World!"
System.out.println(s_buf);
}
}
```

如图 3-30 所示。

```
E:\>javac String_Buffer1.java

E:\>java String_Buffer1
Hello World!
```

图 3-30

② StringBuffer insert()和 int length()。
• 方法 insert()用于在串缓冲区的指定位置 offset 处插入各种类型的数据。

- length()方法的功能是获取字符串对象的长度。

【例 3.34】使用 insert( )方法在 StringBuffer 的内容"Hello World"的"Hello"后面添加"our",然后再使用 insert( )方法在 StringBuffer 内容的结尾添加"!!"。

```
public class String_Buffer2{
public static void main(String args[]){
StringBuffer s_buf = new StringBuffer();//声明 StringBuffer 对象
s_buf.append("Hello World");//添加内容
s_buf.insert(6,"our");//在"Hello"后添加"our"
s_buf.insert(s_buf.length(),"!!");//在最后添加"!!"
System.out.println(s_buf);
}
}
```

如图 3-31 所示。

```
E:\>javac String_Buffer2.java

E:\>java String_Buffer2
Hello our World!!
```

图 3-31

③ StringBuffer delete(int start, int end)。

方法 delete()用来从 StringBuffer 字符串对象中删去从 start 开始到 end-1 结束的子字符串。

【例 3.35】用 delete( )方法删除 StringBuffer 中存放的字符串"aabbcc"中间的"abb"。

```
public class String_Buffer3{
public static void main(String args[]){
StringBuffer s_buf = new StringBuffer("aabbcc");//声明 StringBuffer 对象
System.out.println(s_buf.delete(1,4));//删除"abb",输出结果应为"acc"
}
}
```

如图 3-32 所示。

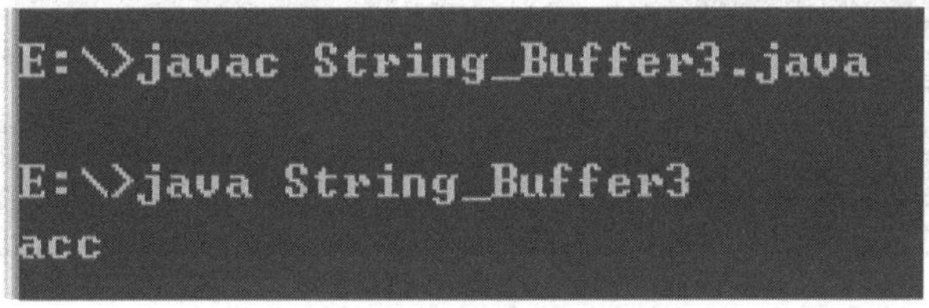

```
E:\>javac String_Buffer3.java

E:\>java String_Buffer3
acc
```

图 3-32

④ deleteCharAt(int index)。

方法 deleteCharAt()用来删除 index 处的字符。

【例 3.36】使用 deleteCharAt()方法删除 StringBuffer 字符串"Hello World!"中的"e"。

```
public class String_Buffer4{
public static void main(String args[]){
StringBuffer s_buf = new StringBuffer("Hello World!");//声明 StringBuffer 对象
System.out.println(s_buf.deleteCharAt(1));//删除字符"e"
}
}
```

如图 3-33 所示。

```
E:\>javac String_Buffer4.java

E:\>java String_Buffer4
Hllo World!
```

图 3-33

⑤ StringBuffer reverse()。

方法 reverse()用于对 StringBuffer 类字符串对象进行颠倒操作。

【例 3.37】将 StringBuffer 中的字符串"Hello World!"翻转后输出。

```
public class String_Buffer5{
public static void main(String args[]){
StringBuffer buf = new StringBuffer("Hello World!");//声明 StringBuffer 对象
System.out.println(buf.reverse());//将内容输出
}
}
```

如图 3-34 所示。

```
E:\>java String_Buffer5
!dlroW olleH

E:\>
```

图 3-34

⑥ void setCharAt(int index, char ch)。

setCharAt()方法的功能是将指定位置 index 处的字符替换为字符 ch。

【例3.38】使用 setCharAt()方法将 StringBuffer 字符串"Hello World!"的"H"替换为"h"。

```
public class String_Buffer6{
public static void main(String args[]){
StringBuffer s_buf = new StringBuffer("Hello World!");//声明 StringBuffer 对象
s_buf.setCharAt(0,'h');//将"H"替换为"h"
System.out.println("替换后的结果为:" + s_buf);//输出内容
}
}
```

如图 3-35 所示。

```
E:\>javac String_Buffer6.java

E:\>java String_Buffer6
替换后的结果为: hello World!
```

图 3-35

⑦ StringBuffer replace(int start, int end, String str)。

该方法的功能是将 StringBuffer 对象字符串从 start 至 end-1 处,用字符串 str 替换。

【例3.39】将 StringBuffer 字符串"＊＊＊＊＊＊＊＊"第 2～5 位的"＊"替换为"Java"。

```
public class String_Buffer7{
public static void main(String args[]){
StringBuffer s_buf = new StringBuffer("＊＊＊＊＊＊＊＊");//声明 StringBuffer 对象
s_buf.replace(2,6,"Java");//将第 2～5 位的"＊"替换为"Java"
System.out.println("替换后的结果为:" + s_buf);//输出内容
}
}
```

如图 3-36 所示。

```
E:\>javac String_Buffer7.java

E:\>java String_Buffer7
替换后的结果为: ＊＊Java＊＊
```

图 3-36

⑧ void getChars(int srcBegin,int srcEnd,char[]dst,int dstBegin)。

该方法的功能是将 StringBuffer 对象字符串中字符复制到目标字符数组中去。复制的字符从 srcBegin 处开始，到 srcEnd－1 处结束，复制字符的个数为 srcEnd－srcBegin，字符被复制到目标数组的 dstBegin 至 dstBegin＋(srcEnd－srcBegin)－1 处。

【例 3.40】 将 StringBuffer 中的字符串"java"中的"va"复制到字符数组 a 的中间位置，字符数组 a 的内容为"＊＊＊＊＊＊＊＊"。

```
public class String_Buffer8{
public static void main(String args[]){
char a[] = {'*','*','*','*','*','*','*','*'};
StringBuffer s_buf = new StringBuffer("java");;//声明 StringBuffer 对象
s_buf.getChars(2,4,a,3);//将"va"复制到字符数组 a 的中间位置
System.out.print("替换后的结果为:");//输出内容
System.out.println(a);
}
}
```

如图 3-37 所示。

```
E:\>javac String_Buffer8.java

E:\>java String_Buffer8
替换后的结果为: ***va***
```

图 3-37

⑨ String substring(int start)。

该方法的功能是获取 StringBuffer 字符串对象从 start 处开始的子串。重载的方法 String substring(int start,int end)用来获取 StringBuffer 字符串对象从 start 至 end－1 处之间的子串。

【例 3.41】 分别使用 substring(int start)和 substring(int start,int end)方法截取 StringBuffer 字符串"Hello World!"的"World!"和"orl"。

```
public class String_Buffer9{
public static void main(String args[]){
StringBuffer s_buf = new StringBuffer("Hello World!");;//声明 StringBuffer 对象
System.out.println(s_buf.substring(6));//截取"World!"并输出
System.out.println(s_buf.substring(7,10));//截取"orl"并输出
}
}
```

如图 3-38 所示。

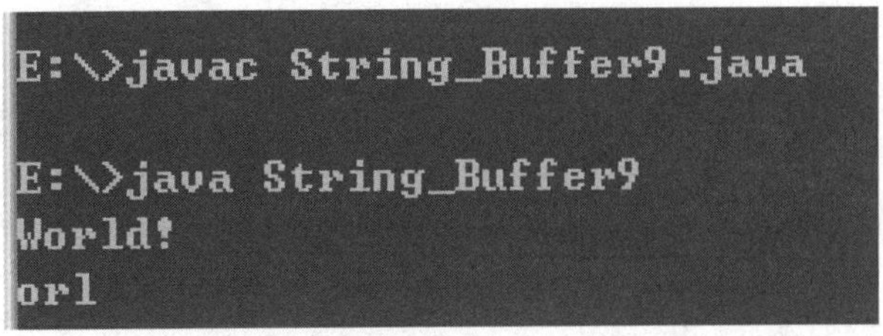

图 3-38

3.2.9 Vector 类

Vector 类似于一个数组,但与数组相比在使用上有以下两个优点。
(1) 使用的时候无须声明上限,随着元素的增加,Vector 的长度会自动增加。
(2) Vector 提供额外的方法来增加、删除元素,比数组操作高效。
Vector 类有 3 个构造函数,分别如下:

```
public Vector();
```

该方法创建一个空的 Vector。

```
public Vector(int initialCapacity);
```

该方法创建一个初始长度为 initialCapacity 的 Vector。

```
public Vector(int initialCapacity, int capacityIncrement);
```

该方法创建一个初始长度为 initialCapacity 的 Vector,当向量需要增长时,增加 capacityIncrement 个元素。
(1) Vector 类中添加、删除对象的方法如下:

```
public void add(int index,Object element)
```

在 index 位置添加对象 element。

```
public boolean add(Object o)
```

在 Vector 的末尾添加对象 o。

```
public Object remove(int index)
```

删除 index 位置的对象,后面的对象依次前提。

(2) Vector 类中访问、修改对象的方法如下：

```
public Object get(int index)
```

返回 index 位置对象。

```
public Object set(int index, Object element)
```

修改 index 位置的对象为 element。

(3) 其他方法：

```
public String toString()
```

将元素转换成字符串。

```
public int size()
```

返回对象的长度。

【例3.42】操作 Vector 对象，进行元素的添加、插入、修改和删除。程序输出结果如图 3-39 所示。源程序代码如下：

```java
//程序文件名为 UseVector.java
import java.util.Vector;//引入 JDK 的 Vector 类
public class UseVector
{
    public static void main(String[]args)
    {
    Vector<String> vScore = new Vector<String>();
    vScore.add("86");    //添加元素
    vScore.add("98");    //添加元素
    vScore.add(1,"99");  //插入元素
    //输出结果
    for(int I = 0;I<vScore.size();I++)
    {
        System.out.print(vScore.get(i)+"   ");
    }
    vScore.set(1,"77");  //修改第二个元素
    vScore.remove(0);//删除第一个元素
    System.out.println("\n修改并删除之后");
    for(int I = 0;I<vScore.size();I++)
```

```
        {
            System.out.print(vScore.get(i)+"  ");
        }
        System.out.println("\n转换成字符串之后的输出\n" + vScore.toString());
    }
};
```

如图 3-39 所示。

```
E:\>java UseVector
86   99   98
修改并删除之后
77   98
转换成字符串之后的输出
[77, 98]

E:\>
```

图 3-39

任务 实施

3.2.10 学生成绩查询实现(类)

```
import java.util.Scanner;
class StudentScore{
    int sno;//定义变量 sno,用于表示学生学号
    String sname;//定义变量 sname,用于表示学生姓名
    char sex;//定义变量 sex,用于表示学生性别
    String cname;//定义变量 Cname,用于表示课程名称
    byte pingShiScore;//定义变量 pingShiScore,用于表示学生平时成绩
    byte guoChengScore;//定义变量 guoChengScore,用于表示学生过程考核成绩
    byte zhongJieScore;//定义变量 zhongJieScore,用于表示学生终结考核成绩
    double zongHeScore;//定义变量 zongHeScore,用于表示学生综合成绩
    //定义变量 zongHeScoreJiBie,用于表示学生综合成绩的等级
    String zongHeScoreJiBie;
}
public class test{
```

```java
public static void main(String[]args){
//定义对象数组,并分配内存,假设最多100名学生
StudentScore a[] = new StudentScore[100];
for(int i = 0;i<a.length;i++){
a[i] = new StudentScore();
}
//键盘输入
Scanner scan = new Scanner(System.in);
//变量n用于统计实际录入的人数并将录入的数据依次存放到对象数组
int n = 0;
//创建并录入学习成绩信息表的相关信息
while(true){
System.out.print("请输入第" + (n+1) + "个学生的学号(输入-1退出):");
int number = scan.nextInt();
if(number == -1){
break;
}
//通过学号,检查是否重复输入学生数据
for(int i = 0;i<a.length;i++){
if(number == a[i].sno){
System.out.println("学号重复!请重新输入:");
break;
}
}
//依次存放学号、姓名等信息到对应的对象
a[n].sno = number;
System.out.print("请输入第" + (n+1) + "个学生的姓名:");
a[n].sname = scan.next();
System.out.print("请输入第" + (n+1) + "个学生的性别:");
a[n].sex = scan.next().charAt(0);
System.out.print("请输入第" + (n+1) + "个学生的所修的课程名称:");
a[n].cname = scan.next();
System.out.print("请输入第" + (n+1) + "个学生的所修的课程的平时成绩:");
a[n].pingShiScore = scan.nextByte();
System.out.print("请输入第" + (n+1) + "个学生的所修的课程的过程考核成绩:");
a[n].guoChengScore = scan.nextByte();
```

```java
            System.out.print("请输入第" + (n + 1) + "个学生的所修的课程的终结考核
成绩:");
            a[n].zhongJieScore = scan.nextByte();
        a[n].zongHeScore =
a[n].pingShiScore * 0.1 + a[n].guoChengScore * 0.4 + a[n].zhongJieScore * 0.5;
            if(a[n].zongHeScore<60){
        a[n].zongHeScoreJiBie = "不及格";
            }else if(a[n].zongHeScore<70){
            a[n].zongHeScoreJiBie = "及格";
            }else if(a[n].zongHeScore<80){
            a[n].zongHeScoreJiBie = "中";
            }else if(a[n].zongHeScore<90){
            a[n].zongHeScoreJiBie = "良";
            }else if(a[n].zongHeScore<=100){
            a[n].zongHeScoreJiBie = "优";
            }
            n++;
            }
            while(true){
            System.out.print("请输入要查找的学生学号:");
            int number = scan.nextInt();
            if(number==-1){
            break;
            }else{
            //调用find方法查找学生的成绩信息
            //参数a表示查询的对象数组
            //参数n表示查询数组a的范围(0到第n个数组对象)
            //参数number表示要查询的学生的学号
            find(a,n,number);
            }
            }
        }
        //按学号查找学生信息
        public static void find(StudentScore a[],int n,int number){
            int j=0;//用于记录是否找到对应的学生
            for(int i=0;i<n;i++){
            if(a[i].sno==number){
```

```
            System.out.println("找到学号为" + number + "的学生,其成绩信息如下:");
            System.out.print("学号:" + a[i].sno + "   ");
            System.out.print("姓名:" + a[i].sname + "   ");
            System.out.print("性别:" + a[i].sex + "   ");
            System.out.print("课程名称:" + a[i].cname + "   ");
            System.out.print("平时成绩:" + a[i].pingShiScore + "   ");
            System.out.print("过程考核成绩:" + a[i].guoChengScore + "   ");
            System.out.print("终结考核成绩:" + a[i].zhongJieScore + "   ");
            System.out.print("综合成绩:" + a[i].zongHeScore + "   ");
            System.out.println("综合成绩:" + a[i].zongHeScoreJiBie + "   ");
            j = 1;
            break;
        }
    }
    if(j = = 0){
        System.out.println("未找到该学生!");
    }
}
```

如图 3 - 40 所示。

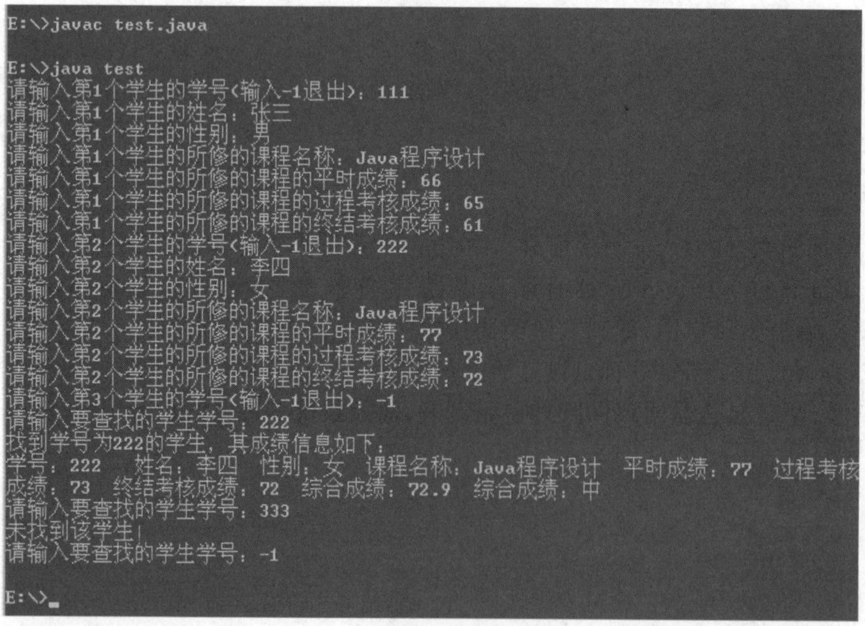

图 3 - 40

任务 拓展

（1）将银行自助服务系统根据需求，修改为继承的方式进行操作实现。
（2）将银行自助服务系统根据需求，修改为多态的方式进行操作实现。
（3）为银行自助服务系统增加一个办理信用卡的业务，用户不一定都办理信用卡。
（4）为银行自助服务系统增加一个办理贷款的业务，用户不一定都有信用卡。
（5）对银行自助服务系统中用户录入的身份证号码进行长度和格式的验证（如下图所示）。

请输入你的身份证号码：
521364383828737834
你录入的身份证信息是：521364383828737834

请输入你的身份证号码：
534324745545
身份证号码的位数不符合要求！！

请输入你的身份证号码：
123471980340734985
你录入的身份证信息是：123471980340734985
身份信息月份的数据不合法！！

请输入你的身份证号码：
123471898340734985
你录入的身份证信息是：123471898340734985
身份信息年份数据不合法！！
身份信息月份的数据不合法！！

请输入你的身份证号码：
123472918937828374
你录入的身份证信息是：123472918937828374
身份信息年份数据不合法！！
身份信息月份的数据不合法！！
身份信息的日期数据不合法！！

请输入你的身份证号码：
132161995032964839
你录入的身份证信息是：132161995032964839
身份信息的年月日均符合规范，可以保存！！

3.3 习题

一、选择题

1. 给定如下 Java 代码,下列哪个选项可以加入 China 类中,并能保证编译正确? (   )

```
public class person{
    public float getNum(){
        return 3.0f;
    }
}
public class China extends Person{
}
```

A. public float getNum(){return 8.0f;}
B. public void getNum(){}
C. public getNum(float f){}
D. public double getNum(float f){return 8.0f;}

2. 在子类的构造方法中,可以使用(   )关键字来调用父类的构造方法。

A. base   B. super
C. this   D. extends

3. 下列选项中关于抽象类和抽象方法的描述正确的是(   )。

A. 抽象类中不可以有非抽象方法
B. 某个非抽象类的父类是抽象类,则这个类必须重载父类的所有抽象方法。
C. 抽象类无法实例化
D. 抽象方法的方法体部分必须使用一对大括号括起来

4. 编译运行如下 Java 代码,输出结果是(   )。

```
public class Base{
    public void method(){System.out.print("Base method");}
}
public class child extends Base{
    public void method(){System.out.print("Child method");}
}
public class Test{
    public static void main(String[ ]args){
        Base base = new Child();
```

```
        Base.method();
    }
}
```

A. Base method

B. Child method

C. Base method Child method

D. 编译错误

5. 下列选项中,关于 Java 接口说法错误的是(　　)。

A. Java 接口是一些方法特征的集合

B. Java 接口中的方法是没有方法实现的

C. Java 接口中的方法在不同的地方进行实现

D. Java 接口中的方法实现必须具有相同的行为

6. 分析如下的 Java 代码,第(　　)行代码不能通过编译。

```
public interface MyInter{
    public final static int numOne = 10;//1
    public int numTwo;//2
    abstract void find();//3
    public void query();//4
}
```

A. 1   B. 2   C. 3   D. 4

7. 在 Java 中,以下关于接口和抽象类的说法错误的是(　　)。

A. 抽象类和接口中都可以定义抽象方法

B. 抽象类也可以实现接口

C. 接口可以继承接口,抽象类不可以继承抽象类

D. 接口中只可以定义常量属性,但是抽象类中既可以定义常量属性也可以定义变量属性

8. 下面的程序中定义了一个接口,其中包含(　　)处错误。

```
publci interface Shape{
    private int MAX_SIZE = 20;
    int MIN_SIZE = 10;
    void use(){System.out.println("use it");};
    private int getSize();
    void setSize(int i);
}
```

A. 1   B. 2   C. 3   D. 编译错误

9. 下面选项中关于 Java 中接口的说法正确的是(   )。
   A. 接口中既可以定义变量,也可以定义常量
   B. 接口中的方法都是抽象方法
   C. 一个类只能实现一个接口
   D. 如果一个类或接口要实现另一个接口,均使用 implements 关键字实现
10. 运行下列程序段,s2 的结果是(   )。

```
String s1 = new String("abc");
String s2 = "ef";
s2 = s1.toUpperCase().concat(s2);
s2 = s2.substring(2,4);
```

   A. Cef   B. cef   C. Ce   D. BCe

11. 运行下面的程序段,输出结果是(   )。

```
String s1 = new String("abc");
StringBuffer s2 = new StringBuffer("abc");
s2.append(s1);
s1 = s2.toString();
s1.concat("abc");
System.out.println(s1);
```

   A. abc   B. abcabc
   C. 编译错误   D. abcabcabc

12. 在 Java 中,以下代码的运行结果是(   )。

```
public class Test{
    public static void main(String[ ]args){
        String str1 = new String("abc");
        String str2 = new String("abc");
        System.out.print(str1 = = str2);
        System.out.print(",");
        System.out.print(str1.equals(str2));
    }
}
```

   A. false,false   B. true,true
   C. true,false   D. false,true

13. 阅读下列 Java 代码片段,输出结果正确的是(   )。

```
String str = "select name,sex from user";
Int num1 = str.indexOf("e");
Int num2 = str.lastIndexOf("e");
System.out.println(num1 + num2);
```

A. 22　　　　　　B. 23　　　　　　C. 24　　　　　　D. 25

14. 下列关于字符串的叙述中错误的是(　　)。
A. 字符串是对象
B. String 对象存储字符串的效率高于 StringBuffer
C. 可以使用 StringBuffer sb＝"这里是字符串"声明并初始化 StringBuffer 对象 sb
D. String 类提供了许多用于操作字符串的方法：连接、提取、查询等

二、简答题

1. 简述 this 和 super 的区别及适应场合。
2. 简述对抽象类和抽象方法的理解。
3. 简述实现动态性多态的条件有哪些。
4. 请指出如下 Java 代码中存在的错误，并解释其原因。

```
public class Other{
    public int I;
}
public class Something{
    public static void main(String[ ]args){
        Other o = new Other();
        new Something().add(o);
    }
    public void add(final Other o){
        o.i++;
        o = new Other();
    }
}
```

5. 请指出如下 Java 代码中存在的错误，并解释原因。

```
public class Person{
    String name;
    int age;
    public void eat(){
        System.out.print("person eating with mouth");
    }
}
```

```java
        public void sleep(){
            System.out.print("sleeping in night");
        }
    }
    public class Chinese extends Person{
        public void eat(){
            System.out.print("Chinese eating rice with mouth by chopsticks");
        }
        public void shadowBoxing(){
            System.out.print("practice shadowboxing every morning");
        }
    }
    public class Test{
        public static void main(String[] args){
            Chinese ch = new Chinese();
            ch.eat();
            ch.sleep();
            ch.shadowboxing();
            Person p = new Chinese();
            p.eat();
            p.sleep();
            p.shadowboxing();
        }
    }
```

6. 编码题：设计 Bird、Fish 类，都继承自抽象类 Animal，实现抽象方法 showInfo()，并输出它们的信息，要求画出类图。

7. 编码题：请使用多态实现以下要求：小明有两个好朋友，一个是中国的小强，喜欢吃粤菜、练太极拳；一个是美国的约翰，喜欢吃披萨、打橄榄球。每当朋友来拜访的时候，小明都会按照各人的喜好招待他们。

8. 阅读如下 Java 代码，指出其中存在的错误，并说明原因。

```java
    publci interface Constants{
        int MAX = 100;
        int MIN = 10;
    }
    public class Test{
        public static void main(String[] args){
            Constants con = new Constants();
```

```
            System.out.println(con.MAX);
            int i = 50;
            if(I>Constants.MAX){
    Constants.MAX = i;
  }
  }
  }
```

9. 阅读下面 Java 代码,并给出运行结果。

```
public interface Animal{
    void shout();
}
public Dog implements Animal{
    public void shout(){
        System.out.println("W W!");
    }
}
public Cat implements Animal{
    public void shout(){
        System.out.println("M M!");
    }
}
public class Store{
    public static Animal get(String choice){
        if(choice.equalsIgnoreCase("dog")){
            return new Dog();
}else{
    return new Cat();
}
}
}
public class AnimateTest{
    public static void main(String[]args){
        Animal a1 = Store.get("dog");
        a1.shout();
}
}
```

10. 在第 9 题的基础上进行功能扩展，要求如下：
(1) 增加一种新的动物类型：Pig(猪)，实现 shout()方法。
(2) 修改 Store 类的 get()方法：如果传入的参数是字符串 dog，则返回一个 Dog 对象；如果传入的参数是字符串 pig，则返回一个 Pig 对象；否则，返回一个 Cat 对象。
(3) 在测试类中加以测试。

11. 编写程序，求柱体的体积：
(1) 为柱体的底面设计一个接口 Geometry，包含计算面积的方法 getArea()；。
(2) 为柱体设计类 pillar，要求：
① 有两个成员变量：底面和高度。底面是任何可以计算面积的几何形状。
② 实现构造方法，对成员变量赋值。
③ 包含成员方法，计算柱体 pillar 的体积。
(3) 编写测试类圆形类、矩形类实现 Geometry 接口，编写测试类 Test，分别用圆形、矩形作为柱体的底面，并计算其体积。

12. 输入 5 种水果的英文名称（如葡萄 grape、橘子 orange、香蕉 banana、苹果 apple、桃 peach），编写一个程序，输出这些水果的名称（按字典里出现的先后顺序输出）。效果如下图所示。

```
请输入第1种水果：grape
请输入第2种水果：apple
请输入第3种水果：banana
请输入第4种水果：peach
请输入第5种水果：orange

这些水果在字典中出现的顺序是：
apple
banana
grape
orange
peach
```

13. 假设中国人的姓都是单个字，请随机输入一个人的姓名，然后分别输出姓和名。效果如下图所示。

```
输入任意一个姓名：张雪怡

姓氏：  张
名字：  雪怡
```

14. 录入用户的 18 位身份证号码,从中提取用户生日的年、月、日。效果如下图所示。

```
Console
<terminated> GetBirthday [Java Application] E:\Program Files\Java\jre7\bin\javaw.exe (2017年8月14日 下午4:29:
请输入用户的身份证号码: 652384199203185371

该用户生日是: 1992年03月18日
```

15. 编写一个程序,当输入一串字符串时,给定要查找的字符内容,显示该字符或字符串在整个字符串中的位置。效果如下图所示。

```
Console
<terminated> StringExplorer [Java Application] E:\Program Fil
请输入一段字符:
我喜欢我的电脑
请输入要查询的字符串:
我
我出现的位置是: 0    3
```

16. 对录入的信息进行有效性验证。录入会员生日时,形式必须是"月/日",如"09/12"。录入的密码位数必须为 6~10 位,允许用户重复录入,直到正确为止。效果如下图所示。

```
Console
<terminated> InputVerify [Java Application] E:\Program Files\Java\jre7\bin\javaw.exe (2017年8月14
请输入会员生日<月/日: 00/00>:  35.9
生日形式输入错误!

请输入会员生日<月/日: 00/00>:  9/15
生日形式输入错误!

请输入会员生日<月/日: 00/00>:  09/15
该会员生日是: 09/15

请输入会员密码<6~10位>:  wang
密码形式输入错误!

请输入会员密码<6~10位>:  wang0915
该会员的密码是: wang0915
```

学生成绩信息的异常处理

项目 4 首先介绍了异常的概念、异常的处理机制、异常类的层次和主要子类、异常类的方法和属性,其次讲解了异常的处理,尤其是用户自定义的异常,为后续 Java 面向对象程序设计中程序的异常处理奠定基础。

工作 任务

学生成绩输入异常的处理(类)。

学习 目标

掌握异常的概念,理解异常的处理机制,掌握异常处理的方法,能够根据实际情况独立设计用户自定义异常。

任务 学生成绩输入异常的处理(类)

任务描述 及分析

小明在录入学生成绩时(假设要录入 10 人的综合成绩,以便计算平均成绩),不小心将第 5 个学生的成绩输入错误,将 89.7 输入为 897。此时,由于 897 是合法的数据,所以系统没有进行提示,小明就继续录入成绩,导致最终计算出来的平均成绩错误。要求小明针对此种情况,完成用户自定义异常程序设计,实现成绩输入的异常检测。要完成这个工作任务,第一步,要掌握 Java 对异常的处理机制及异常类的方法和属性;第二步,要掌握 Java 语言中,不同类型异常的处理方式;第三步,根据要解决的实际问题,选取合适的异常处理方式,然后具体设计实现。

相关知识

4.1.1 异常概述

异常是导致程序中断运行的一种指令流。如果不对异常进行正确的处理,则可能导致程序的中断执行,造成不必要的损失,所以在程序的设计中必须要考虑各种异常的发生,并正确作好相应的处理,这样才能保证程序正常的执行。在 Java 语言中,Java 异常处理机制提供了去检查及处理产生各种错误、异常事件的方法。

1. 异常的概念

在 Java 程序运行过程中发生的、会打断程序正常执行的事件称为异常。程序运行过程中可能会有许多意料之外的事情发生,例如,零用作了除数(产生算术异常 ArithmeticException),在指定的磁盘上没有要打开的文件(产生文件未找到异常 FileNotFoundException),数组下标越界(产生数组下标越界异常 ArrayIndexOutOfBoundsException),等等。对于一个实用的程序来说,处理异常的能力是一个不可缺少的部分,它的目的是保证程序在出现任何异常的情况下仍能按照计划执行。

【例 4.1】Java 系统对除数为 0 的异常处理演示。

```
public class Exception1{
public static void main(String args[]){
System.out.println("* * * * * * * * * *除数为0异常演示开始* * * * * *
* * * * *");
int i = 10;//定义整型变量
int j = 0;//定义整型变量
int temp = i/j;//此处产生了异常
System.out.println("两个数字相除的结果:" + temp);
System.out.println("* * * * * * * * * * * *演示结束* * * * * * * * * *
*");
}
}
```

如图 4-1 所示。

```
E:\>java Exception1
********** 除数为0异常演示开始 **********
Exception in thread "main" java.lang.ArithmeticException: / by zero
        at Exception1.main(Exception1.java:6)
```

图 4-1

屏幕显示的信息指明了异常的类型:ArithmeticException:/by zero(算术异常/用0除)。在该例中,未进行程序的异常处理。这是因为除数为零是算术异常,它属于运行时异常(RunTimeException),通常运行时异常在程序中不作处理,Java运行时系统能对它们进行处理。Java语言本身提供的Exception类已考虑了多种异常的处理,如果要想处理异常,则必须掌握异常的处理机制。

2. Java 的异常处理机制

Java异常处理机制提供了一种统一和相对简单的方法来检查及处理可能的错误,处理的步骤如下:

(1)一旦产生异常,则首先会产生一个异常类的实例化对象。

(2)异常处理程序对此异常对象进行捕捉。

(3)产生的异常对象与异常处理程序中的各个异常类型进行匹配。如果匹配成功,则执行相应的异常处理代码;如果系统找不到相应的异常处理程序,则由 Java 默认的异常处理程序来处理,即在输出设备上输出异常信息,同时程序停止运行。

3. 异常类的层次和主要子类

Java用面向对象的方法处理异常,所有的异常都是 Throwable 类的子类生成的对象,即所有的异常类都是 Throwable 类的子类。Throwable 类的父类是 Java 的基类 Object,它有两个直接子类:Error 类和 Exception 类,运行时异常 RuntimeException 类是 Exception 类的子类。如图 4-2 所示。

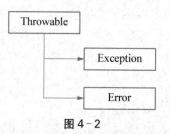

图 4-2

Exception 类一般表示的是程序中出现的问题,可以直接使用 try-catch 处理;Error 类一般指的是 JVM 错误,程序中无法处理;只有 Throwable 类的后代才可以作为一个异常被抛出。Java 语言的异常处理的主要子类如表 4-1~表 4-3 所示。

表 4-1

子类名	子类名说明
AbstractMethodError	调用抽象方法错误
ClassFormatError	类文件格式错误
IllegalAccessError	非法访问错误
IncompatibleClassChangeError	非法改变一个类错误
InstantiationError	实例化一个接口或抽象类错误
InternalError	Java 内部错误
LinkageError	连接失败错误

续 表

子类名	子类名说明
NoClassDefFoundError	找不到类定义错误
NoSuchFieldError	域未找到错误
NoSuchMethodError	调用不存在的方法错误
OutOfMemoryError	内存溢出错误
StackOverflowError	堆栈溢出错误
ThreadDeadError	线程死亡错误
UnknownError	未知错误
UnstatisfiedLinkError	链接不满足错误
VerifyError	校验失败错误
VirtualMachineError	虚拟机错误

表 4-2

子类名	子类名说明
ClassNotFoundException	类未找到异常
DataFormatException	数据格式异常
IllegalAceessException	非法存取异常
InstantiationException	实例化异常
InterruptedException	中断异常
NoSuchMethodException	调用不存在方法异常
RuntimeException	运行时异常

表 4-3

子类名	子类名说明
ArithmeticException	算术异常
ArrayIndexOutOfBoundException	数组越界异常
ArrayStoreException	数组存储异常
ClassCastException	类强制转换异常
IllegalArgumentException	非法参数异常
IllegalThreadstateException	非法线程状态异常
IndexOutOfBoundException	索引越界异常
NegativeArraySizeException	负值数组大小异常
NullPointerException	空引用异常

续 表

子类名	子类名说明
NumberFormatException	数值格式异常
SecurityException	安全异常
StringIndexOutOfBoundException	字符串越界异常

4. 异常类的方法和属性

(1) 异常类的构造方法。

Exception 类有 4 个重载的构造方法,常用的两个构造方法为:

① public Exception()创建新异常;

② public Exception(String message)用字符串参数 message 描述异常信息,创建新异常。

(2) 异常类的方法。

Exception 类常用的方法有:

① public String toString()返回描述当前异常对象信息的字符串。

② public String getMessage()返回描述当前异常对象的详细信息。

③ public void printStackTrace()在屏幕上输出当前异常对象使用堆栈的轨迹,即程序中先后调用了哪些方法,使得运行过程中产生了这个异常对象。

注意:

在输出异常信息的时候,不仅可以使用 public void printStackTrace(),还可直接使用 System. out. println()打印异常对象。

4.1.2 异常处理

在 Java 语言中,异常有以下 3 种处理方式:

(1) 可以不处理运行时异常,由 Java 虚拟机自动进行处理。

(2) 使用 try-catch-finally 语句捕获异常。

(3) 通过 throws 子句声明抛出异常,还可以自定义异常,用 throw 语句来抛出它。

1. 运行时异常

运行时异常是系统在程序运行中检测到的,可能在程序的任何部分发生,而且数量可能较多,如果逐个处理,工作量很大,有可能影响程序的可读性及执行效率。因此,Java 编译器允许程序不对运行时异常进行处理,而将它交给默认的异常处理程序,一般的处理方法是在屏幕上输出异常的内容以及异常的发生位置。必要的时候,也可以声明、抛出、捕获运行时异常。

【例 4.2】运行时异常演示。

```
public class Exception2{
public static void main(String[]args){
String s = "12.3";
Integer i = new Integer(s);
```

```
        System.out.println(i);
    }
}
```

如图 4-3 所示。

```
E:\>javac Exception2.java

E:\>java Exception2
Exception in thread "main" java.lang.NumberFormatException: For input string: "1
2.3"
        at java.lang.NumberFormatException.forInputString(NumberFormatException.
java:48)
        at java.lang.Integer.parseInt(Integer.java:458)
        at java.lang.Integer.<init>(Integer.java:660)
        at Exception2.main(Exception2.java:4)
```

图 4-3

本程序运行结果说明，程序运行时产生一个 NumberFormatException 数值格式异常。在用 Integer 将一个字符串转换为 Integer 数据时，参数字符串格式不对，所以产生了这个运行时异常。Java 系统调用方法 printStackTrace()将调用堆栈的轨迹打印了出来，输出的第一行信息也是 toString()方法输出的结果，对这个异常对象进行简单说明。其余各行显示的信息表示了异常产生过程中调用的方法，最终是在调用 Integer.parseInt()方法时产生的异常，调用的出发点在 main()方法中。

2. try-catch-finally 语句

在 Java 语言中，允许自己来处理异常。Java 语言提供 try-catch-finally 语句来捕获和处理异常，该语句的格式如下：

```
try{
statements//可能产生异常的语句
}catch(Throwable-subclass e){//异常参数
statements//异常处理程序
}catch(Throwable-subclass e){//异常参数
statements//异常处理程序
}...
finally{//finally 将作为异常的统一出口,不管是否有异常都会执行此语句,即使
在之前有 return 语句一样要执行 finally
statements
}
```

try 语句块中是可能产生异常对象的语句，一旦其中的语句产生了异常，程序即跳到紧跟其后的第一个 catch 子句。try 程序块之后的 catch 子句可以有多个，也可以没有。

catch 子句中都有一个代表异常类型的形式参数，这个参数指明了这个 catch 程序块可以捕获并处理的异常类型，若产生的异常类型与 catch 子句中声明的异常参数类型匹配，则

执行这个 catch 程序块中的异常处理程序。"匹配"是指异常参数类型与实际产生的异常类型一致或是其父类。若不匹配,则顺序寻找下一个 catch 子句,因此 catch 语句的排列顺序应该从特殊到一般,否则,放在后面的 catch 语句将永远执行不到。

此外,还可以用一个 catch 语句处理多个异常类型,这时它的异常类型参数应该是这多个异常类型的父类。

若所有 catch 参数类型与实际产生的异常类型都不匹配,则标准异常处理程序将被调用,即在输出设备上输出异常信息,同时程序停止运行。

【例 4.3】 对除数为 0 的异常进行捕获处理。

```java
public class Exception3{
    public static void main(String args[]){
        System.out.println("＊＊＊＊＊＊＊＊＊＊除数为0异常演示开始＊＊＊＊＊＊＊＊＊＊＊");
        int i=10;//定义整型变量
        int j=0;//定义整型变量
        try{
            int temp=i/j;//此处产生了异常
            System.out.println("两个数字相除的结果:"+temp);
            System.out.println("--------------------------");
        }catch(ArithmeticException e){
            System.out.println("出现异常了:"+e);
        }
        System.out.println("＊＊＊＊＊＊＊＊＊＊＊演示结束＊＊＊＊＊＊＊＊＊＊＊");
    }
}
```

如图 4-4 所示。

```
E:\>java Exception3
********** 除数为0异常演示开始 **********
出现异常了: java.lang.ArithmeticException: / by zero
********** 演示结束 **********
```

图 4-4

上例中只使用了基本的异常处理格式:try-catch。try 中捕获异常,出现异常之后的代码将不再被执行,而是跳转到相应的 catch 语句中执行,用于处理异常。此外,对于异常也可以使用 finally 设置其统一的出口。

【例4.4】使用 finally 设置除数为 0 的异常处理的统一出口。

```
public class Exception4{
public static void main(String args[]){
System.out.println("＊＊＊＊＊＊＊＊＊＊除数为0异常演示开始＊＊＊＊＊
＊＊＊＊＊");
int i = 10;//定义整型变量
int j = 0;//定义整型变量
try{
int temp = i/j;//此处产生了异常
System.out.println("两个数字相除的结果:"+temp);
System.out.println("---------------------
--");
}catch(ArithmeticException e){//捕获算术异常
System.out.println("出现异常了:"+e);
}finally{//作为异常的统一出口
System.out.println("不管是否出现异常,都执行此代码");
}
System.out.println("＊＊＊＊＊＊＊＊＊＊演示结束＊＊＊＊＊＊＊＊＊
＊");
}
}
```

如图 4-5 所示。

```
E:\>java Exception4
＊＊＊＊＊＊＊＊＊ 除数为0异常演示开始 ＊＊＊＊＊＊＊＊＊＊
出现异常了: java.lang.ArithmeticException: / by zero
不管是否出现异常, 都执行此代码
＊＊＊＊＊＊＊＊＊ 演示结束 ＊＊＊＊＊＊＊＊＊＊
```

图 4-5

上述两个示例只是针对一个异常进行处理,如果有多个异常,则需要对应的多个 catch 语句进行处理。

【例4.5】多 catch 子句示例。

```
public class Exception5{
public static void main(String args[]){
try{
```

```
        int a = args.length;
        System.out.println("a = " + a);
        int b = 42/a;
        int c[] = {1};
        c[4] = 99;
    }catch(ArithmeticException e){//捕获算术运算异常
        System.out.println("Divide by 0:" + e);
    }catch(ArrayIndexOutOfBoundsException e){//捕获数组下标越界异常
        System.out.println("Array index oob:" + e);
    }
    System.out.println("After try/catch blocks.");
    }
}
```

如图 4-6 所示。

```
E:\>java Exception5
a = 0
Divide by 0: java.lang.ArithmeticException: / by zero
After try/catch blocks.
```

图 4-6

3. throw 语句和 throws 子句

throw 语句可使得用户自己根据需要抛出异常。throw 语句以一个异常类的实例对象作为参数,其一般的形式是:

```
MethodName(arguments)throws 异常类{
    ……
    throw new 异常类();
}
```

new 运算符被用来生成一个异常类的实例对象。例如:

```
throw new ArithmeticException();
```

包含 throw 语句的方法要在方法头参数表后书写 throws 子句,它的作用是通知所有要调用此方法的其他方法,必须要处理该方法抛出的那些异常。若方法中的 throw 语句不止一个,throws 子句应指明抛出的所有可能产生的异常。

注意:

(1) 通常使用 throws 子句的方法本身不处理本方法中产生的异常,而是由调用该方法的方法进行处理,并且应该在调用方法的 try 块中调用包含 throw 语句的方法,才能捕获 throw 抛出的异常。

(2) throw 语句一般应放入分支语句中,表示仅在满足一定条件后才被执行,而且 throw 语句后不允许有其他语句,否则将出现编译错误信息 unreachable statement。

(3) 主方法 main()也可以使用 throws 关键字,但主方法是程序的起点,所以主方法向上抛异常,就只能将异常抛给 JVM 处理。此外,使用 throws 关键字后,在 main()主方法中不对异常进行处理。

【例 4.6】throw 关键字示例。

```
public class throw1{
public static void main(String args[]){
try{
throw new Exception("throw 演示");//用户自己根据需要,抛出异常的实例化对象
}catch(Exception e){
System.out.println(e);
}
}
}
```

如图 4-7 所示。

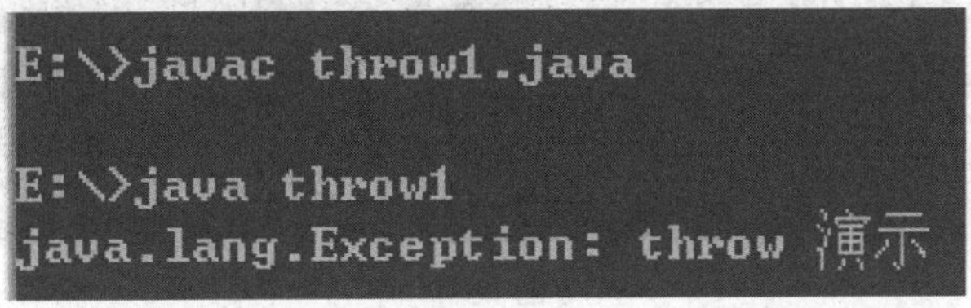

```
E:\>javac throw1.java

E:\>java throw1
java.lang.Exception: throw 演示
```

图 4-7

【例 4.7】throws 关键字示例。

```
class Math{
public int div(int i,int j)throws Exception{//定义除法操作,
//如果有异常,则交给被调用处处理(即由程序根据实际情况决定是否抛出异常)
System.out.println("*****除法操作计算开始*****");
int temp=0;//定义局部变量
try{
temp=i/j;//计算,但是此处有可能出现异常
}catch(Exception e){
throw e;
}finally{//不管是否有异常,都要执行统一出口
```

```
System.out.println("*****除法操作计算结束*****");
}
return temp;
}
};
public class throws1{
public static void main(String args[]){
Math x = new Math();
try{
System.out.println("除法操作:" + x.div(10,0));
}catch(Exception e){
System.out.println("异常产生:" + e);
}
}
}
```

如图 4-8 所示。

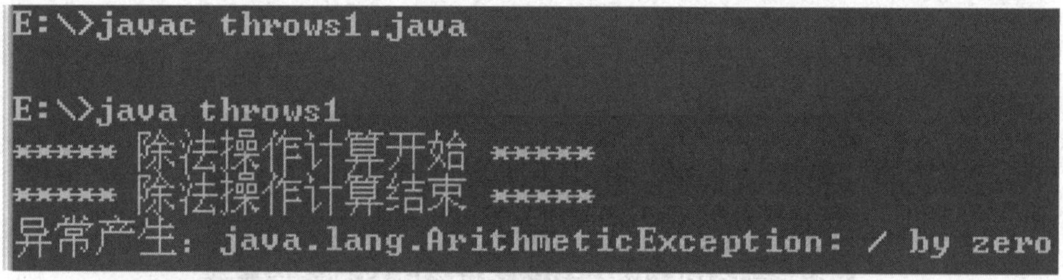

```
E:\>javac throws1.java

E:\>java throws1
***** 除法操作计算开始 *****
***** 除法操作计算结束 *****
异常产生：java.lang.ArithmeticException: / by zero
```

图 4-8

4. 用户自定义异常

Java 语言还允许定义用户自己的异常类,从而实现用户自己的异常处理机制。定义自己的异常类要继承 Throwable 类或它的子类,通常是继承 Exception 类。

【例 4.8】用户自定义异常示例。

```
class U_Exception extends Exception{//自定义异常类,继承 Exception 类
public U_Exception(String str){//定义构造函数
super(str);//调用 Exception 类中有一个参数的构造方法,传递错误信息
}
};
public class User_Exception{
public static void main(String args[]){
try{
```

```
throw new U_Exception("用户自定义异常演示.");//抛出异常
}catch(Exception e){
System.out.println(e);
}
}
}
```

如图 4-9 所示。

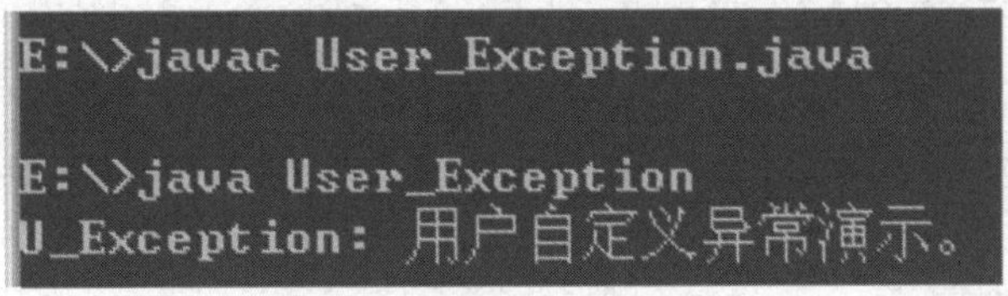

图 4-9

任务 实施

4.1.3 学生成绩输入异常处理实现(类)

```
import java.util.Scanner;
//用户自定义异常
class MyException extends Exception{
public String inputTest()
{
return"没有输入 0—100 的实数";
}
}
//录入学生成绩
public class ScoreExceptionTest{
public static void main(String[]args){
double s[] = new double[10];//存放10个学生的综合成绩
double sum = 0.0,average = 0.0;//变量 sum 存放总成绩,average 存放平均成绩
Scanner scan = new Scanner(System.in);//键盘输入
try{
System.out.println("请输入 10 个学生的综合成绩:");
for(int i = 0;i<10;i + + )
{
```

```
System.out.println("第" + (i+1) + "个学生的综合成绩:");
s[i] = scan.nextDouble();
if(s[i]<0||s[i]>100){
throw new MyException();
}
sum = sum + s[i];
}
}
catch(MyException e)
{
System.out.println(e.inputTest());
}
finally
{
average = sum/10;
System.out.println("该门课程的平均成绩为:" + average);
}
}
}
```

如图 4-10、图 4-11 所示。

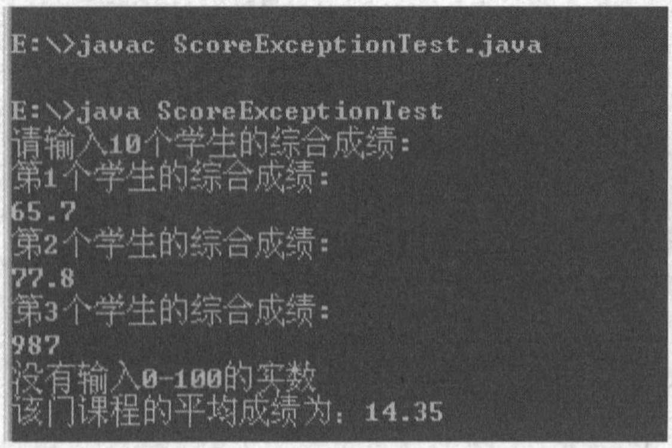

图 4-10

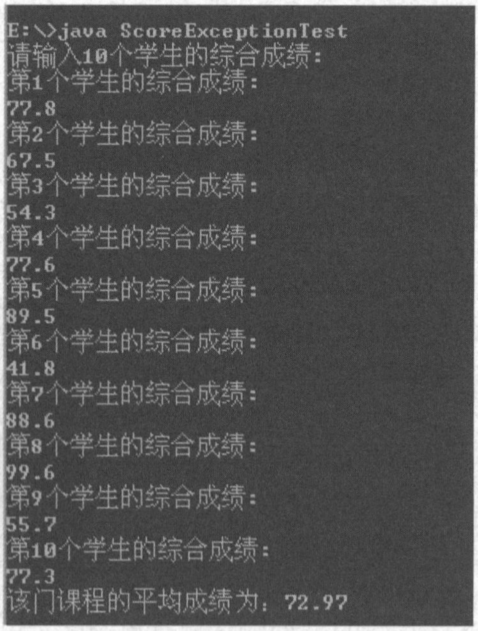

图 4-11

任务拓展

定义一个银行类模拟银行账户,通过银行卡的号码完成识别,可以进行存钱和取钱的操作,然后自定义异常类用于抛出当取出钱多于余额时的错误提示。

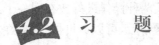

4.2 习题

一、选择题

1. 对下面的代码段:

```
try{
run();
}catch(IOException e){
System.out.println("Exception1");
return;
}catch(Exception e1){
System.out.println("Exception2");
return;
}finally{
System.out.println("finally");
}
```

若 run()方法抛出一个空指针异常 NullPointerException,显示器上将显示(　　)。

A. 无输出

B. Exception1
　　finally

C. Exception2
　　finally

D. Exception1

E. Exception2

2. 处理异常用到 3 个保留字,除了 try 外,还有(　　)。

A. catch　　　　　B. class　　　　　C. throw　　　　　D. return

3. catch 一般放在其他 catch 子句的后面,该子句的作用是(　　)。

A. 抛出异常　　　　　　　　　　　B. 捕获所有类型的异常

C. 检测并处理异常　　　　　　　　D. 有语法错误

4. 关于异常的描述中,错误的是(　　)。

A. 异常既可以被硬件引发,又可以被软件引发

B. 运行异常可以预料,但不能避免,它是由系统运行环境造成的
C. 异常是指从发生问题的代码区域传递到处理问题的代码区域的一个对象
D. 在程序运行中,一旦发生异常,程序立即中断运行
5. 下列说法中错误的是(　　)。
A. 引发异常后,首先在引发异常的函数内部寻找异常处理过程
B. 抛出异常是没有任何危险的
C. "抛出异常"和"捕捉异常"两种操作最好放在同一个函数中
D. 异常处理过程在处理完异常后,可以通过带有参数的 throw 继续传播异常

二、简答题

1. 什么是异常?试列出 3 个系统定义的运行时异常类。
2. try-catch-finally 语句的执行顺序是怎样的?

学生成绩信息保存到文件

项目 5 首先总体介绍了 Java 程序的输入、输出处理,其次讲解了文件的顺序访问、文件的随机访问、字节—字符转换流,然后讲解了目录和文件管理,最后介绍了其他常用的流处理,为后续 Java 网络程序设计奠定基础。

工作 任务

学生成绩信息保存到文件。

学习 目标

(1) 掌握 File 类的作用,能够使用 File 类中的方法对文件进行操作。

(2) 掌握流的概念,掌握字节流与字符流的作用,掌握文件的标准操作步骤,掌握字节与字符操作的区别,能够用 InputStream 和 OutputStream 的子类或 Reader 和 Writer 的子类实现文件的顺序访问。

(3) 掌握 OutputStreamWriter 类的作用,掌握 InputStreamReader 类的作用,能够实现字节流—字符流转换。

(4) 掌握 RandomAccessFile 类的作用,能够使用 RandomAccessFile 类中的方法读取指定位置的数据。

(5) 理解内存操作流、管道流、合并流、压缩流、BufferedReader 类等常用的 Java IO 流处理,能够根据工作任务需要选取合适的 IO 操作。

5.1 任务 学生成绩信息保存到文件

小明在之前的程序设计中发现,无论是用结构化方法还是面向对象方法来实现学生成绩管理系统,都不能永久保存学生的成绩信息,每次运行程序都要先输入学生成绩等基本信息,比较麻烦。要求小明针对此种情况,完成将学生成绩信息保存到文件中的程序设计。要完成这个工作任务,第一步,要对 Java 程序的输入、输出操作有整体的认知;第二步,需要在总体认知的基础上,掌握对 Java 程序实现顺序输入输出操作的最核心的 5 个类(File、InputStream、OutputStream、Reader、Writer),同时还要掌握字节流—字符流的转换;第三步,在掌握文件顺序访问的基础上,进一步学习文件的随机访问;第四步,了解并理解部分常用的 Java IO 流处理;第五步,根据工作任务要求结合已掌握的 Java IO 程序设计知识,选取适当的 Java IO 流实现学生成绩保存到文件中。

相关知识

5.1.1 输入/输出流概念

为了使得一个 Java 程序能与外界交流数据信息,Java 语言必须提供输入/输出的功能。例如,从键盘读取数据,从文件中读数据或向文件中写数据,将数据输出到打印机以及在一个网络连接上进行读写操作等。输入/输出时,数据在通信通道中流动。所谓"数据流 stream"指的是所有数据通信通道之中数据的起点和终点。例如,执行程序通常会输出各种信息到显示器,用户可以随时了解程序的状态信息,而这些信息的通道就是一个数据流,其中的数据就是要显示的信息,数据的源(起点)就是执行的程序,而数据的终点就是显示器。又如,一个程序在打开某一文件时,程序和文件之间就建立起一个数据流,文件的内容就是数据流中的数据。若这个文件是程序所要读取的,那么数据流的源就是文件,而目的就是程序;若要对文件进行写入操作,则情况相反。总之,只要是数据从一个地方"流"到另外一个地方,这种数据流动的通道都可以称为数据流。

从程序设计的角度看,用数据流的概念编写程序也是比较简单的。当程序是数据流的源时,一旦建立起数据流,便可以不去理会数据流的目的是什么(可能是显示器、打印机、网络系统中的远端客户等),可以将对方看成是一个会接受数据的"黑匣子",程序只负责提供数据就可以了。而程序若是数据流的终点目的,那么等到数据流建立完成后,也同样不必关心数据流的起点是什么,只要索取自己想要使用的数据就可以了。

输入/输出是相对于程序来说的,程序在使用数据时所扮演的角色有两个:一个是源,一个是目的。若程序是数据流的源,即数据的提供者,这个数据流对程序来说就是一个"输出数据流"(数据从程序流出);若程序是数据流的终点,这个数据流对程序而言就是一个"输入数据流"(数据从程序外流向程序)。

5.1.2 输入/输出类

在 java.io 包中有 60 多个类,用于处理 Java IO 流。这些类从功能上分为两大类:输入流和输出流。输入数据用输入流,输出数据用输出流。从流结构上可分为字节流(以字节为处理单位)和字符流(以字符为处理单位)。字节流的输入流和输出流基础是 InputStream 和 OutputStream 这两个抽象类,字节流的输入输出操作由这两个类的子类实现;字符流是 Java 1.1 版后新增加的以字符为单位进行输入输出处理的流,字符流输入输出的基础是抽象类 Reader 和 Writer。InputStream 和 OutputStream、Reader 和 Writer 主要用于文件的顺序访问,对文件进行随机访问用 RandomAccessFile 类,它允许对文件进行随机访问,还可以同时对文件进行输入读或输出写操作。

上述输入/输出类主要用于 Java 程序中数据的输入、输出操作,而在整个 java.io 包中,唯一与文件本身有关的类就是 File 类,使用 File 类可以进行创建或删除文件等常用操作。

5.1.3 目录和文件管理

java.io 包中的 File 类提供了与具体平台无关的方式来描述目录和文件对象的属性功能,其中包含大量的方法可用来获取路径、目录和文件的相关信息,并对它们进行创建、删除、改名等管理工作。因为不同的系统平台对文件路径的描述不尽相同,为做到与平台无关,Java 语言中使用抽象路径等概念。Java 自动进行不同系统平台的文件路径描述与抽象文件路径之间的转换。

File 类的直接父类是 Object 类,使用 File 类进行创建或删除文件等常用操作的方法如下。

1. 目录管理

File 类用于目录操作的主要方法有:

(1) public boolean mkdir()根据抽象路径名创建目录。

(2) public String[]list()返回抽象路径名表示路径中的文件名和目录名。

2. 文件管理

File 类还可以对文件进行新建、删除、属性修改等操作。此外,在进行文件操作时,常需要知道一个文件的相关信息,Java 的 File 类提供了相应的成员方法用于获得文件的信息。

(1) 新建文件对象。

可用 File 类的构造方法来生成 File 对象,File 类的构造方法有以下 3 种:

① File(String pathname)通过给定的路径名变换的抽象路径创建文件对象。

② File(File parent,String child)从父抽象路径(目录)和子路径字符串创建文件对象。

③ File(String parent,String child)从父路径和子路径字符串创建文件对象。

这些构造方法取决于访问文件的方式。例如,若在应用程序里只用一个文件,第一种创建文件的结构是最容易的。但若需要在同一目录里打开数个文件,则后两种方法更好一些。

【例5.1】在 E 盘根目录创建一个记事本文件,其名称为 ceshi.txt。

```
import java.io.File;
import java.io.IOException;
public class File1{
public static void main(String args[]){
File f1 = new File("E:\\ceshi.txt");//实例化 File 类的对象
try{
f1.createNewFile();//创建文件,根据给定的路径创建
}catch(IOException e){
e.printStackTrace();//输出异常信息
}
}
}
```

如图 5-1 所示。

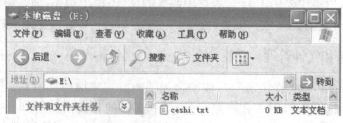

图 5-1

上例已经成功创建了记事本文档 ceshi.txt。但是，在各种操作系统中，路径的分隔符是不一样的。例如：

在 Windows 中使用反斜杠："\"；

在 Linux 中使用正斜杠："/"。

如果想让 Java 程序的可移植性继续保持，最好根据所在的操作系统来自动使用分隔符，即使用常量 File.pathSeparator 和 File.separator，如下例所示。

【例 5.2】在 E 盘的 file 文件夹下创建记事本文档，其名称为 ceshi.txt。

```
import java.io.File;
import java.io.IOException;
public class File2{
public static void main(String args[]){
File f1 = new File("E:" + File.separator + "file" + File.separator + "ceshi.txt");//实例化File类的对象
try{
f1.createNewFile();//创建文件,根据给定的路径创建
}catch(IOException e){
e.printStackTrace();//输出异常信息
}
}
}
```

如图 5-2 所示。

图 5-2

(2) 文件测试和管理。

文件对象创建成功后,便可使用下述方法来获得文件相关信息或管理文件:

① 获得文件名。
- public String getName()得到一个文件名。
- public String getParent()得到一个抽象路径的父路径名。
- public String getPath()得到一个文件的路径名。
- public String getAbsolutePath()得到一个抽象路径的绝对路径名。

② 获得一般文件信息。
- public long lastModified()得到抽象路径表示的文件最近一次修改的时间。
- public long length()得到抽象路径表示的文件的长度。

③ 文件重命名。
- public boolean renameTo(File dest)将抽象路径文件名重命名为给定的新文件名。

④ 文件删除。
- public boolean delete()删除抽象路径表示的文件或目录。

⑤ 文件测试。
- public boolean exists()检查抽象路径表示的文件是否存在。
- public boolean canWrite()检查抽象路径表示的文件是否可写。
- public boolean canRead()检查抽象路径表示的文件是否可读。
- public boolean isFile()检查抽象路径表示的文件是否为正常文件(非目录)。
- public boolean isDirectory()检查抽象路径表示的是否为目录。
- public boolean isAbsolute()检查抽象路径表示的是否为绝对路径。

【例 5.3】删除 E 盘根目录下的记事本文档 ceshi.txt。

```
import java.io.File;
import java.io.IOException;
public class File3{
public static void main(String args[]){
File f1 = new File("e:" + File.separator + "ceshi.txt");//实例化File类的对象
f1.delete();//删除文件
System.out.println("e盘根目录下的ceshi.txt已删除");
}
}
```

如图 5-3 所示。

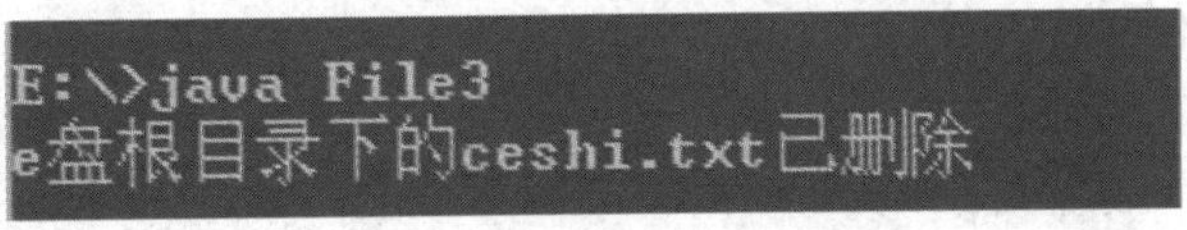

```
E:\>java File3
e盘根目录下的ceshi.txt已删除
```

图 5-3

上例成功删除了 E 盘根目录下的记事本文件 ceshi.txt。但是,为了使程序更加完善,应该在删除文件前判断文件是否存在。在 File 类中可以使用 exists() 方法判断一个文件是否存在,如下例所示。

【例 5.4】删除 E 盘 file 文件夹里名称为"ceshi.txt"的记事本文档,在删除前判断文件是否存在。

```java
import java.io.File;
import java.io.IOException;
public class File4{
public static void main(String args[]){
File f1 = new File("E:" + File.separator + "file" + File.separator + "ceshi.txt");//实例化 File 类的对象
if(f1.exists()){//如果文件存在则删除
f1.delete();//删除文件
System.out.println("e 盘 file 文件夹里的记事本文档 ceshi.txt,已经删除");
}
}
}
```

如图 5-4 所示。

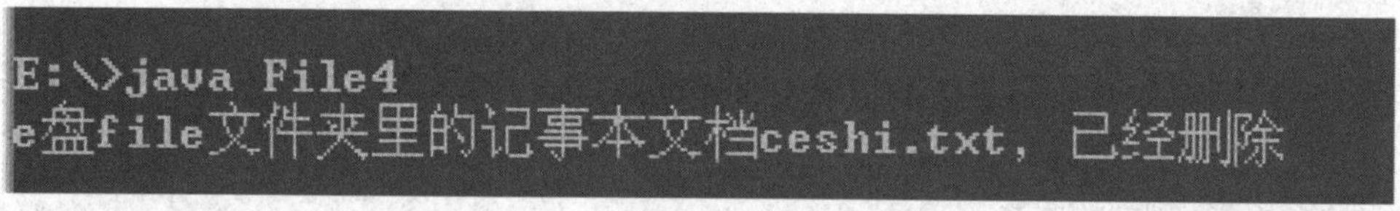

```
E:\>java File4
e盘file文件夹里的记事本文档ceshi.txt, 已经删除
```

图 5-4

【例 5.5】判断 E 盘根目录是否存在记事本文档 ceshi.txt。如果存在就将其删除;如果不存在就创建该文档。

```java
import java.io.File;
import java.io.IOException;
public class File5{
public static void main(String args[]){
File f1 = new File("e:" + File.separator + "ceshi.txt");//实例化 File 类的对象
if(f1.exists()){//如果文件存在则删除
f1.delete();//删除文件
}else{
try{
f1.createNewFile();//创建文件,根据给定的路径创建
```

```
        }catch(IOException e){
            e.printStackTrace();//输出异常信息
        }
    }
  }
}
```

上例是对文件的删除、创建、检测的综合应用。如果要创建文件夹，可以使用 mkdir() 方法，如下列所示。

【例 5.6】在 E 盘根目录创建一个文件夹，将其命名为"dir"。

```
import java.io.File;
import java.io.IOException;
public class File6{
    public static void main(String args[]){
        File f1 = new File("e:" + File.separator + "dir");//实例化 File 类的对象
        f1.mkdir();//创建文件夹
        System.out.println("dir 文件夹创建成功");
    }
}
```

如图 5-5 所示。

```
E:\>java File6
dir文件夹创建成功
```

图 5-5

上例已经成功创建了一个文件夹。如果要列出指定目录的全部文件，在 File 类中存在如下两个方法：
- 以字符串数组的形式返回：public String[]list()
- 以 File 数组的形式返回：public File[]listFile()

【例 5.7】使用 list()方法列出 E 盘根目录下的全部内容。

```
import java.io.File;
import java.io.IOException;
public class File7{
    public static void main(String args[]){
        File f1 = new File("e:" + File.separator);//实例化 File 类的对象
        String str[] = f1.list();//列出给定目录中的内容
```

```
for(int i = 0;i<str.length;i++){
System.out.println(str[i]);
}
}
}
```

如图 5-6 所示。

图 5-6

【例 5.8】使用 listFiles( )方法列出 E 盘根目录下的全部内容。

```
import java.io.File;
import java.io.IOException;
public class File8{
public static void main(String args[]){
File f = new File("e:" + File.separator);//实例化 File 类的对象
File files[] = f.listFiles();//列出全部内容
for(int i = 0;i<files.length;i++){
System.out.println(files[i]);
}
}
}
```

如图 5-7 所示。

图 5-7

上述例题中,文件和文件夹均有路径,如何判断一个路径是文件还是文件夹呢?可以通过 File 类的 public boolean isDirectory()方法判断给定路径是不是目录,如下例所示。

【例 5.9】判断一个给定的路径是不是目录。

```
import java.io.File;
import java.io.IOException;
public class File9{
public static void main(String args[]){
File f = new File("e:" + File.separator);//实例化 File 类的对象
if(f.isDirectory()){//判断是不是目录
System.out.println(f.getPath() + "是目录。");
}else{
System.out.println(f.getPath() + "不是目录。");
}
}
}
```

如图 5-8 所示。

图 5-8

5.1.4 文件的顺序访问

在 Java 程序中要实现文件的顺序访问,可以通过字节流或字符流实现。字节流或字符流操作的流程如下:

(1) 使用引入语句引入 java.io 包:"import java.io.*;"。
(2) 根据不同数据源和输入输出任务,建立字节流或字符流对象。
(3) 若需要对字节或字符流信息组织加工为数据,在已建字节流或字符流对象上构建数据流对象。
(4) 用输入输出流对象类的成员方法进行读写操作,需要时设置读写位置指针。
(5) 关闭流对象。

其中步骤(2)~(5)要考虑异常处理。由于在程序中实例化、写、关闭时都可能发生异常,为方便可直接在主方法上使用 throws 关键字抛出异常,以减少 try-catch 语句。

1. 字节流(InputStream 和 OutputStream 类)

字节流主要是操作 byte 类型数据,主要操作类为 InputStream 和 OutputStream。

(1) InputStream 和 FileInputStream。

InputStream 中包含一系列字节输入流需要的方法,可以完成最基本的从输入流读入数据的功能。当 Java 程序需要接收外设的数据时,可根据数据的不同形式,创建一个适当的 InputStream 子类类型的对象来完成与该外设的连接,然后再调用执行这个流类对象的特定输入方法,如 read(),来实现对相应外设的输入操作。InputStream 类的类层次如下所示:

InputStream
├ FileInputStream
├ ByteArrayInputStream
├ PipedInputStream
├ SequenceInputStream
├ StringBufferInputStream
└ FilterInputStream
　├ DataInputStream(实现 DataInput 接口)
　├ LineNumberInputStream
　├ BufferedInputStream
　└ PushbackInputStream

InputStream 类的常用方法有读数据的方法 read(),获取输入流字节数的方法 available(),定位输入位置指针的方法 skip()、reset()、mark()等,通过 InputStream 可以从文件中把内容读取进来。

由于 InputStream 类是抽象类,不能直接生成对象,因此,InputStream 类如果要实现从文件中读取数据,必须依靠其子类 FileInputStream。类 FileInputStream 重写或实现了 InputStream 父类中的一些方法以顺序访问本地文件,是字节流操作的基础类,其常用方法如下:

① 创建字节输入文件流 FileInputStream 类对象。

若需要以字节为单位顺序读出一个已存在文件的数据,可使用字节输入流 FileInputStream。可以用文件名文件对象或文件描述符建立字节文件流对象,FileInputStream 类的构造方法有:

• FileInputStream(String name)用文件名 name 建立流对象。

例如:

```
FileInputStream fis = new FileInputStream("c:/config.sys");
```

- FileInputStream(File file)用文件对象 file 建立流对象。

例如：

```
File myFile = new File("c:/config.sys");
FileInputStream fis = new FileInputStream(myFile);
```

若创建 FileInputStream 输入流对象成功,就相应地打开了该对象对应的文件,接着就可以从文件读取信息了。若创建对象失败,将产生异常 FileNotFoundException,这是一个非运行时异常,必须捕获和抛出,否则编译会出错。

② 读取文件信息。

从 FileInputStream 流中读取字节信息,一般用 read()成员方法,该方法有重载：

- int read()读流中一个字节,若流结束则返回-1。
- int read(byte b[])从流中读字节填满字节数组 b,返回所读字节数,若流结束则返回-1。
- int read(byte b[],int off,int len)从流中读字节填入 b[off]开始处,返回所读字节数,若流结束则返回-1。

【例 5.10】使用 InputStream 读取 E 盘根目录中的记事本文档 ceshi.txt。

```
import java.io.File;
import java.io.InputStream;
import java.io.FileInputStream;
public class InputStream1{
    public static void main(String args[])throws Exception{//异常抛出,不处理
        //第 1 步:使用 File 类找到一个文件
        File f1 = new File("e:" + File.separator + "ceshi.txt");//声明 File 对象
        //第 2 步:通过子类实例化父类对象
        InputStream input1 = null;//准备好一个输入的对象
        input1 = new FileInputStream(f1);//通过对象多态性,进行实例化
        //第 3 步:进行读操作
        byte b[] = new byte[1024];//所有的内容都读到此数组之中
        input1.read(b);//读取内容
        //第 4 步:关闭输出流
        input1.close();//关闭输出流
        System.out.println("内容为:" + new String(b));//把 byte 数组变为字符串输出
    }
}
```

如图 5-9 所示。

图 5-9

上例实现了记事本文档的内容读取,但是在内容显示后面有多行空白,这是由于我们设置的数组大小为 1024 字节,但记事本的实际内容只有 21 个字节,导致没有用到的后续的 1003 个字节转为空格字符串。鉴于此,我们可以通过 read()方法的返回值(该返回值记录了从文件中实际读取的数据个数)解决该问题,如下例所示。

【例 5.11】使用 InputStream 读取 E 盘根目录中的记事本文档 ceshi.txt,只显示该文档的实际内容。

```
import java.io.File;
import java.io.InputStream;
import java.io.FileInputStream;
public class InputStream2{
public static void main(String args[])throws Exception{//异常抛出,不处理
//第 1 步:使用 File 类找到一个文件
File f1 = new File("e:" + File.separator + "ceshi.txt");//声明 File 对象
//第 2 步:通过子类实例化父类对象
InputStream input1 = null;//准备好一个输入的对象
input1 = new FileInputStream(f1);//通过对象多态性,进行实例化
//第 3 步:进行读操作
byte b[] = new byte[1024];//所有的内容都读到此数组之中
int len = input1.read(b);//读取内容
//第 4 步:关闭输出流
```

```
    input1.close();//关闭输出流
    System.out.println("读入数据的长度:"+len);
    System.out.println("内容为:"+new String(b,0,len));//把 byte 数组变为字符
串输出
  }
}
```

如图 5-10 所示。

```
E:\>java InputStream2
读入数据的长度: 21
内容为: hello, 我是测试文档!

E:\>
```

图 5-10

上例成功解决了多余数组空间转换为空白字符串问题,但还是造成大量的资源浪费(总计 1024 个字节,有 1003 个字节没有用到)。如果想不浪费资源,我们可以使用 File 类的 public long length()方法获取文件的数据量,然后根据文件大小来限定数组开辟空间的大小。

【例 5.12】使用 InputStream 读取 E 盘根目录中的记事本文档 ceshi.txt,只显示该文档的实际内容,且用于存放文档数据的数组大小根据文件的实际数据量创建。

```
import java.io.File;
import java.io.InputStream;
import java.io.FileInputStream;
public class InputStream3{
public static void main(String args[])throws Exception{//异常抛出,不处理
//第 1 步:使用 File 类找到一个文件
File f1 = new File("e:"+File.separator+"ceshi.txt");//声明 File 对象
//第 2 步:通过子类实例化父类对象
InputStream input1 = null;//准备好一个输入的对象
input1 = new FileInputStream(f1);//通过对象多态性,进行实例化
//第 3 步:进行读操作
byte b[] = new byte[(int)f1.length()];//数组大小由文件决定
int len = input1.read(b);//读取内容
//第 4 步:关闭输出流
input1.close();//关闭输出流
System.out.println("读入数据的长度:"+len);
```

```
        System.out.println("内容为:"+new String(b));//把 byte 数组变为字符串输出
    }
}
```

如图 5-11 所示。

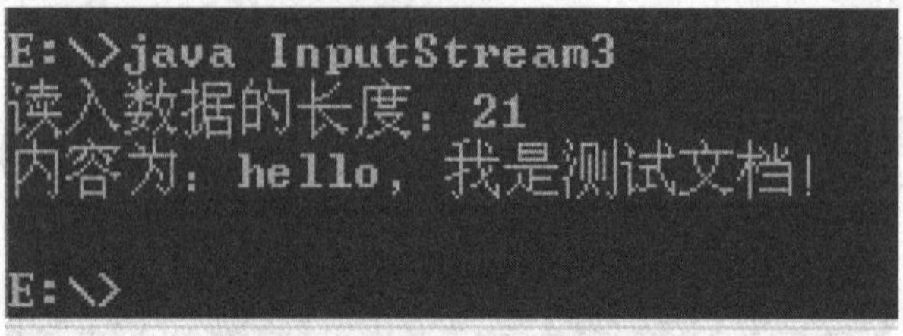

```
E:\>java InputStream3
读入数据的长度: 21
内容为: hello, 我是测试文档!

E:\>
```

图 5-11

以上例题都是将文件内容从文档中全部读取完毕后再显示。如果想一边读取,一边显示(即读取一个显示一个),可以采用循环的方式,使用 read()读取,然后显示,如下例所示。

【例 5.13】使用 InputStream 读取 E 盘根目录中的记事本文档 ceshi.txt,要求边读取边显示。

```
import java.io.File;
import java.io.InputStream;
import java.io.FileInputStream;
public class InputStream4{
public static void main(String args[])throws Exception{//异常抛出,不处理
//第 1 步:使用 File 类找到一个文件
File f1 = new File("e:"+File.separator+"ceshi.txt");//声明 File 对象
//第 2 步:通过子类实例化父类对象
InputStream input1 = null;//准备好一个输入的对象
input1 = new FileInputStream(f1);//通过对象多态性,进行实例化
//第 3 步:进行读操作
byte b;//数组大小由文件决定
for(int i = 0;i<f1.length();i++){
b = (byte)input1.read();//读取内容
System.out.print((char)(b));//把 byte 变为字符输出
}
//第 4 步:关闭输出流
input1.close();//关闭输出流
    }
}
```

如图 5-12 所示。

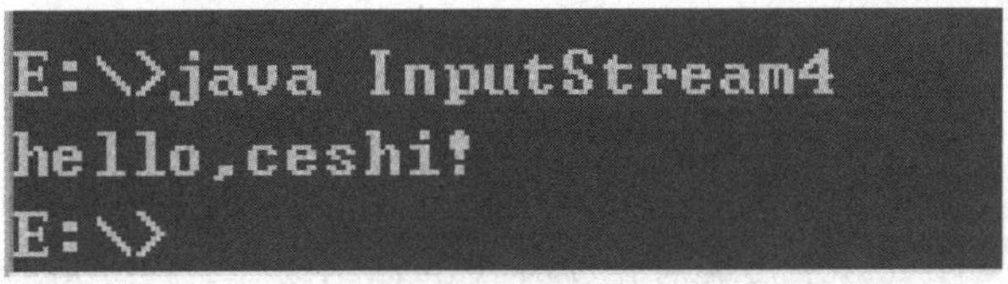

```
E:\>java InputStream4
hello,ceshi!
E:\>
```

图 5-12

上例是通过 File 类的 length()方法获取了文件的大小,即知道了输入流的大小,然后通过循环依次读取。但有时候,无法得知文件大小,又如何实现文件内容的依次读取呢?可以通过判断是否读取到文件末尾来实现(文件末尾用-1 表示),如下例所示。

【例 5.14】使用 InputStream 读取 E 盘根目录中的记事本文档 ceshi.txt,要求边读取边显示,且使用判断是否读取到文件末尾的方法来实现。

```
import java.io.File;
import java.io.InputStream;
import java.io.FileInputStream;
public class InputStream5{
public static void main(String args[])throws Exception{//异常抛出,不处理
//第 1 步:使用 File 类找到一个文件
File f1 = new File("e:" + File.separator + "ceshi.txt");//声明 File 对象
//第 2 步:通过子类实例化父类对象
InputStream input1 = null;//准备好一个输入的对象
input1 = new FileInputStream(f1);//通过对象多态性,进行实例化
//第 3 步:进行读操作
int b = 0;
while((b = input1.read())! = -1){//文件的结尾是用-1 作为标志
System.out.print((char)(b));//把 b 变为字符输出
}
//第 4 步:关闭输出流
input1.close();//关闭输出流
}
}
```

如图 5-13 所示。

```
E:\>java InputStream5
hello,ceshi!
E:\>
```

图 5-13

(2) OutputStream 和 FileOutputStream。

OutputStream 中包含一系列字节输出流需要的方法,可以完成最基本的输出数据到输出流的功能。当 Java 程序需要将数据输出到外设时,可根据数据的不同形式,创建一个适当的 OutputStream 子类类型的对象来完成与该外设的连接,然后再调用执行这个流类对象的特定输出方法,如 write(),来实现对相应外设的输出操作。OutputStream 类的类层次如下所示:

OutputStream
├ ObjectOutputStream
├ ByteArrayOutputStream
├ FileOutputStream
├ PipedOutputStream
└ FilterOutputStream
　　├ DataOutputStream(实现 DataOutput 接口)
　　├ BufferedOutputStream
　　└ PrintStream

OutputStream 类常用的方法有:写数据的方法 write()、关闭流方法 close()、刷新数据缓冲区方法 flush()等。由于 OutputStream 类也是抽象类,不能直接生成对象,因此,OutputStream 类如果要实现写入数据到文件中,必须依靠其子类 FileOutputStream。

FileOutputStream 类的常用方法如下:

① 创建字节输出文件流 FileOutputStream 类对象。

FileOutputStream 可表示一种创建并顺序写的文件。在构造此类对象时,若指定路径的文件不存在,会自动创建一个新文件;若指定路径已有一个同名文件,该文件的内容将被保留或删除。FileOutputStream 对象用于向一个文件写数据。像输入文件一样,也要先打开这个文件后,才能写这个文件。要打开一个 FileOutputStream 对象,像打开一个输入流一样,可以将字符串或文件对象作为参数。FileOutputStream 类的构造方法有:

● FileOutputStream(String name)用文件名 name 创建流对象。

例如:

```
FileOutputStream fos = new FileOutputStream("d:/out.dat");
```

● FileOutputStream(File file)用文件对象 file 建立流对象。

例如:

```
File myFile = new File("d:/out.dat");
FileOutputStream fos = new FileOutputStream(myFile);
```

上述两种格式的构造方法还允许使用第二个参数:boolean append。若这个参数的值为 true,则向文件尾输出字节流,即为添加数据方式使用字节流。

② 向输出流写信息。

向 FileOutputStream 中写入信息,一般用 write()方法,该方法有重载:

- void write(int b)将整型数据的低字节写入输出流。
- void write(byte b[])将字节数组 b 中的数据写入输出流。
- void write(byte b[],int off,int len)将字节数组 b 中从 off 开始的 len 个字节数据写入输出流。

③ 关闭 FileInputStream。

当完成一个文件的操作后,可用两种方法关闭它:显式关闭和隐式关闭。隐式关闭是让系统自动关闭它,Java 有自动垃圾回收的功能。显式关闭是使用 close()方法,例如:

```
fos.close();
```

[例 5.15] 将"Hello ceshi!"写入 E 盘根目录下的记事本文档 ceshi.txt。

```
import java.io.File;
import java.io.OutputStream;
import java.io.FileOutputStream;
public class OutputStream1{
public static void main(String args[])throws Exception{//异常抛出,不处理
//第 1 步:使用 File 类找到一个文件
File f1 = new File("e:" + File.separator + "ceshi.txt");//声明 File 对象
//第 2 步:通过子类实例化父类对象
OutputStream out1 = null;//准备好一个输出的对象
out1 = new FileOutputStream(f1);//通过对象多态性,进行实例化
//第 3 步:进行写操作
String str = "Hello ceshi!";//准备一个字符串
byte b[] = str.getBytes();//只能输出 byte 数组,所以将字符串变为 byte 数组
out1.write(b);//将内容输出,保存文件
//第 4 步:关闭输出流
out1.close();//关闭输出流
}
}
```

如图 5-14 所示。

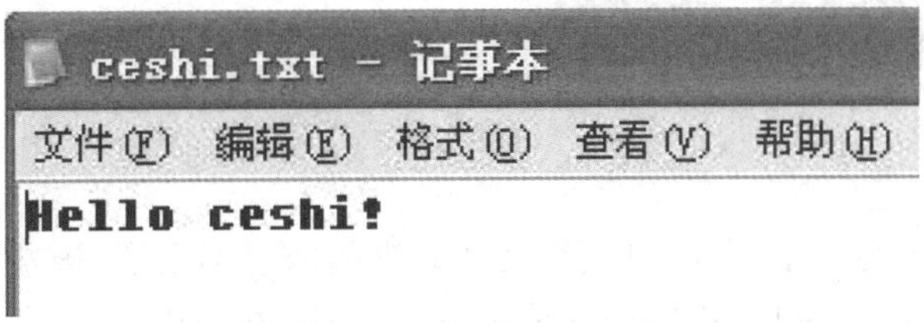

图 5-14

上例是将字符数组变为 byte 数组后,再一次性写入记事本文档,如果想将字符串的内容直接依次写入记事本文档,可使用 write(int i)方法通过循环按字节一个一个地写入记事本文档,如下例所示。

【例 5.16】将"Hello ceshi!"写入 E 盘根目录下的记事本文档 ceshi.txt,要求使用 write(int i)方法实现。

```java
import java.io.File;
import java.io.OutputStream;
import java.io.FileOutputStream;
public class OutputStream2{
public static void main(String args[])throws Exception{//异常抛出,不处理
//第 1 步:使用 File 类找到一个文件
File f1 = new File("e:" + File.separator + "ceshi.txt");//声明 File 对象
//第 2 步:通过子类实例化父类对象
OutputStream out1 = null;//准备好一个输出的对象
out1 = new FileOutputStream(f1);//通过对象多态性,进行实例化
//第 3 步:进行写操作
String str = "Hello ceshi!";//准备一个字符串
for(int i = 0;i<str.length();i++){//采用循环方式写入
byte temp = (byte)str.charAt(i);
out1.write(temp);//每次只写入一个内容
}
//第 4 步:关闭输出流
out1.close();//关闭输出流
}
};
```

如图 5-15 所示。

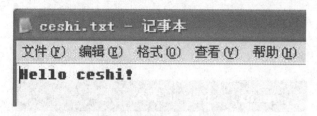

图 5-15

以上例题,将数据写入记事本文档后,文档原来的内容不见了(被覆盖了)。如果想在原来的内容后面追加数据,可以在 OutputStream 的构造方法中将参数 append 的值设置为 true,如下例所示。

【例 5.17】 在 E 盘根目录下的记事本文档 ceshi.txt 中追加内容"new ceshi!"。

```java
import java.io.File;
import java.io.OutputStream;
import java.io.FileOutputStream;
public class OutputStream3{
    public static void main(String args[])throws Exception{//异常抛出,不处理
        //第1步:使用File类找到一个文件
        File f1 = new File("e:" + File.separator + "ceshi.txt");//声明File对象
        //第2步:通过子类实例化父类对象
        OutputStream out1 = null;//准备好一个输出的对象
        out1 = new FileOutputStream(f1,true);//此处表示在文件末尾追加内容
        //第3步:进行写操作
        String str = "new ceshi!";//准备一个字符串
        for(int i = 0;i<str.length();i++){//采用循环方式写入
            out1.write((byte)str.charAt(i));//每次只写入一个内容
        }
        //第4步:关闭输出流
        out1.close();//关闭输出流
    }
}
```

如图 5-16 所示。

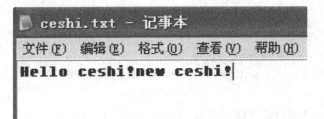

图 5-16

（3）DataInputStream 和 DataOutputStream 类。

字节文件流 FileInputStream 和 FileOutputStream 只能提供纯字节或字节数组的输入/输出，如果要进行其他数据类型数据的输入/输出（如整数和浮点数的输入/输出），则要用到过滤流类的子类二进制数据文件流 DataInputStream 和 DataOutputStream。数据输入流 DataInputStream 和数据输出流 DataOutputStream 与平台无关，其主要特点是在输入/输出数据的同时，能对所传输的数据作指定类型或格式的转换，即可实现对二进制字节数据的理解和编码转换，方便数据处理。

注意：

这两个类的对象必须和一个输入类或输出类联系起来，而不能直接用文件名或文件对象建立。

使用数据文件流的一般做法是分 3 步：首先创建字节文件流对象（即 OutputStream 或 InputStream 对象），然后基于字节文件流对象建立数据文件流对象（即 DataOutputStream 或 DataInputStream 对象），再用此对象的方法对基本类型的数据进行输入/输出操作（DataOutputStream 或 DataInputStream 的常用方法见表 5-1）。

表 5-1

数据类型	DataInputStream	DataOutputStream
byte	readByte	writeByte
short	readShort	writeShort
int	readInt	writeInt
long	readLong	writeLong
float	readFloat	writeFloat
double	readDouble	writeDouble
boolean	readBoolean	writeBoolean
char	readChar	writeChar
String	readUTF	writeUTF
byte[]	readFully	

DataInputStream 类的构造方法如下：

DataInputStream(InputStream in)创建过滤流 FilterInputStream 对象并为以后的使用保存 InputStream 参数 in。

DataOutputStream 类的构造方法如下：

DataOutputStream(OutputStream out)创建输出数据流对象写数据到指定的 OutputStream。

【例 5.18】在 E 盘根目录下建立文件 file.dat，存储 1～10 共计 10 个整数。

```
import java.io.OutputStream;
import java.io.FileOutputStream;
import java.io.File;
import java.io.DataOutputStream;
public class DataOutputStream1{
public static void main(String args[])throws Exception{
//创建字节文件输出流
OutputStream fos = new FileOutputStream("e:" + File.separator + "file.dat");
DataOutputStream dos = new DataOutputStream(fos);//创建数据输出流
```

```
//写1～10到文件中
for(int i = 1;i< = 10;i + + ){
dos.writeInt(i);
}
fos.close();//关闭文件输出流
System.out.println("文件创建成功");
}
}
```

如图 5-17、图 5-18 所示。

图 5-17

图 5-18

【例5.19】使用 DataInputStream 读取上例创建的文件 file.dat 的内容。

```
import java.io.InputStream;
import java.io.FileInputStream;
import java.io.File;
import java.io.DataInputStream;
import java.io.EOFException;
public class DataInputStream1{
public static void main(String args[])throws Exception{
//创建文件输入流
InputStream fis = new FileInputStream("e:" + File.separator + "file.dat");
DataInputStream dis = new DataInputStream(fis);//创建数据输入流
int readCount = dis.available();//获取要读取的总字节数,用于计算循环读取的控制条件
for(int i = 0;i<readCount/4;i + + ){
```

```
        System.out.print(dis.readInt() + "  ");
    }
    //关闭文件输入流
    dis.close();
    fis.close();
    }
}
```

如图 5-19 所示。

```
E:\>java DataInputStream1
1  2  3  4  5  6  7  8  9  10
E:\>
```

图 5-19

上述两例中使用了整数的输入方法 readInt() 和输出方法 writeInt(),其他可用的输入/输出方法见表 5-1。

(4) BufferedInputStream 和 BufferedOutputStream 类。

若处理的数据量较多,为避免每个字节的读写都对流进行,可以使用过滤流类的子类缓冲流。缓冲流建立一个内部缓冲区,输入输出数据先读写到缓冲区中进行操作,这样可以提高文件流的操作效率。

BufferedInputStream 缓冲输入流与 BufferedOutputStream 缓冲输出流分别是 FilterInputStream 类和 FilterOutputStream 类的子类。缓冲输出流 BufferedOutputStream 类提供和 FileOutputStream 类同样的写操作方法,但所有输出全部写入缓冲区中。当写满缓冲区或关闭输出流时,它再一次性输出到流,或者用 flush() 方法主动将缓冲区输出到流。同理,缓冲输入流 BufferedInputStream 也继承了 FilterInputStream 类的所有方法,当创建缓冲输入流 BufferedInputStream 时,一个输入缓冲区数组被创建,来自流的数据填入缓冲区,一次可填入许多字节。

① 创建 BufferedOutputStream 流对象。

若要创建一个 BufferedOutputStream 流对象,首先需要一个 FileOutputStream 流对象,然后基于这个流对象创建缓冲流对象。

BufferedOutputStream 类的构造方法如下:

• BufferedOutputStream(OutputStream out) 创建缓冲输出流,写数据到参数指定的输出流,缓冲区设为默认 512 字节大小。

• BufferedOutputStream(OutputStream out, int size) 创建缓冲输出流,写数据到参数指定的输出流,缓冲区设为指定的 size 字节大小。

例如,下面的代码可创建一个缓冲输出流 bos:

```
FileOutputStream fos = new FileOutputStream("e:/file.dat");
BufferedOutputStream bos = new BufferedOutputStream(fos);
```

② BufferedOutputStream 类的其他常用方法。
- flush()方法。

如果想在程序结束之前将缓冲区里的数据写入磁盘,除了填满缓冲区或关闭输出流外,还可以显式调用 flush()方法。flush()方法的声明为:

```
public void flush()throws IOException
```

例如:

```
bos.flush();
```

- public void write(int b)向输出流中输出一个字节。
- public void write(byte[]b,int off,int len)将指定 byte 数组中从偏移量 off 开始的 len 个字节写入此缓冲的输出流。

【例 5.20】 使用 BufferedOutputStream 在 E 盘根目录下建立文件 file.dat,存储 1～10 共计 10 个整数。

```java
import java.io.FileOutputStream;
import java.io.File;
import java.io.BufferedOutputStream;
public class BufferedOutputStream1{
public static void main(String args[])throws Exception{
//创建字节文件输出流
FileOutputStream fos = new FileOutputStream("e:" + File.separator + "file.dat");
BufferedOutputStream bos = new BufferedOutputStream(fos);//创建缓冲输出流
//写1～10 到文件中
for(int i = 1;i <= 10;i + +){
bos.write(i);
}
bos.flush();//刷新输出
bos.close();
fos.close();//关闭文件输出流
System.out.println("文件创建成功");
}
}
```

如图 5-20、图 5-21 所示。

```
E:\>java BufferedOutputStream1
文件创建成功

E:\>
```

图 5-20

图 5-21

③ 创建 BufferedInputStream 流对象。

BufferedInputStream 类的构造方法如下：

- BufferedInputStream(InputStream in)创建一个 BufferedInputStream 对象并保存其参数，即输入流 in，以便将来使用。同时，自动创建一个内部缓冲区数组，该缓冲区的大小默认为 8192。
- BufferedInputStream(InputStream in, int size)创建具有指定缓冲区大小（size 用于指定缓冲区大小）的 BufferedInputStream 对象并保存其参数，即输入流 in，以便将来使用。

④ BufferedInputStream 类的其他常用方法。

- public int read()从输入流中读取一个字节。
- public int read(byte[]b, int off, int len)从此字节输入流中给定偏移量处开始将各字节读取到指定的 byte 数组中。
- int available()返回底层流对应的源中有效可供读取的字节数。
- void close()关闭此流、释放与此流有关的所有资源。

【例 5.21】使用 BufferedInputStream 读取上例创建的文件 file.dat 的内容。

```
import java.io.BufferedInputStream;
import java.io.FileInputStream;
import java.io.File;
public class BufferedInputStream1{
    public static void main(String[]args)throws Exception{
        //TODO Auto-generated method stub
        //创建 FileInputStream 对象
        FileInputStream fis = new FileInputStream("e:" + File.separator + "file.dat");
        //创建 BufferedInputStream 对象
        BufferedInputStream bis = new BufferedInputStream(fis);
```

```java
                //自己定义一个缓冲区,缓冲区大小为6个字节(即一次可读取6个字节数据)
                //由于缓冲区较小,所以一个文件要分多次才能读取完毕
                byte[]buffer = new byte[6];
                //count用于记录每次从文件中读取的字节数
                int count = 0;
                int j = 1;//j用于记录这是第几次从文件中读取数据
                //判断是否从文件中读取到数据,将读取到的数据的字节数保存到变量count
                while((count = bis.read(buffer))! = -1){
                //输出提示信息"第j次读取到的数据:"
                System.out.print("第"+j+"次读取到的数据:");
                //依次输出本次从文件中读取到的内容
                    for(int i = 0;i<count;i++){
                        System.out.print(buffer[i]+"");
                    }
                    System.out.print("\n");//换行,为下次读取内容的输出作准备
                    j++;//j的值加一,用于记录下一次读取是第几次
                }
            bis.close();
            fis.close();
        }
    }
```

如图 5-22 所示。

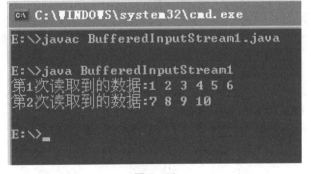

图 5-22

(5) PrintStream 类。

过滤流类的子类 PrintStream 类提供了一系列的 print 和 println 方法,可以将 Java 的任何类型转换为字符串类型输出,使用 PrintStream 可以更加方便地输出数据。

创建 PrintStream 流也需要 OutputStream 流对象,PrintStream 类的构造方法有:

- public PrintStream(OutputStream out)创建一个新的打印流对象。
- public PrintStream(OutputStream out, boolean autoFlush)创建一个新的打印流对象。布尔值的参数 autoFlush 为 true 时,当写一个字节数组、引用 println( )方法或写 newline 字符或写字节('\n')时,缓冲区内容将被写到输出流。

【例5.22】通过 PrintStream 向 E 盘根目录下的记事本文件 ceshi.txt 中打印信息"Hello World!"。

```java
import java.io.PrintStream;
import java.io.FileOutputStream;
import java.io.File;
public class PrintStream1{
public static void main(String arg[])throws Exception{
//声明 FileOuputStream 对象(OutputStream 的子类对象),意味着所有的输出是向文件之中
FileOutputStream fos = new FileOutputStream(new File("e:" + File.separator + "ceshi.txt"));
//声明打印流对象
PrintStream ps = new PrintStream(fos);
//使用 print 方法向文件输出"Hello  "
ps.print("Hello  ");
//使用 println 方法向文件输出"World!"
ps.println("World!");
ps.close();
}
}
```

如图 5-23 所示。

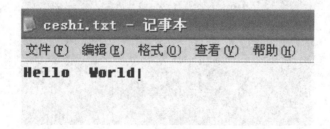

图 5-23

2. 字符流(Reader 类和 Writer 类)

由于 Java 采用 16 位的 Unicode 字符,因此需要基于字符的输入/输出操作。从 Java 1.1 版开始加入了专门处理字符流的抽象类 Reader 和 Writer,前者用于处理输入,后者用于处理输出。这两个类类似于 InputStream 和 OuputStream,也只是提供一些用于字符流的接

口,本身不能用来生成对象。

(1) Reader。

Reader 中包含一套字符输入流需要的方法,可以完成最基本的从输入流读入数据的功能。当 Java 程序需要外设的数据时,可根据数据的不同形式,创建一个适当的 Reader 子类类型的对象来完成与该外设的连接,然后再调用执行这个流类对象的特定输入方法,如 read(),来实现对相应外设的输入操作。Reader 类的类层次如下所示:

```
Reader
├ BufferedReader
│   └ LineNumberReader
├ CharArrayReader
├ FilterReader
│   └ PushbackInputStream
├ InputStreamReader
│   └ FileReader
├ PipedReader
└ StringReader
```

(2) Writer。

Writer 中包含一套字符输出流需要的方法,可以完成最基本的输出数据到输出流的功能。当 Java 程序需要将数据输出到外设时,可根据数据的不同形式,创建一个适当的 Writer 子类类型的对象来完成与该外设的连接,然后再调用执行这个流类对象的特定输出方法,如 write(),来实现对相应外设的输出操作。Writer 类的类层次如下所示:

```
Writer
    ├ BufferedWriter
    ├ CharArrayWriter
    ├ FilterWriter
    ├ OutputStreamWriter
    │   └ FileWriter
    ├ PipedWriter
    ├ StringWriter
    └ PrintWriter
```

Reader 和 Writer 类也有各自的子类,与字节流类似,它们用来创建具体的字符流对象进行 IO 操作。字符流的读写等方法与字节流的相应方法都很类似,但读写对象使用的是字符。

(3) InputSteamReader 和 OutputStreamWriter 类。

Java 支持字节流和字符流,有时需要在字节流和字符流之间转换。InputSteamReader 和 OutputStreamWriter 类是 java.io 包中用于处理字符流的基本类,用来在字节流和字符流之间搭一座"桥"。InputStreamReader 可以将一个字节流中的字节解码成字符,OuputStreamWriter 将写入的字符编码成字节后写入一个字节流。InputSteamReader 和 OutputStreamWriter 类处理的字节流的编码规范与具体的平台有关,可以在构造流对象时指定规范,也可以使用当前平台的默认规范。

InputSteamReader 和 OutputStreamWriter 类的构造方法如下:
- public InputSteamReader(InputSteam in)
- public InputSteamReader(InputSteam in,String enc)
- public OutputStreamWriter(OutputStream out)
- public OutputStreamWriter(OutputStream out,String enc)

其中 in 和 out 分别为输入和输出字节流对象,enc 为指定的编码规范(若无此参数,表示使用当前平台的默认规范,可用 getEncoding()方法得到当前字符流所用的编码方式)。

读写字符的方法 read()、write(),关闭流的方法 close()等,与 Reader 和 Writer 类的同名方法用法都是类似的。

【例 5.23】将字节的文件输出流,以字符的形式输出。

```
import java.io.File;
import java.io.Writer;
import java.io.OutputStreamWriter;
import java.io.FileOutputStream;
public class OutputStreamWriter1{
public static void main(String args[])throws Exception{//所有异常抛出
File f = new File("e:" + File.separator + "ceshi.txt");
Writer out = null;//字符输出流
out = new OutputStreamWriter(new FileOutputStream(f));//字符流变为字节流
out.write("hello OutputStreamWriter!");//使用字符流输出
out.close();
}
}
```

如图 5-24 所示。

图 5-24

【例 5.24】将字节输入流变为字符输入流。

```
import java.io.File;
```

```java
import java.io.Reader;
import java.io.InputStreamReader;
import java.io.FileInputStream;
public class InputStreamReader1{
public static void main(String args[])throws Exception{
File f = new File("e:" + File.separator + "ceshi.txt");
Reader reader = null;
reader = new InputStreamReader(new FileInputStream(f));//将字符流变为字节流
char c[] = new char[1024];
int len = reader.read(c);//读取
reader.close();//关闭
System.out.println(new String(c,0,len));
}
}
```

如图 5-25 所示。

图 5-25

以上两个示例程序都是以文件操作为例,将字节流的操作类以字符流的形式进行输入和输出,即 OutputStreamWriter() 中接收的类型是 OutputStream,只要是字节输出流都可以使用字符的形式操作;对应的 InputStreamReader() 中接受的类型是 InputStream,只要是字节的输入流都可以使用字符的输入流操作。

(4) FileReader 和 FileWriter 类。

FileReader 和 FileWriter 类是 InputSteamReader 和 OutputStreamWriter 类的子类,利用它们可方便地进行字符输入/输出操作。

FileReader 类的构造方法有:
- FileReader(File file)对指定要读的 file 创建 FileReader 对象。
- FileReader(String fileName)对指定要读的 fileName 创建 FileReader 对象。

FileWriter 类的构造方法有:
- FileWriter(File file)对指定的 file 创建 FileWriter 对象。
- FileWriter(String fileName)对指定的 fileName 创建 FileWriter 对象。

这里列出的 FileWriter 类的两个构造方法都可带第二个布尔值的参数 append,当 append 为 true 时,为添加到输出流。

FileReader 类中可用的方法有:read()返回输入字符、read(char[]buffer)输入字符到字符数组中等。

FileWriter 类中常用的方法有:write(String str)和 write(char[ ]buffer)输出字符串、write(int char)输出字符、flush( )输出缓冲字符、close( )在执行 flush 后关闭输出流、getEncoding( )获得文件流字符的编码等。

【例 5.25】使用 FileWriter 类向 E 盘根目录下的记事本 ceshi.txt 中输入"FileWriter and FileReader!",然后使用 FileReader 类从 ceshi.txt 中读取出来打印到显示器。

```java
import java.io.File;
import java.io.Writer;
import java.io.FileWriter;
import java.io.Reader;
import java.io.FileReader;
public class FileWriter_FileReader
{
public static void main(String args[])throws Exception
{
File f = new File("e:" + File.separator + "ceshi.txt");
Writer out = null;
out = new FileWriter(f);
//声明一个 String 类型对象
String str = "FileWriter and FileReader!";
//将 str 内容写入文件之中
out.write(str);
out.close();
//以下为读文件操作
Reader in = null;
in = new FileReader(f);
//开辟一个空间用于接收文件读进来的数据
char c1[ ] = new char[1024];
int i = 0;
//将 c1 的引用传递到 read( )方法之中,同时此方法返回读入数据的个数
i = in.read(c1);
in.close();
//将字符数组转换为字符串输出
System.out.println(new String(c1,0,i));
}
}
```

如图 5-26、图 5-27 所示。

图 5-26

图 5-27

注意:

字节流中的 FileOutputStream、FileInputStream 类分别是 OutputStream、InputStream 的直接子类,但字符流中的 FileWriter 并不直接是 Writer 的子类,而是 OutputStreamWriter(转换流)的直接子类,同理 FileReader 也不直接是 Reader 的子类,而是 InputStreamReader(转换流)的直接子类。因此,不管是使用字节流还是字符流,实际上最终都是以字节的形式操作输入输出流的,在传输或者是从文件中读取数据的时候,文件里面真正保存的数据永远是字节。

(5) BufferedReader 和 BufferedWriter 类。

InputStreamReader 和 OutputStreamWriter 虽然可以实现字节流和字符流之间的转换,但为了避免频繁地进行字符与字节间的相互转换,达到最高的效率,最好不要直接使用这两个类来进行读写,应尽量使用 BufferedWriter 类包装 OutputStreamWriter 类,用 BufferedReader 类包装 InputStreamReader 类。

缓冲字符流类 BufferedReader 和 BufferedWriter 的使用的构造方法如下:

- public BufferedReader(Reader in)
- public BufferedReader(Reader in, int sz)
- public BufferedWriter(Writer out)
- public BufferedWriter(Writer out, int sz)

其中 in 和 out 分别为字符流对象,sz 为缓冲区大小,从上述构造方法的声明可以看出,缓冲流的构造方法是基于字符流创建相应的缓冲流。

在 BufferedReader 和 BufferedWriter 类中,除了 Reader 和 Writer 中提供的基本读写方法外,增加了对整行字符的处理方法 readLine() 和 newLine()。前者从输入流中读取一行字符,行结束标志为回车符和换行符;后者向字符输出流中写入一个行结束标记,该标记是由系统定义的属性 line.separator。

【例 5.26】 使用 BufferedWriter 和 BufferedReader 类实现向 E 盘根目录的记事本 ceshi.txt 写入"BufferedWriter and BufferedReader Test!",并读取显示到屏幕。

```java
import java.io.BufferedReader;
import java.io.BufferedWriter;
import java.io.File;
import java.io.FileReader;
import java.io.FileWriter;
import java.io.Reader;
import java.io.Writer;
public class BufferedReader_BufferedWriter{
    public static void main(String[]args)throws Exception{
        File file = new File("e:" + File.separator + "ceshi.txt");
        Writer writer = new FileWriter(file);
        BufferedWriter bw = new BufferedWriter(writer);
        bw.write("BufferedWriter and BufferedReader Test!");
        bw.close();
        writer.close();
        if (file.exists()){
            Reader reader = new FileReader(file);
            BufferedReader br = new BufferedReader(reader);
            String str = null;
            while((str = br.readLine())! = null){
                System.out.println(str);
            }
            reader.close();
            br.close();
        }
    }
}
```

如图 5-28、图 5-29 所示。

```
E:\>javac BufferedReader_BufferedWriter.java
E:\>java BufferedReader_BufferedWriter
BufferedWriter and BufferedReader Test!
```

图 5-28

图 5-29

(6) PrintWriter 类。

类似 PrintStream 类，PrintWriter 类提供字符流的输出处理。由于该类的对象可基于字节流或字符流来创建，写字符的方法 print()、println()可直接将 Java 基本类型的数据转换为字符串输出，用起来很方便。

PrintWriter 类的构造方法如下：

- PrintWriter(OutputStream out)
- PrintWriter(OutputStream out, boolean autoFlush)
- PrintWriter(Writer out)
- PrintWriter(Writer out, boolean autoFlush)

例如，为文件 test.txt 创建 PrintWriter 对象 pw 的语句可为：

```
PrintWriter pw = new PrintWriter(new FileOutputStream("test.txt"));
```

或

```
PrintWriter pw = new PrintWriter(new FileWriter("test.txt"));
```

【例 5.27】 通过 PrintWriter 向文件中打印信息。

```
import java.io.File;
import java.io.PrintWriter;
import java.io.FileWriter;
public class PrintWriter1{
    public static void main(String args[])throws Exception{
        PrintWriter out = null;
        File f = new File("e:" + File.separator + "ceshi.txt");
        out = new PrintWriter(new FileWriter(f));
        //由 FileWriter 实例化,则向文件中输出
        out.print("Hello PrintWrite");
        out.close();
    }
}
```

如图 5-30 所示。

图 5-30

5.1.5 文件的随机访问

有时读文件不是从头至尾顺序读的,也可能想将一个文本文件当作一个数据库,读完一个记录后跳到另一个记录,这些记录在文件的不同地方;或者对一个文件进行又读又写的操作。Java 提供的 RandomAccessFile 类可进行这种类型的输入输出。

RandomAccessFile 类直接继承于 Object,但由于实现了 DataInput 和 DataOutput 接口,因而与同样实现该接口的 DataInputStream 和 DataOutputStream 类方法很相似。

1. 建立随机访问文件流对象

建立 RandomAccessFile 类对象类似于建立其他流对象,RandomAccessFile 类的构造方法如下:

- RandomAccessFile(File file, String mode)
- RandomAccessFile(String name, String mode)

其中,name 为文件名字符串,file 为 File 类的对象,mode 为访问文件的方式,有"r"或"rw"两种形式。若 mode 为"r",则文件只能读出,对这个对象的任何写操作将抛出 IOException 异常;若 mode 为"rw"并且文件不存在,则该文件将被创建。若 name 为目录名,也将抛出 IOException 异常。

例如,打开一个数据库后更新数据:

```
RandomAccessFile rf = new RandomAccessFile("/usr/db/stock.dbf","rw");
```

2. 访问随机访问文件

RandomAccessFile 对象的读写操作和 DataInput/DataOutput 对象的操作方式一样,可以使用在 DataInputStream 和 DataOutputStream 里出现的所有 read()和 write()方法。

3. 移动文件指针

随机访问文件的任意位置的数据记录读写是通过移动文件指针指定文件读写位置来实现的。与文件指针有关的常用方法有:

- public long getFilePointer() throws IOException 返回文件指针的当前字节位置。
- public void seek(long pos) throws IOException 将文件指针定位到一个绝对地址 pos。

pos 参数指明相对于文件头的偏移量,地址 0 表示文件的开头。例如,将文件 rf 的文件指针移到文件尾,可用语句:

```
rf.seek(rf.length());
```

其中,public long length( )throws IOException 返回文件的长度,地址 length( )表示文件的结尾。

- public int skipBytes(int n)throws IOException 将文件指针向文件尾方向移动 n 个字节。

4. 向随机访问文件增加信息

可以用访问方式"rw"打开随机访问文件后,向随机访问文件增加信息。例如:

```
rf = new RandomAccessFile("c:/config.sys","rw");
rf.seek(rf.length());//任何顺序写将添加到文件
```

【例 5.28】 使用 RandomAccessFile 类将表 5-2 所示的信息写入 E 盘根目录的记事本文件 ceshi.txt。

表 5-2

序号	学号	姓名
1	111	zhangsan
2	112	lisi
3	113	wangwu
4	114	zhaoliu

```
import java.io.File;
import java.io.RandomAccessFile;
public class RandomAccessFile1{
//所有的异常直接抛出,程序中不再进行处理
public static void main(String args[])throws Exception{
File f = new File("E:" + File.separator + "ceshi.txt");//指定要操作的文件
//声明 RandomAccessFile 类的对象
//读写模式,如果文件不存在,会自动创建
RandomAccessFile rdf = new RandomAccessFile(f,"rw");
String name = null;
int xuehao = 0;
xuehao = 111;//数字的长度为 4
name = "zhangsan";//字符串长度为 8
rdf.writeInt(xuehao);//将学号写入文件之中
rdf.writeBytes(name);//将姓名写入文件之中
xuehao = 112;//数字的长度为 4
name = "lisi";//字符串长度为 4
rdf.writeInt(xuehao);//将学号写入文件之中
rdf.writeBytes(name);//将姓名写入文件之中
```

```
xuehao = 113;//数字的长度为4
name = "wangwu";//字符串长度为6
rdf.writeInt(xuehao);//将学号写入文件之中
rdf.writeBytes(name);//将姓名写入文件之中
xuehao = 114;//数字的长度为4
name = "zhaoliu";//字符串长度为7
rdf.writeInt(xuehao);//将学号写入文件之中
rdf.writeBytes(name);//将姓名写入文件之中
rdf.close();//关闭
    }
}
```

如图5-31所示。

图5-31

【例5.29】将上例中写入记事本ceshi.txt的4个人的信息按照4、3、2、1的顺序读取并显示。

```
import java.io.File;
import java.io.RandomAccessFile;
public class RandomAccessFile2{
//所有的异常直接抛出,程序中不再进行处理
public static void main(String args[])throws Exception{
File f = new File("E:" + File.separator + "ceshi.txt");//指定要操作的文件
//声明RandomAccessFile类的对象
//以只读的方式打开文件
RandomAccessFile rdf = new RandomAccessFile(f,"r");
String name = null;
int xuehao = 0;
byte b[] = new byte[8];//开辟byte数组
//读取第四个人的信息,意味着要空出前三个人的信息
rdf.skipBytes(30);//跳过前三个人的信息
xuehao = rdf.readInt();//读取数字,即读取第四个人的学号
for(int i = 0;i<7;i + +){//读取第四个人的姓名
```

```
        b[i] = rdf.readByte();//读取一个字节
    }
    name = new String(b);//将读取出来的 byte 数组变为字符串
    System.out.println("第四个人的信息-->学号:" + xuehao + ";姓名:" + name);
    //读取第三个人的信息
    rdf.seek(20);//指针回到第三个人的信息起始位置
    xuehao = rdf.readInt();//读取数字,即读取第三个人的学号
    for(int i = 0;i<6;i++){//读取第三个人的姓名
        b[i] = rdf.readByte();//读取一个字节
    }
    name = new String(b);//将读取出来的 byte 数组变为字符串
    System.out.println("第三个人的信息-->学号:" + xuehao + ";姓名:" + name);
    //读取第二个人的信息
    rdf.seek(12);//指针回到第二个人的信息起始位置
    xuehao = rdf.readInt();//读取数字,即读取第二个人的学号
    for(int i = 0;i<4;i++){//读取第二个人的姓名
        b[i] = rdf.readByte();//读取一个字节
    }
    name = new String(b);//将读取出来的 byte 数组变为字符串
    System.out.println("第二个人的信息-->学号:" + xuehao + ";姓名:" + name);
    //读取第一个人的信息
    rdf.seek(0);//指针回到第一个人的信息起始位置
    xuehao = rdf.readInt();//读取数字,即读取第一个人的学号
    for(int i = 0;i<8;i++){//读取第一个人的姓名
        b[i] = rdf.readByte();//读取一个字节
    }
    name = new String(b);//将读取出来的 byte 数组变为字符串
    System.out.println("第一个人的信息-->学号:" + xuehao + ";姓名:" + name);
    rdf.close();//关闭
    }
}
```

如图 5-32 所示。

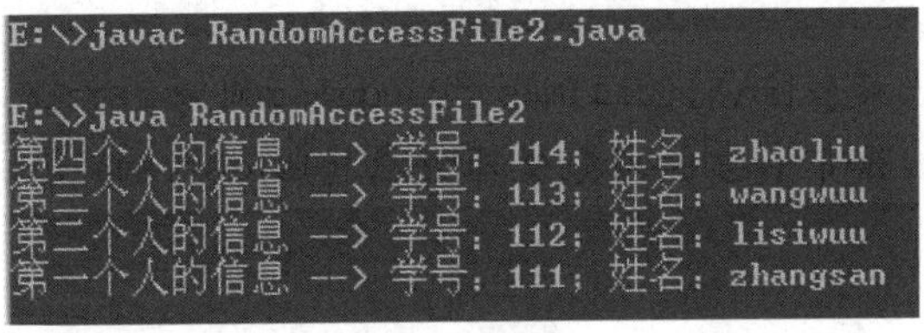

图 5-32

字节流与字符流的使用虽然非常相似,但两者的操作代码是不同的。此外,字节流在操作的时候本身是不会用到缓冲区(内存)的,是对文件本身直接操作的,而字符流在操作的时候是使用到缓冲区的。

注意:

由于字节流没有用到缓冲区,因此在使用字节流的操作中,即使最后没有关闭字节流,最终也是可以输出的。但字符流用到了缓冲区,在最后必须关闭字符流或手工强制性调用刷新方法,才能把内容输出。

在所有的硬盘上保存文件或是进行传输的时候都是以字节的方式进行的,包括图片也是按字节完成的。而字符是只有在内存中才会形成的,因此使用字节的操作是最多的。

5.1.6 其他常用的流

1. 内存操作流

在 Java 语言中,为了保证安全性而禁止直接操作内存。但 Java 语言提供了 ByteArrayInputStream 和 ByteArrayOutputStream 类来利用内存,可将字节数组中的数据视为内存数据进行操作,通过 ByteArrayInputStream 类将数组中的数据以流方式从内存读出,或将数据通过 ByteArrayOutputStream 类以流方式写入内存中暂时保存。这两个类的构造方法如下:

• public ByteArrayInputStream(byte[]buf)创建一个字节数组输入流,buf 用作缓冲区数组。

• public ByteArrayInputStream(byte[]buf, int offset, int length)创建一个字节数组输入流,buf 用作缓冲区数组,offset 为数组读出开始位置,length 为读出字节数。

• public ByteArrayOutputStream()创建字节数组输出流,缓冲区容量初始化为 32 字节。

• public ByteArrayOutputStream(int size)创建字节数组输出流,缓冲区容量指定为 size 个字节。

【例 5.30】 使用内存操作流实现将内存中存放的字符串"Hello ByteArray"中的大写字母转换为小写字母并显示。

```
import java.io.ByteArrayInputStream;
import java.io.ByteArrayOutputStream;
public class BAIS_BAOS{
public static void main(String args[])throws Exception{
String str = "Hello ByteArray";//定义一个字符串
ByteArrayInputStream bis = null;//内存输入流
ByteArrayOutputStream bos = null;//内存输出流
bis = new ByteArrayInputStream(str.getBytes());//向内存中输出内容
bos = new ByteArrayOutputStream();//准备从内存 ByteArrayInputStream 中读取内容
int temp = 0;
```

```
while((temp = bis.read())! = -1){
char c = (char)temp;//读取的数字变为字符
bos.write(Character.toLowerCase(c));//将字符变为小写
}
//所有的数据就都在 ByteArrayOutputStream 中
String newStr = bos.toString();//取出内容
bis.close();
bos.close();

System.out.println(newStr);
}
}
```

如图 5-33 所示。

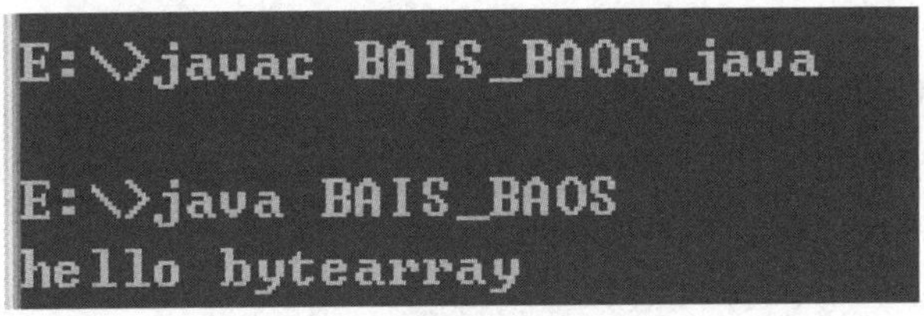

```
E:\>javac BAIS_BAOS.java

E:\>java BAIS_BAOS
hello bytearray
```

图 5-33

2. 管道流

管道是一种数据流的形式,是线程之间传输数据的通道,分为管道输出流和管道输入流。在 java.io 包中,PipedInputStream 和 PipedOutputStream 类描述了管道的输入和输出。

管道流的构造方法如下:

- PipedInputStream()创建未连接管道输出流的管道输入流对象。
- PipedInputStream(PipedOutputStream src)创建连接到管道输出流 src 的管道输入流对象。
- PipedOutputStream()创建未连接管道输入流的管道输出流对象。
- PipedOutputStream(PipedInputStream snk)创建连接到管道输入流 snk 的管道输出流对象。

管道流的主要作用是可以进行两个线程间的通讯,如果要进行管道输出,将一个线程中的管道输出流连接到另一个线程的管道输入流中就可以通过管道在两个线程中传送数据了。在 PipedOutputStream 类上有如下的一个方法用于连接管道:

- public void connect(PipedInputStream snk)throws IOException

注意:

由于可能引起系统死锁的原因,在单线程的程序中一般不使用管道操作。

3. 合并流

合并流可以将两个或多个输入流合并为一个输入流,主要用于将两个文件的内容合并为一个文件。例如,将多个文件的内容一次读到内存中,就可以使用合并流完成。要实现合并流,必须使用 SequenceInputStream 类。SequenceInputStream 类的构造方法为:

- public SequenceInputStream(Enumeration e)用枚举类的对象来创建顺序输入流。
- public SequenceInputStream(InputStream s1,InputStream s2)用两个输入流来创建顺序输入流。

【例5.31】使用合并流将 E 盘根目录的记事本文件 ceshi1.txt(内容为:ceshi1)和 ceshi2.txt(内容为:ceshi2)合并为一个文件 ceshi.txt。

```java
import java.io.File;
import java.io.SequenceInputStream;
import java.io.FileInputStream;
import java.io.InputStream;
import java.io.FileOutputStream;
import java.io.OutputStream;
public class test{
public static void main(String args[])throws Exception{//所有异常抛出
InputStream is1 = null;//输入流 1
InputStream is2 = null;//输入流 2
OutputStream os = null;//输出流
SequenceInputStream sis = null;//合并流
is1 = new FileInputStream("e:" + File.separator + "ceshi1.txt");
is2 = new FileInputStream("e:" + File.separator + "ceshi2.txt");
os = new FileOutputStream("e:" + File.separator + "ceshi.txt");
sis = new SequenceInputStream(is1,is2);//实例化合并流
int temp = 0;//接收内容
while((temp = sis.read())! = -1){//循环输出
os.write(temp);//保存内容
}
sis.close();//关闭合并流
is1.close();//关闭输入流 1
is2.close();//关闭输入流 2
os.close();//关闭输出流
}
}
```

如图 5-34 至图 5-36 所示。

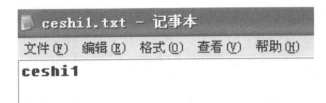

图 5-34

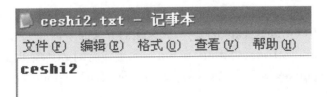

图 5-35

图 5-36

4. 压缩流

在日常中,我们经常会使用到像 WinRAR 或 WinZIP 这样的压缩文件,通过这些软件可以把一个很大的文件进行压缩以方便传输。在 Java 中为了减少传输时的数据量,也提供了专门的压缩流,可以将文件或文件夹压缩成 ZIP、JAR、RAR 等文件形式。

(1) Java ZIP 压缩流简介。

ZIP 是一种较为常见的压缩形式,在 Java 中想实现 ZIP 的压缩需要导入 java.util.zip 包,可以使用此包中的 ZipFile、ZipOutputStream、ZipInputStream 和 ZipEntry 几个类完成操作。在 Java 10 中,不仅可以实现 ZIP 压缩格式的输入和输出,还可以实现 JAR 及 GZIP 文件格式的压缩。

JAR 压缩的支持类保存在 java.util.jar 包中,常用类有如下 4 个:

- JAR 压缩输出流:JarOutputStream。
- JAR 压缩输入流:JarInputStream。
- JAR 文件:JARFile。
- JAR 实体:JAREntry。

GZIP 是用于 Unix 系统的文件压缩,在 Linux 中经常会使用到"*.gz"的文件,就是 GZIP 格式。GZIP 压缩的支持类保存在 java.util.zip 包中,常用类有如下两个:

- GZIP 压缩输出流:GZIPOutputStream。
- GZIP 压缩输入流:GZIPInputStream。

在每一个压缩文件中都会存在多个子文件,每个子文件在 Java 中就用 ZipEntry 表示。

ZipEntry 类的常用方法如下：
- public ZipEntry(String name)创建对象并指定要创建的 ZipEntry 名称。
- public boolean isDirectory()判断此 ZipEntry 是否是目录。

注意：

压缩的输入流和输出流也属于 InputStream、OutputStream 的子类，但却没有定义在 java.io 包中，而是以一种工具类的形式提供的，在操作时还需要 java.io 包的支持。

（2）ZipOutputStream。

如果想完成一个文件或文件夹的压缩，就要使用 ZipOutputStream 类。ZipOutputStream 是 OutputStream 的子类，其常用的方法如下：
- public ZipOutputStream(OutputStream out)创建新的 ZIP 输出流。
- public void putNextEntry(ZipEntry e) throws IOException 设置每一个 ZipEntry 对象。
- public void setComment(String comment)设置 ZIP 文件的注释。

【例 5.32】使用 ZipOutputStream 将 E 盘根目录下的记事本文件 ceshi.txt 压缩到文件 ceshi.zip(该文件位于 E 盘根目录)。

```java
import java.io.File;
import java.io.FileInputStream;
import java.io.InputStream;
import java.util.zip.ZipEntry;
import java.util.zip.ZipOutputStream;
import java.io.FileOutputStream;
public class ZipOutputStream1{
public static void main(String args[])throws Exception{//所有异常抛出
File file = new File("e:" + File.separator + "ceshi.txt");//定义要压缩的文件
File zipFile = new File("e:" + File.separator + "ceshi.zip");//定义压缩文件名称
InputStream input = new FileInputStream(file);//定义文件的输入流
ZipOutputStream zipOut = null;//声明压缩流对象
zipOut = new ZipOutputStream(new FileOutputStream(zipFile));
zipOut.putNextEntry(new ZipEntry(file.getName()));//设置 ZipEntry 对象
zipOut.setComment("test ZipOutputStream");//设置注释
int temp = 0;
while((temp = input.read())! = -1){//读取内容
zipOut.write(temp);//压缩输出
}
input.close();//关闭输入流
zipOut.close();//关闭输出流
}
}
```

如图 5-37 所示。

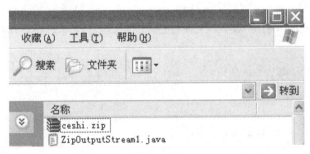

图 5-37

(3) ZipFile 类。

ZipFile 是一个专门表示压缩文件的类。在 Java 中,每一个压缩文件都可以使用 ZipFile 表示,还可以使用 ZipFile 根据压缩后的文件名称找到每一个压缩文件中的 ZipEntry 并将其进行解压缩操作。ZipFile 类的常用方法如下:

• public ZipFile(File file) throws ZipException,IOException 根据 File 类实例化 ZipFile 对象。

• public ZipEntry getEntry(String name)根据名称找到其对应的 ZipEntry。

• public InputStream getInputStream(ZipEntry entry) throws IOException 根据 ZipEntry 取得 InputStream 实例。

• public String getName()得到压缩文件的路径名称。

ZipFile 在实例化的时候必须接受 File 类的实例,此 File 类的实例是指向一个压缩的 "*.zip"文件。

【例 5.33】使用 ZipFile 将上例生成的压缩文件,解压缩到 F 盘根目录(其文件名为 ceshi.txt)。

```
import java.io.File;
import java.io.FileInputStream;
import java.io.InputStream;
import java.io.OutputStream;
import java.util.zip.ZipEntry;
import java.util.zip.ZipOutputStream;
import java.util.zip.ZipFile;
import java.io.FileOutputStream;
public class ZipFile1{
public static void main(String args[])throws Exception{//所有异常抛出
File file = new File("e:" + File.separator + "ceshi.zip");//找到压缩文件
File outputFile = new File("f:" + File.separator + "ceshi.txt");//定义解压缩的文件名称
ZipFile zipFile = new ZipFile(file);//实例化 ZipFile 对象
```

```
ZipEntry entry = zipFile.getEntry("ceshi.txt");//得到一个压缩实体
OutputStream out = new FileOutputStream(outputFile);//实例化输出流
InputStream input = zipFile.getInputStream(entry);//得到一个压缩实体的输入流
int temp = 0;
while((temp = input.read())! = -1){
  out.write(temp);
}
input.close();//关闭输入流
out.close();//关闭输出流
}
}
```

如图 5-38 所示。

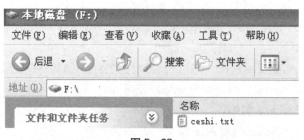

图 5-38

(4) ZipInputStream。

ZipInputStream 是 InputStream 的子类,通过这个类可以方便地读取 ZIP 格式的压缩文件,其常用方法如下:
- public ZipInputStream(InputStream in)实例化 ZipInputStream 对象。
- public ZipEntry getNextEntry()throws IOException 取得下一个 ZipEntry。

使用 ZipInputStream 读取 ZIP 格式的文件,可以通过 getNextEntry()方法依次读取压缩文件中每一个 ZipEntry 对象的名称,然后通过 ZipFile 类取得每一个 ZipEntry 的输入流对象,从而完成文件的解压缩。

注意:

ZipInputStream 解压 ZIP 文件时,在输出文件前,必须先判断其输出文件夹和文件是否存在。如果不存在,必须先创建。

【例 5.34】使用 ZipInputStream 解压缩 E 盘根目录下的压缩文件 ceshi.zip。

```
import java.io.File;
import java.io.OutputStream;
import java.io.InputStream;
import java.util.zip.ZipEntry;
```

```java
import java.util.zip.ZipFile;
import java.util.zip.ZipInputStream;
import java.io.FileInputStream;
import java.io.FileOutputStream;
public class ZipInputStream1{
public static void main(String args[])throws Exception{//所有异常抛出
File file = new File("e:" + File.separator + "ceshi.zip");//定义压缩文件名称
File outFile = null;//输出文件的时候要有文件夹的操作
ZipFile zipFile = new ZipFile(file);//实例化 ZipFile 对象
ZipInputStream zipInput = null;//定义压缩输入流
OutputStream out = null;//定义输出流,用于输出每一个实体内容
InputStream input = null;//定义输入流,读取每一个 ZipEntry
ZipEntry entry = null;//每一个压缩实体
zipInput = new ZipInputStream(new FileInputStream(file));//实例化 ZipInputStream
while((entry = zipInput.getNextEntry())! = null){//得到一个压缩实体
System.out.println("解压缩" + entry.getName() + "文件。");
outFile = new File("e:" + File.separator + entry.getName());//定义输出的文件路径
if(!outFile.getParentFile().exists()){//如果输出文件夹不存在
outFile.getParentFile().mkdir();//创建文件夹
}
if(!outFile.exists()){//判断输出文件是否存在
outFile.createNewFile();//创建文件
}
input = zipFile.getInputStream(entry);//得到每一个实体的输入流
out = new FileOutputStream(outFile);//实例化文件输出流
int temp = 0;
while((temp = input.read())! = -1){
out.write(temp);
}
input.close();//关闭输入流
out.close();//关闭输出流
}
input.close();
}
}
```

如图 5-39、图 5-40 所示。

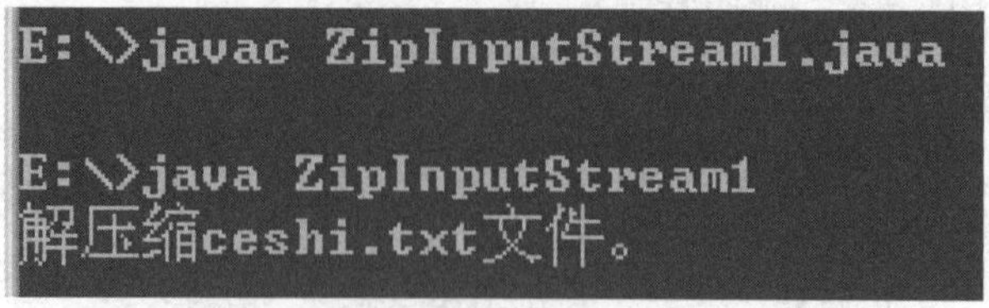

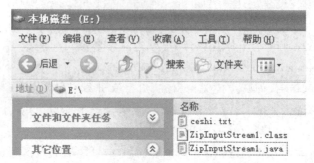

图 5-39

图 5-40

任务 实施

5.1.7 学生成绩信息保存到文件实现(类)

```
import java.io.File;
import java.io.FileInputStream;
import java.io.FileWriter;
import java.util.Scanner;
import java.util.ArrayList;

public class ScoreSave2{
    public static void main(String[]args)throws Exception{
        ArrayList<StudentScore>arr = new ArrayList<StudentScore>();//存储学生的成绩信息
        Scanner sc = new Scanner(System.in);//键盘输入
        while(true){
        System.out.print("请输入学号(输入-1退出):");
        int sno = sc.nextInt();
        if(sno = = -1){
           break;
```

```java
            }
            for(int i = 0;i<arr.size();i++){
                StudentScore stu = (StudentScore)arr.get(i);
                if(sno = = stu.sno){
                    System.out.println("学号重复！请重新输入:");
                    continue;
                }
            }
            System.out.print("请输入姓名:");
            String sname = sc.next();
            System.out.print("请输入性别:");
            char sex = sc.next().charAt(0);
            System.out.print("请输入课程名称:");
            String cname = sc.next();
            System.out.print("请输入平时成绩:");
            byte pingShiScore = sc.nextByte();
            System.out.print("请输入过程考核成绩:");
            byte guoChengScore = sc.nextByte();
            System.out.print("请输入终结考核成绩:");
            byte zhongJieScore = sc.nextByte();
            double zongHeScore = pingShiScore * 0.1 + guoChengScore * 0.4 + zhongJieScore * 0.5;
            String zongHeScoreJiBie = null;
            if(zongHeScore<60){
                zongHeScoreJiBie = "不及格";
            }else if(zongHeScore<70){
                zongHeScoreJiBie = "及格";
            }else if(zongHeScore<80){
                zongHeScoreJiBie = "中";
            }
            else if(zongHeScore<90){
                zongHeScoreJiBie = "良";
            }
            else if(zongHeScore< = 100){
                zongHeScoreJiBie = "优";
            }
```

```java
                StudentScore student = new StudentScore(sno,sname,
sex,cname,pingShiScore,guoChengScore,zhongJieScore,zongHeScore,zongHeScoreJiBie);
                arr.add(student);
            }
            FileWriter wr = new FileWriter("studentScore.txt");
            for(int i = 0;i<arr.size();i++){

    wr.write(arr.get(i).sno + "\t" + arr.get(i).sname + "\t" + arr.get(i).sex + "\
t" + arr.get(i).cname + "\t" + arr.get(i).pingShiScore + "\t" + arr.get(i).
guoChengScore + "\t" + arr.get(i).zhongJieScore + "\t" + arr.get(i).zongHeScore
 + "\t" + arr.get(i).zongHeScoreJiBie + "\t" + "\n");
                wr.flush();
            }
            wr.close();
            System.out.println("文件保存成功!");

        }
    }

class StudentScore{
        int sno;//定义变量sno,用于表示学生学号
        String sname;//定义变量sname,用于表示学生姓名
        char sex;//定义变量sex,用于表示学生性别
        String cname;//定义变量Cname,用于表示课程名称
        byte pingShiScore;//定义变量pingShiScore,用于表示学生平时成绩
        byte guoChengScore;//定义变量guoChengScore,用于表示学生过程考核成绩
        byte zhongJieScore;//定义变量zhongJieScore,用于表示学生终结考核成绩
        double zongHeScore;//定义变量zongHeScore,用于表示学生综合成绩
        //定义变量zongHeScoreJiBie,用于表示学生综合成绩的等级
        String zongHeScoreJiBie;

    public StudentScore(){//空构造函数
        }
    public StudentScore(int sno, String sname, char sex, String cname, byte
pingShiScore, byte guoChengScore, byte zhongJieScore, double zongHeScore, String
zongHeScoreJiBie){//带参数的构造函数
            this.sno = sno;
            this.sname = sname;
```

```java
        this.sex = sex;
        this.cname = cname;
        this.pingShiScore = pingShiScore;
        this.guoChengScore = guoChengScore;
        this.zhongJieScore = zhongJieScore;
        this.zongHeScore = zongHeScore;
        this.zongHeScoreJiBie = zongHeScoreJiBie;
    }
    public int getSno(){
        return sno;
    }

    public void setSno(int sno){
        this.sno = sno;
    }
    public String getSname(){
        return sname;
    }
    public void setSname(String sname){
        this.sname = sname;
    }
    public char getSex(){
        return sex;
    }
    public void setSex(char sex){
        this.sex = sex;
    }
    public String getCname(){
        return cname;
    }
    public void setCname(String cname){
        this.cname = cname;
    }
    public byte getPingShiScore(){
        return pingShiScore;
    }
    public void setPingShiScore(byte pingShiScore){
        this.pingShiScore = pingShiScore;
```

```java
        }
        public byte getGuoChengScore(){
            return guoChengScore;
        }
        public void setGuoChengScore(byte guoChengScore){
            this.guoChengScore = guoChengScore;
        }
        public byte getZhongJieScore(){
            return zhongJieScore;
        }
        public void setZhongJieScore(byte zhongJieScore){
            this.zhongJieScore = zhongJieScore;
        }
        public double getZongHeScore(){
            return zongHeScore;
        }
        public void setZongHeScore(double zongHeScore){
            this.zongHeScore = zongHeScore;
        }
        public String getZongHeScoreJiBie(){
            return zongHeScoreJiBie;
        }
        public void setZongHeScoreJiBie(String zongHeScoreJiBie){
            this.zongHeScoreJiBie = zongHeScoreJiBie;
        }
    }
```

如图 5-41 至图 5-43 所示。

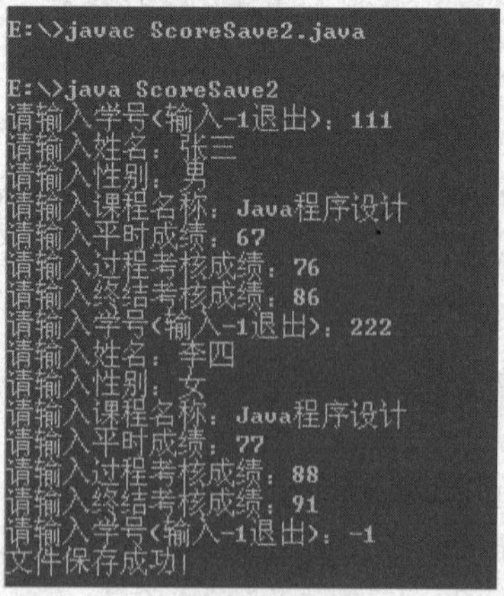

图 5-41

名称	大小	类型	修改日期
studentScore.txt	1 KB	文本文档	2018-11-5 09:08
StudentScore.class	2 KB	CLASS 文件	2018-11-5 09:07
ScoreSave2.class	3 KB	CLASS 文件	2018-11-5 09:07
ScoreSave2.java	5 KB	JAVA 文件	2018-11-5 09:05

图 5-42

```
studentScore.txt - 记事本
文件(F) 编辑(E) 格式(O) 查看(V) 帮助(H)
111   张三   男   Java程序设计   67   76   86   88.1   良
222   李四   女   Java程序设计   77   88   91   88.4   良
```

图 5-43

任务 拓展

（1）实现自助银行服务系统用户账户信息的保存功能，要求将如图 5-44 所示的用户信息存入记事本文档 BankInfo.txt。

用户名	账号	密码	身份证号码	余额
王丽丽	179708064356	1234	210050619890808185	1000.0
张颖颖	179708064359	4321	210050619891231127	2000.0
刘华	179708064368	4567	410207198904051271	3000.0

图 5-44

（2）实现 BankInfo.txt 中存储的用户信息的查询、删除、修改操作，要求如下：
① 查询账号为 179708064356 的账户信息。
② 删除账号为 179708064368 的账户信息。
③ 将账号为 179708064359 的账户的余额修改为 5000 元。

 5.2 习 题

一、选择题

1. 请选出正确的文件操作语句。（　　）
   A. File f1＝new File("TEST.bat");
   B. DataInputStream DT＝new DataInputStream(System.in);
   C. OutputStreamWriter OS＝new OutputStreamWriter(System.out);
   D. RandomAccessFile RA＝new RandomAccessFile("OF1");

2. 请选出正确的建立随机访问文件流对象语句。（　　）
   A. RandomAccessFile RAF1＝new RandomAccessFile("TEST.txt","rw");

B. RandomAccessFile RAF1=new RandomAccessFile(new DataInputStream());

C. RandomAccessFile RAF1=new RandomAccessFile("TEST.txt");

D. RandomAccessFile RAF1=new RandomAccessFile(new File("TEST.txt"));

3. 为大型文件的读取选取合适的创建对象方法。（   ）

A. new FileInputStream("FILENAME");

B. new InputStreamReader(new FileInputStream("FILENAME"));

C. new BufferedReader（new InputStreamReader（new FileInputStream（"FILENAME"）））;

D. new RandomAccessFile RAF=new RandomAccessFile("FILENAME","+rw");

4. 选取正确的创建 InputStreamReader 对象的方法。（   ）

A. new InputStreamReader(new FileInputStream("FILE"));

B. new InputStreamReader(new FileReader("FILE"));

C. new InputStreamReader(new BufferedReader("FILE"));

D. new InputStreamReader("FILE");

E. new InputStreamReader(System.in);

5. 过滤流类 FilterOutputStream 是 BufferedOutputStream、DataOutputStream 和 PrintStream 类的父类，下列哪些是 FilterOutputStream 类构造方法有效的参数？（   ）

A. InputStream

B. OutputStream

C. File

D. RandomAccessFile

二、简答题

1. 将 100～300 之间的所有素数保存到文件 sushu.txt 中。

2. 将下表所示的学生信息存入二进制数据文件 STUDENTINFO.DAT 中。

学号	姓名	性别	专业	系别	民族	政治面貌
111	张三	男	计算机应用技术	信息工程系	汉	团员
222	李四	女	软件与信息服务	信息工程系	汉	党员
333	王五	男	物联网应用技术	信息工程系	汉	群众
444	赵六	男	计算机应用技术	信息工程系	汉	团员
555	燕七	女	软件与信息服务	信息工程系	汉	党员
666	钱八	女	物联网应用技术	信息工程系	汉	群众

3. 读取第 2 题中的文件 STUDENTINFO.DAT 里的数据，然后查找并显示学号为 222 的学生信息。

4. 读取第 2 题中的文件 STUDENTINFO.DAT 里的数据，然后将学号为 333 的学生的政治面貌改为党员，再删除学号为 666 的学生信息。

创建学生成绩管理系统的图形界面

项目6首先分析了学生成绩管理系统的基本功能,并总体介绍了Java图形用户界面程序设计的相关知识(如awt、swing等);其次,讲解了窗口、菜单条、菜单、菜单项,并设计实现了程序主窗口;然后,讲解了标签、文本框、按钮、布局管理器、事件处理机制,并设计实现了程序的登录窗口;接下来,讲解了复选框和单选按钮组件、组合框、对话框,并设计实现了信息管理(添加、修改、删除)窗口;最后,讲解了表格等的相关知识,并设计实现了程序的查询窗口。

工作 任务

(1) 学生成绩管理系统功能分析。
(2) 学生成绩管理系统主界面设计。
(3) 学生成绩管理系统登录界面设计。
(4) 学生成绩管理系统信息管理窗口设计。
(5) 学生成绩管理系统信息查询窗口设计。

学习 目标

(1) 了解学生成绩管理系统的基本功能,熟悉GUI组件的相关知识,能够根据工作任务需要完成系统的基本功能的分析。

(2) 掌握界面应用程序的创建步骤,掌握窗口的属性、菜单的组成、菜单的建立,能够设计实现程序的主窗口。

(3) 掌握标签、文本框、按钮的常用属性和方法,掌握窗口组件的添加方法,理解并掌握文本框、按钮事件的处理,能够设计实现程序的登录窗口。

(4) 掌握部分常用组件(复选框、单选按钮组件、组合框、文本区TextArea、列表框等)的属性和方法,掌握对话框的建立和使用方法,掌握Java组件的常用事件的处理方法,能够设计实现程序的信息管理(添加、删除、修改)窗口。

(5) 掌握容器、表格的使用,掌握组件的布局设计,能够设计实现程序的查询窗口。

任务1　学生成绩管理系统功能分析

任务描述及分析

小明要完成学生成绩管理系统的图形界面设计,必须先完成学生成绩管理系统的功能分析,这样才能根据系统的不同功能确定相应的图形界面。要完成这个工作任务,第一步,需要完成系统的功能分析,并根据系统的功能需要确定要完成的系统界面;第二步,在明确了要设计完成哪些图形界面的基础上,应该对Java图形用户界面设计的相关知识有总体认知,为后续具体界面的设计奠定基础。

相关知识

6.1.1　系统功能分析

1. 需求分析

根据调研客户需求,明确客户要求设计学生成绩管理系统,实现数据录入、数据修改、数据删除、数据查询功能,同时还要有基本的用户安全管理功能。由于数据录入、数据修改、数据删除对相应的一整条数据信息(记录)进行操作后,数据信息会发生变化,而且三者操作的数据信息的格式一致,因此可以合并为一个功能模块(信息管理)。

2. 系统总体结构设计

通过对学生成绩管理系统的功能分析,可以定义出系统的总体结构模块图,如图6-1所示。

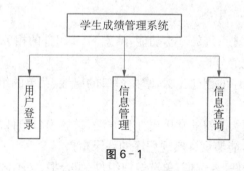

图6-1

根据功能分析,我们可以按照功能模块需求,设计4个图形界面:

(1) 学生成绩管理系统主界面,通过主界面可以调用其他功能模块,因此主界面应具有操作性好、界面清晰的特点,使用户能够很方便地找到所需功能,可以通过菜单栏(信息添加、信息修改、信息删除、信息查询)选项进入相应的功能模块,其中信息添加、信息修改、信息删除都在信息管理模块下。

(2) 学生成绩管理系统登录界面,需要用户输入用户名、密码进行验证,以确保用户安全管理。

(3) 学生成绩管理系统信息管理窗口,包括信息添加、信息修改、信息删除功能,当点击对应的菜单项,就进入对应的功能界面。

(4) 学生成绩管理系统信息查询窗口,通过查询窗口用户可以按照学号、班级等不同的关键字进行成绩信息的查询。

6.1.2 图形用户界面设计概述

在图形用户界面应用程序中,各种图形用户界面元素有机结合在一起,它们不但提供了漂亮的外观,而且提供了与用户交互的各种手段。在 Java 语言中,这些元素主要通过 java.awt 包和 javax.swing 包中的类来进行控制和操作。

1. java.awt 包

Java 语言在 java.awt 包中提供了大量的进行图形用户界面设计所使用的类和接口,包括绘制图形、设置字体和颜色、控制组件处理事件等内容,AWT 是 Java 语言进行图形用户界面程序设计的基础。

java.awt 包中主要的类及其层次关系如下所示:

```
javalangObject java 所有类的超类
  ├─ Font 字体类
  ├─ Color 颜色类
  ├─ Graphics 几何绘图类
  ├─ Component 组件类
  │   ├─ Label 标签类
  │   ├─ Button 按钮类
  │   ├─ TextComponent 文本组件类
  │   │   ├─ TextField 单行文本框类
  │   │   └─ TextArea 多行文本框类
  │   ├─ List 列表类
  │   ├─ Container 容器类
  │   │   ├─ Panel 面板类
  │   │   │   └─ Applet 小程序类
  │   │   └─ Window 窗口类
  │   │       ├─ Frame 框架类
  │   │       └─ Dialog 对话框类
  │   └─ Checkbox 单选按钮与复选按钮类
  ├─ CheckboxGroup 按钮组合类
  ├─ MenuComponent 菜单组件类
  │   ├─ MenuBar 菜单条类
  │   └─ MenuItem 菜单项类
  ├─ FlowLayout 流式布局管理类
  └─ BorderLayout 边界布局管理类
```

2. javax.swing 包

Swing 包是 Java 基础类库(Java Foundation Classes，JFC)的一部分，Swing 提供了从按钮到可分拆面板和表格的所有组件。Swing 组件是 Java 语言提供的第二代 GUI(图形用户界面)设计工具包，它以 AWT 为基础，在 AWT 内容的基础上新增或改进了一些 GUI 组件，使得 GUI 程序功能更强大、设计更容易、更方便。"Swing"是开发新组件的项目代码名，现在，这个名字常用来引用新组件和相关的 API。

Swing 包首先出现在 JDK 11 中，以前的版本中使用的是 AWT 组件，在新的 Java 版本中仍然支持 AWT 组件，但几乎所有的 AWT 组件都有对应的新的、功能更强的 Swing 组件，所以现在开发 GUI 程序时，一般建议用 Swing 组件代替 AWT 组件。鉴于此，本书的图形用户界面均采用 Swing 组件设计。AWT 组件和对应的 Swing 组件，从名称上很容易记忆和区别。例如，AWT 的框架类、面板类、按钮类和菜单类被命名为 Frame、Panel、Button 和 Menu，而 Swing 对应的组件类被命名为 JFrame、JPanel、JButton 和 JMenu。与 AWT 组件相比，Swing 组件的名前多一个"J"字母。另外，AWT 组件在 java.awt 包中，而 Swing 组件在 javax.swing 包中。

3. 组件、容器、布局

一般地，从一个 GUI 程序的外貌可以见到一些对界面起到装饰美化作用的圆、矩形等几何图形和图像，也可以见到如按钮、列表等一些可进行人机交互的组件。在 Java GUI 程序中，这些界面元素应处于一个容器中，其中组件在容器中的摆放位置和大小由容器的布局管理器决定。

(1) 组件和容器。

一个 Java 的图形用户界面的最基本元素是组件，组件是可以以图形化的方式显示在屏幕上并能与用户进行交互的对象，如一个按钮、一个文本框等。Component 类是 AWT 包中的一个抽象类，Java 中的图形组件大多数都是 Component 类的子类，Swing 包中的常用组件，如 JLabel(标签)、JButton(按钮)等都是 JComponent 类的子类，而 JComponent 类又是 Component 类的子类，其层次结构如图 6-2 所示：

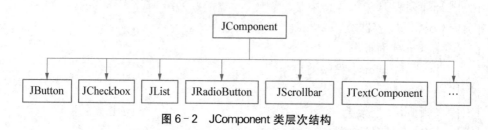

图 6-2 JComponent 类层次结构

在 Java 语言中，通常将组件放在一定的容器内使用。容器实际上是一种具有容纳其他组件和容器的功能的组件，抽象类 Container 是所有容器的父类，其中包含了很多有关容器的功能和方法，而类 Container 又是 Java 语言的组件类 Component 的子类。在 Swing 中依然存在容器的概念，Swing 中所有的容器类都是继承自 AWT 组件包。每一个使用 Swing 的 GUI 应用程序都必须包含至少一个顶层 Swing 容器组件，这样的容器有 3 种：JFrame、JDialog 和 JApplet(用于 Java 小程序)。每一个 JFrame 对象实现一个主窗口，每一个 JDialog 对象实现一个第二窗口(依赖另一窗口的窗口)，每一个 JApplet 对象实现一个 Java

小程序的显示区域。容器类中包含了将组件加入容器的方法 add()、将组件移出容器的方法 remove()和方法 removeAll()、获得组件的方法 getComponent()等。

GUI 的应用程序一般包含多个组件(如标签、文本框、按钮等),这些组件的排列形式由容器所用的布局管理器决定,各种容器都有自己默认的布局管理器。

(2) 布局管理器 Layout Manager。

为了使得图形用户界面具有良好的平台无关性,Java 语言提供了布局管理器这个工具来管理组件在容器中的布局,而不使用直接设置组件位置和大小的方式。容器中的组件定位由布局管理器决定,每个容器都有一个默认的布局管理器。当容器需要对某个组件进行定位或判断其大小尺寸时,就会调用其相应的布局管理器。但也可以不用默认的布局管理器,而是在程序中指定其新的布局管理器。

Java 平台提供多种布局管理器,常用的有 FlowLayout、BorderLayout、GridLayout、CardLayout、BoxLayout 和 GridBagLayout 等。如图 6-3 所示。

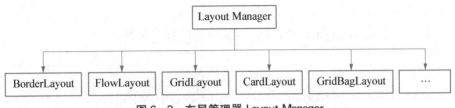

图 6-3 布局管理器 Layout Manager

使用不同的布局管理器,组件在容器上的位置和大小都是很不一样的。在程序中安排组件的位置和大小时,应该注意:

① 容器中的布局管理器负责各个组件的大小和位置,因此用户无法在这种情况下直接设置这些属性。若试图使用 Java 语言提供的 setLocation()、setSize()、setBounds()等方法则都会被布局管理器覆盖。

② 若用户确实需要亲自设置组件的位置和大小,则应取消该容器的布局管理器,方法为:

```
setLayout(null);
```

随后,用户必须使用 setLocation()、setSize()、setBounds()等方法为组件设置大小和位置,但这种方法将会导致程序的系统相关。在一个 GUI 应用程序的界面上除了可以见到上述的容器、组件等标准 GUI 元素外,还可以见到一些几何图形、图案、图像等内容,这些内容通常是不能被系统识别进行人机交互的,在界面上只起到装饰、美化界面的作用。

任务 2 学生成绩管理系统主界面设计

任务描述 及分析

小明要设计实现学生成绩管理系统主界面,在主界面中用不同菜单项(信息添加、信息修改、信息删除、信息查询)实现不同功能模块的调用。要完成这个工作任务,第一步,我们需要掌握 Swing 的基本容器类 JFrame,并利用 JFrame 创建窗口;第二步,我们需要掌握菜单条、菜单、菜单项相关知识,并在创建的窗口上创建对应的菜单。

相关知识

6.2.1 JFrame 类

Swing 的 JFrame 类可以用来创建窗口,该类创建的窗口有边界、标题、关闭按钮等。JFrame 类继承于 Frame 类,其常用操作方法如下:

- public JFrame()throws HeadlessException 创建一个普通的窗体对象。
- public JFrame(String title)throws HeadlessException 创建一个窗体对象,并指定标题。
- public void setSize(int width, int height)设置窗体大小。
- public void setBackground(Color c)设置窗体的背景颜色。
- public void setLocation(int x, int y)设置组件的显示位置。
- public void setVisible(boolean b)显示或隐藏组件。
- public Component add(Component comp)向容器中增加组件。
- public void setLayout(LayoutManager mgr)设置布局管理器,如果设置为 null 表示不适用。
- public void pack()调整窗口大小,以适合其子组件首选大小和布局。
- public Container getContentPane()返回此窗体的容器对象。
- setDefaultCloseOperation(int operation):设置用户在此窗体上发起"close"时默认执行的操作,若参数值为 3 或 EXIT_ON_CLOSE(在 JFrame 中定义),应用程序将使用 System.exit 方法退出。

(1) 创建带标题"Java GUI 应用程序"的窗口,可用语句:

```
JFrame frame = new JFrame("Java GUI 应用程序");
```

(2) 要显示窗口,可使用方法 setVisible()语句:

```
frame.setVisible(true);//使 JFrame 类对象 frame 表示的窗口显示到屏幕上
```

(3) 一般在显示窗口前,应设置窗口的初始显示大小,可使用 setSize()方法或 pack()方法。例如:

```
frame.setSize(200,150);//设置框架窗口初始大小为 200 * 150 点
frame.pack();//设置框架窗口初始大小为刚好只显示出所有的组件
```

(4) 若要设置窗口在计算机显示器屏幕上的显示位置,可用 setLocation() 或 setLocationRelativeTo(null)方法。例如:

frame.setLocation(300,300);//设置窗口在显示器的显示位置
frame.setLocationRelativeTo(null);//设置窗口在显示器的中央显示

(5) 向窗口添加组件(如标签、按钮等)时,并不直接添加组件到窗口,而是添加到内容窗格(content pane),可通过getContentPane()方法获得窗体的容器对象,然后通过add()方法将组件添加到内容窗格。若希望用自己的容器(如JPanel)替换掉内容窗格,可以使用setContentPane()方法。例如:

frame.getContentPane().add(b1);//将组件b1添加到窗口容器

或

Container cont = frame.getContentPane();//得到窗体容器
cont.add(b1);//将组件b1添加到窗口容器

(6) 设置窗口背景色,可用setBackground()方法。例如:

frame.getContentPane().setBackground(Colorblue);//将窗口背景设为蓝色

注意:
使用setBackground()方法,必须导入头文件java.awt.Color。

【例6.1】创建第一个Java窗体。

```
import javax.swing.JFrame;
import java.awt.Color;
public class JFrame1{
public static void main(String args[]){
JFrame f = new JFrame("第一个Swing窗体");
f.setSize(230,80);//设置组件的大小
f.setBackground(Color.WHITE);//将背景设置成白色
f.setLocation(300,200);//设置组件的显示位置
f.setVisible(true);//让组件可见
}
}
```

如图6-4所示。

图6-4

【例 6.2】将上例创建的窗体显示在屏幕中央,并且创建窗体时继承 **JFrame** 类。

```
import javax.swing.JFrame;
import java.awt.Color;
public class JFrame2 extends JFrame{//继承 JFrame 类
public JFrame2(){//构造方法
super("第二个 Swing 窗体");//设置标题为"第二个 Swing 窗体"
this.setSize(300,200);//设置组件的大小
this.setBackground(Color.WHITE);//将背景设置成白色
this.setLocationRelativeTo(null);//设置组件的显示位置在屏幕中央
this.setVisible(true);//让组件可见
}
public static void main(String args[]){
JFrame2 JF2 = new JFrame2();//创建窗体
}
}
```

如图 6-5 所示。

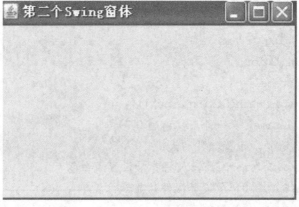

图 6-5

6.2.2 菜单

菜单将一个应用程序的命令按层次化管理并组织在一起,是一种常用的 GUI 组件。常见的菜单为下拉式菜单和弹出式菜单快捷菜单等。下拉式菜单包含有一个菜单条(也称为菜单栏,MenuBar),在菜单条上安排有若干个菜单(Menu),每个菜单又包含若干菜单项(MenuItem),每个菜单项对应了一个命令或子菜单项,它们构成一个应用程序的菜单系统。用鼠标或键盘选择对应一个命令的菜单项与选择一个按钮类似,使用菜单系统可方便地向程序发布命令。

在构建一个自己的菜单系统时,可按照菜单系统的层次一步一步地进行:

(1) 用类 JMenuBar 创建菜单条。

可简单地用 JMenuBar()构造方法类创建一个新菜单条。例如:

```
JMenuBar aMenuBar = new JMenuBar();
```

(2) 用类 JMenu 创建菜单。

用类 JMenu 的构造方法来创建菜单,其构造方法有:
- public JMenu()构造一个无文本的菜单。
- public JMenu(String s)用字符串 s 作为文本来构造一个菜单。

例如:

```
JMenu aMenu = new JMenu("文件");//创建文件菜单
```

(3) 用 JMenuItem 类创建菜单项。

JMenuItem 类的构造方法如下:
- public JMenuItem()创建一个菜单项但不设置文本和图标。
- public JMenuItem(Icon icon)创建一个带图标的菜单项。
- public JMenuItem(String text)创建一个具有指定文本的菜单项。
- public JMenuItem(String text,Icon icon)创建具有文本和图标的菜单项。
- public JMenuItem(String text,int mnemonic)创建具有文本和快捷字母的菜单项。

例如:

```
JMenuItem aMenuItem = new JMenuItem("新建");//创建"新建"菜单项
```

(4) 将菜单项加入菜单中,将菜单加入菜单条中。

可用 JMenuBar 类和 JMenu 类的 add()方法完成添加工作,例如:

```
aMenuBaradd(aMenu);
aMenuadd(aMenuItem);
```

另外,可用 addSeparator()方法向菜单添加分割线。

(5) 将菜单条加入容器中。

可向实现了 MenuContainer 接口的容器(如窗口)加入菜单系统。在 JFrame 类中有方法:

```
public void setJMenuBar(JMenuBar menubar)
```

可为窗口设置菜单条,例如:

```
JFrame aFrame = new JFrame();
aFrame.setJMenuBar(aMenuBar);
```

(6) 处理菜单项选择事件。

为了检测对菜单项作出的选择,要监听菜单项的 ActionEvent 事件。选择一个菜单项如同选择了一个 JButton 按钮一样,事件处理详见工作任务 6.3。

【例6.3】创建菜单栏,显示"文件""编辑"菜单,"文件"菜单包括"新建""打开""保存""打印""退出"菜单项,"编辑"菜单包括"查找""替换""全选"菜单项,而且需要用分割线分割相关菜单项。

```java
import javax.swing.JFrame;
import javax.swing.JMenuBar;
import javax.swing.JMenu;
import javax.swing.JMenuItem;
public class JMenu1{
    public static void main(String args[]){
        JFrame f = new JFrame("第一个菜单");//创建窗体f,标题为"第一个菜单"
        JMenuBar MenuBar1 = new JMenuBar();//创建菜单栏
        JMenu Menu1 = new JMenu("文件");//创建菜单1,显示文件
        JMenu Menu2 = new JMenu("编辑");//创建菜单2,显示编辑
        JMenuItem MenuItem11 = new JMenuItem("新建");//创建菜单项11,显示新建
        JMenuItem MenuItem12 = new JMenuItem("打开");//创建菜单项12,显示打开
        JMenuItem MenuItem13 = new JMenuItem("保存");//创建菜单项13,显示保存
        JMenuItem MenuItem14 = new JMenuItem("打印");//创建菜单项14,显示打印
        JMenuItem MenuItem15 = new JMenuItem("退出");//创建菜单项15,显示退出
        JMenuItem MenuItem21 = new JMenuItem("查找");//创建菜单项21,显示查找
        JMenuItem MenuItem22 = new JMenuItem("替换");//创建菜单项22,显示替换
        JMenuItem MenuItem23 = new JMenuItem("全选");//创建菜单项23,显示全选
        //将菜单项11~15加入Menu1,并在不同位置添加分割线
        Menu1.add(MenuItem11);
        Menu1.add(MenuItem12);
        Menu1.add(MenuItem13);
        Menu1.addSeparator();//添加分割线
        Menu1.add(MenuItem14);
        Menu1.addSeparator();
        Menu1.add(MenuItem15);
        //将菜单项21~23加入Menu2,并添加分割线
        Menu2.add(MenuItem21);
        Menu2.add(MenuItem22);
        Menu2.addSeparator();//添加分割线
        Menu2.add(MenuItem23);
        //将菜单Menu1、Menu2加入菜单栏MenuBar1
```

```
MenuBar1.add(Menu1);
MenuBar1.add(Menu2);
f.setJMenuBar(MenuBar1);//将菜单条 MenuBar1 加入容器中
f.setSize(300,200);//设置窗体的大小
f.setLocationRelativeTo(null);//设置窗体的显示位置
f.setVisible(true);//让组件可见
    }
}
```

如图 6-6、图 6-7 所示。

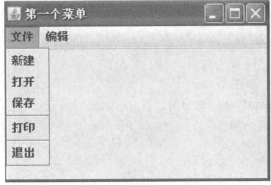

图 6-6

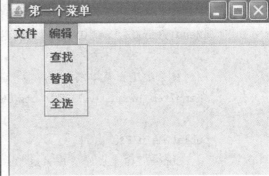

图 6-7

任务 实施

6.2.3 学生成绩管理系统主界面设计实现

```
import javax.swing.JFrame;
import javax.swing.JMenu;
import javax.swing.JMenuBar;
import javax.swing.JMenuItem;

public class MainJFrame extends JFrame  {
    //创建菜单栏
    JMenuBar jMenuBar = new JMenuBar();

    //创建信息管理菜单
    JMenu jMenu01 = new JMenu("信息管理");
```

```java
//创建信息管理菜单项
JMenuItem jMenuItem01_00 = new JMenuItem("信息添加");
JMenuItem jMenuItem01_01 = new JMenuItem("信息修改");
JMenuItem jMenuItem01_02 = new JMenuItem("信息删除");

//创建信息查询菜单
JMenu jMenu02 = new JMenu("信息查询");

//创建信息查询菜单项
JMenuItem jMenuItem02_00 = new JMenuItem("成绩查询");

//创建系统退出菜单
JMenu jMenu03 = new JMenu("系统退出");

//创建系统退出菜单项
JMenuItem jMenuItem03_00 = new JMenuItem("退出");

public MainJFrame(){
    super("学生成绩管理系统");

    //菜单栏
    jMenu01.add(jMenuItem01_00);
    jMenu01.addSeparator();
    jMenu01.add(jMenuItem01_01);
    jMenu01.addSeparator();
    jMenu01.add(jMenuItem01_02);
    jMenu02.add(jMenuItem02_00);
    jMenu03.add(jMenuItem03_00);

    jMenuBar.add(jMenu01);
    jMenuBar.add(jMenu02);
    jMenuBar.add(jMenu03);

    //在窗口上设置菜单栏
    this.setJMenuBar(jMenuBar);

    //设置窗口
    this.setDefaultCloseOperation(JFrame.EXIT_ON_CLOSE);
```

```
            this.setSize(700,580);
            this.setLocationRelativeTo(null);
            this.setVisible(true);
        }

        public static void main(String args[]){
            new MainJFrame();
        }

    }
```

如图 6-8 至图 6-10 所示。

图 6-8

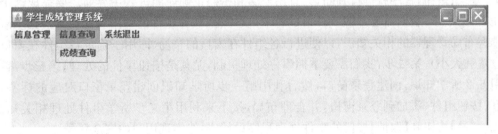

图 6-9

图 6-10

任务 拓展

完成如图 6-11 所示的自助银行服务系统主界面设计。

图 6-11

6.3 任务 3 学生成绩管理系统登录界面设计

任务描述 及分析

小明要设计实现学生成绩管理系统登录界面,在登录界面中要输入用户名和密码,然后点击【登录】按钮,程序就进行判断,只有用户名和密码正确才能进入主界面。要完成这个工作任务,第一步,我们需要掌握 Swing 的标签、文本框、按钮组件的常用属性和方法,此外,还要掌握布局管理器的相关知识,以便进行各组件在窗口的布局(即用于确定各组件在窗口中的位置和大小);第二步,我们需要掌握事件处理机制,尤其是按钮事件的处理;第三步,首先利用 6.2 所学知识,创建登录窗口,然后利用第一步所学知识创建登录窗口对应的标签、文本框、按钮组件,添加到登录窗口,并合理布局,接下来利用第二步所学事件处理相关知识,完成【登录】按钮和【取消】按钮的事件处理程序设计,最终实现用户的安全登录。

相关 知识

6.3.1 标签 JLabel

JLabel 标签常用来在界面上输出信息,JLabel 类的构造方法有:
- JLabel()创建一个空标签。
- JLabel(Icon image)创建一个带指定图像的标签。
- JLabel(Icon image, int horizontalAlignment)创建一个带指定图像和水平对齐方式的标签。

- JLabel(String text)创建一个带文字的标签。
- JLabel(String text, Icon icon, int horizontalAlignment)创建一个带文字图像和指定水平对齐方式的标签。
- JLabel(String text, int horizontalAlignment)创建一个带文字和指定水平对齐方式的标签。

其中,horizontalAlignment 水平对齐方式可以使用表示左对齐、右对齐、居中对齐的常量 JLabel.LEFT、JLabel.RIGHT 和 JLabel.CENTER。

【例6.4】用 JLabel(String text)创建一个带文字的标签。

```
import javax.swing.JFrame;
import javax.swing.JLabel;
public class JLabel1{
public static void main(String args[]){
JFrame f = new JFrame("第一个标签");//创建窗体 f,标题为"第一个标签"
JLabel JL1 = new JLabel("标签一");//实例化标签对象 JL1,其文本默认左对齐
f.getContentPane().add(JL1);
f.setSize(300,200);//设置窗体的大小
f.setLocationRelativeTo(null);//设置窗体的显示位置
f.setVisible(true);//让组件可见
}
}
```

如图 6-12 所示。

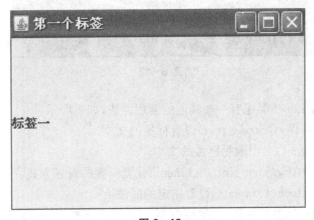

图 6-12

【例6.5】用 JLabel(Icon image)创建一个带指定图像的标签。

```
import javax.swing.JFrame;
import javax.swing.JLabel;
```

```
import javax.swing.Icon;
import javax.swing.ImageIcon;
public class JLabel2{
public static void main(String args[]){
JFrame f = new JFrame("第二个标签");//创建窗体f,标题为"第二个标签"
Icon icon = new ImageIcon("images/Java.jpg");//获取图像
JLabel JL1 = new JLabel(icon);//实例化标签对象JL1,并指定图像
f.getContentPane().add(JL1);
f.setSize(300,200);//设置窗体的大小
f.setLocationRelativeTo(null);//设置窗体的显示位置
f.setVisible(true);//让组件可见
}
}
```

如图6-13所示。

图6-13

除了构造方法,JLabel类还有一些其他的常用方法,如下所示:
- public void setText(String text)设置标签的文本。
- public String getText()取得标签的文本。
- public void setAlignment(int alignment)设置标签的对齐方式。
- public void setIcon(Icon icon)设置指定的图像。

此外,还可以设置标签的字体,若要更改JLabel的文字样式,可以直接使用Component类中定义的以下方法:
- public void setFont(Font f)

注意:

(1)使用setFont()方法设置字体,必须导入头文件java.awt.Font。

```
import java.awt.Font;
```

（2）setFont(Font f)方法设置字体时,要用 Font 类的实例化对象作为实参,因此,在设置字体之前,必须用 Font 类的构造方法实例化 Font 类对象。Font 类的构造方法如下:

public Font(String name,int style,int size)实例化 Font 对象,指定字体、显示风格及字的大小。name 用于指定字体,显示风格 style 可为 BOLD(粗体)、ITALIC(斜体)、PLAIN(普通样式),size 用于指定字的大小。例如:

```
Font f = new Font("宋体",Font.ITALIC + Font.BOLD,28);//实例化 Font 类对象 f,
将其字体设置为宋体,文字显示风格设置为斜体并加粗,文字的大小为 28
JLab.setFont(f);//设置标签 JLab 的字体
```

【例 6.6】设置标签的字体。

```
import javax.swing.JFrame;
import javax.swing.JLabel;
import java.awt.Font;
public class JLabel3{
public static void main(String args[]){
JFrame f = new JFrame("第三个标签");//创建窗体 f,标题为"第三个标签"
JLabel JL3 = new JLabel("第三个标签");//实例化标签对象 JL3
Font f1 = new Font("宋体",Font.ITALIC + Font.BOLD,28);//实例化 Font 类对象
f1,将其字体设置为宋体,文字显示风格设置为斜体并加粗,文字的大小为 28
JL3.setFont(f1);//设置标签 JL3 的字体
f.getContentPane().add(JL3);
f.setSize(300,200);//设置窗体的大小
f.setLocationRelativeTo(null);//设置窗体的显示位置
f.setVisible(true);//让组件可见
  }
}
```

如图 6-14 所示。

图 6-14

6.3.2 文本框

Java语言提供了单行文本输入框、密码文本输入框和多行文本输入框等文本框形式，它们都是人机交互的主要组件。

1. 单行文本输入框 JTextField

单行文本输入框一般用来让用户输入如姓名、地址这样的信息，它是一个能够接收用户的键盘输入的单行文本区域。JTextField类提供对单行文本框的支持，它有如下的几种构造方法：

- JTextField()创建一个新的单行文本框。
- JTextField(int columns)创建具有指定长度的空单行文本框。
- JTextField(String text)创建带初始文本内容的单行文本框。
- JTextField(String text,int columns)创建带初始文本内容并具有指定长度的单行文本框。

JTextField类的常用方法有：

- public void setText(String s)在文本框中显示字符串s。
- public String getText()获得文本框中的字符串。
- public void setEditable(boolean b)设置此文本框是否可编辑。

【例6.7】创建单行文本框。

```java
import javax.swing.JFrame;
import javax.swing.JTextField;
import javax.swing.JLabel;
import java.awt.GridLayout;
public class JTextField1{
public static void main(String args[]){
JFrame frame = new JFrame("第一个单行文本框");
JTextField JT1 = new JTextField(30);
JTextField JT2 = new JTextField("欢迎!",10);
JLabel JL1 = new JLabel("输入用户姓名:");
JLabel JL2 = new JLabel("不可编辑文本:");
JT2.setEnabled(false);//表示不可编辑
frame.setLayout(new GridLayout(2,2));
frame.add(JL1);
frame.add(JT1);
frame.add(JL2);
frame.add(JT2);
frame.setSize(300,100);
frame.setLocationRelativeTo(null);
frame.setVisible(true);
```

        }
    }

如图 6-15 所示。

图 6-15

2. 密码文本输入框 JPasswordField

密码文本输入框 JPasswordField 类是 JTextField 类的子类，在 JPasswordField 对象中输入的文字会被其他字符替代，这个组件常用来在 Java 程序中输入口令。

JPasswordField 类的构造方法及常用的其他方法有：

public JPasswordField()构造默认的 JPasswordField 对象。

public JPasswordField(String text)构造指定内容的 JPasswordField 对象。

void setEchoChar(char c)设置回显字符，默认为"＊"。

public char getEchoChar()得到回显的字符，默认为"＊"。

public char[]getPassword()得到此文本框的所有内容，即返回输入的口令。

【例 6.8】创建密码框

```java
import java.awt.GridLayout;
import javax.swing.JFrame;
import javax.swing.JPasswordField;
import javax.swing.JLabel;
public class JPasswordField1{
    public static void main(String args[]){
        JFrame frame = new JFrame("创建密码框");
        JPasswordField JPF1 = new JPasswordField();
        JPasswordField JPF2 = new JPasswordField();
        JPF2.setEchoChar('#');//设置回显
        JLabel JL1 = new JLabel("默认回显:");
        JLabel JL2 = new JLabel("设置回显为"#":");
        frame.setLayout(new GridLayout(2,2));
        frame.add(JL1);
        frame.add(JPF1);
        frame.add(JL2);
```

```
        frame.add(JPF2);
        frame.setSize(300,100);
        frame.setLocationRelativeTo(null);
        frame.setVisible(true);
    }
}
```

如图 6-16 所示。

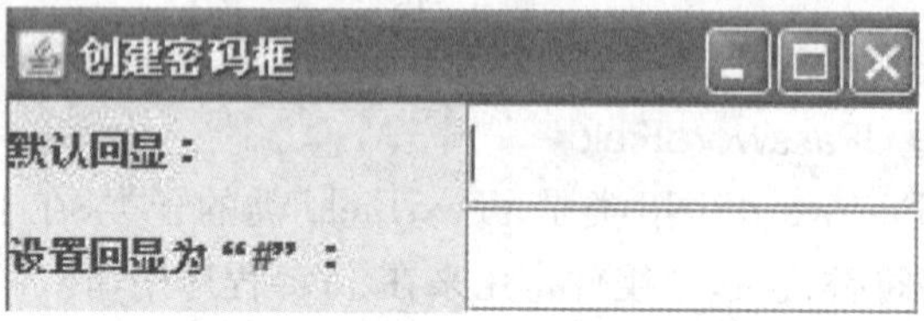

图 6-16

3. 多行文本输入框 JTextArea

JTextField 是单行文本框,不能显示多行文本,如果想要显示大段的多行文本,可以使用类 JTextArea 支持的多行文本框。JTextArea 有 6 个构造方法,常用的有 4 个:
- JTextArea()创建空多行文本框。
- JTextArea(int rows, int columns)创建指定行列数的多行文本框。
- JTextArea(String text)创建带初始文本内容的多行文本框。
- JTextArea(String text, int rows, int columns)创建带初始文本内容和指定大小的多行文本框。

其中,text 为 JTextArea 的初始化文本内容;rows 为 JTextArea 的高度,以行为单位;columns 为 JTextArea 的宽度,以字符为单位。例如构造一个高 5 行、宽 15 个字符的多行文本框的语句为:

```
JTextArea textArea = new JTextArea(5,15);
```

JTextArea 类的常用方法如下:
- public void append(String str)在文本域中追加内容。
- public void replaceRange(String str, int start, int end)替换文本域中指定范围的内容。
- public void insert(String str, int pos)在指定位置插入文本。
- public void setLineWrap(boolean wrap)设置换行策略。

多行文本框默认是不会自动折行的(但可以输入回车符换行),可以使用类 JTextArea 的 setLineWrap(boolean wrap)方法设置是否允许自动折行。wrap 为 true 时允许自动折行,多行文本框会根据用户输入的内容自动扩展大小。若不自动折行,那么多行文本框的宽度是由最长的一行文字确定的。若数据行数超过了预设的行数,则多行文本框会扩展自身的高度去适应,但多行文本框不会自动产生滚动条,可用滚动窗格 JScrollPane 来为多行文本框增加滚动条。

滚动窗格 JScrollPane 是一个能够自己产生滚动条的容器,通常只包容一个组件,并且根据这个组件的大小自动产生滚动条,这种性质正是 JTextArea 组件所需要的,即当数据行数超过包含在 JScrollPane 中的 JTextArea 预设区域大小时,JTextArea 的扩展就会反映到滚动窗格的滚动条上。

此外,多行文本框里文本内容的获得和设置,同样可以使用 getText() 和 setText() 两个方法来完成,还可以用 setEditable(boolean) 确定是否可对多行文本框的内容进行编辑。

【例 6.9】

```java
import java.awt.GridLayout;
import javax.swing.JFrame;
import javax.swing.JTextArea;
import javax.swing.JLabel;
public class JTextArea1{
public static void main(String args[]){
JFrame frame = new JFrame("第一个文本域");
JTextArea JTA1 = new JTextArea(3,10);//设置大小
JLabel JL1 = new JLabel("多行文本域:");
frame.setLayout(new GridLayout(1,1));
frame.add(JL1);
frame.add(JTA1);
frame.setSize(300,150);
frame.setLocationRelativeTo(null);
frame.setVisible(true);
}
}
```

如图 6-17 所示。

图 6-17

【例 6.10】加滚动条的文本域。

```java
import java.awt.GridLayout;
import javax.swing.JFrame;
import javax.swing.JTextArea;
import javax.swing.JLabel;
import javax.swing.JScrollPane;
public class JTextArea2{
    public static void main(String args[]){
        JFrame frame = new JFrame("第二个文本域");
        JTextArea JTA1 = new JTextArea(3,10);//设置大小
        JScrollPane JSP1 = new JScrollPane(JTA1);
        JLabel JL1 = new JLabel("多行文本域:");
        frame.setLayout(new GridLayout(1,1));
        frame.add(JL1);
        frame.add(JSP1);
        frame.setSize(300,150);
        frame.setLocationRelativeTo(null);
        frame.setVisible(true);
    }
}
```

如图 6-18 所示。

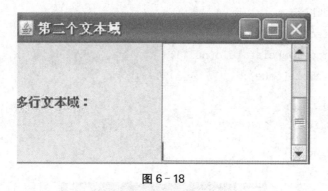

图 6-18

6.3.3 按钮 JButton

按钮是 GUI 中非常重要的一种基本组件,一般对应一个事先定义好的事件、执行功能、一段程序。当使用者单击按钮时,系统自动执行与该按钮联系的程序,从而完成预定的功能。JButton 类有如下的构造方法:

- JButton()创建空按钮。

- JButton(Icon icon)创建带图标的按钮。
- JButton(String text)创建带文字的按钮。
- JButton(String text,Icon icon)创建带文字和图标的按钮。

JButton组件与AWT的Button组件相比,增加了显示文本中可用HTML标记、可带图标等功能。

JButton类的常用方法如下:
- public void setLabel(String label)设置Button的显示内容。
- public String getLabel()得到Button的显示内容。
- public void setBounds(int x,int y,int width,int height)设置组件的大小及显示方式。
- public void setMnemonic(int mnemonic)设置按钮的快捷键。

【例6.11】创建带文字的按钮。

```
import java.awt.GridLayout;
import javax.swing.JFrame;
import javax.swing.JButton;
import javax.swing.JLabel;
public class JButton1{
public static void main(String args[]){
JFrame frame = new JFrame("第一个按钮");//实例化窗体对象
JLabel JL1 = new JLabel("第一个按钮:");
JButton JB1 = new JButton("按钮");
frame.setLayout(new GridLayout(1,1));
frame.add(JL1);
frame.add(JB1);
frame.setSize(200,70);
frame.setLocationRelativeTo(null);
frame.setVisible(true);
}
}
```

如图6-19所示。

图6-19

【例 6.12】 创建带文字和图片的按钮。

```java
import java.awt.GridLayout;
import javax.swing.JFrame;
import javax.swing.JButton;
import javax.swing.JLabel;
import javax.swing.Icon;
import javax.swing.ImageIcon;
public class JButton2{
public static void main(String args[]){
JFrame frame = new JFrame("第二个按钮");//实例化窗体对象
JLabel JL1 = new JLabel("第二个按钮:");
Icon icon = new ImageIcon("images/Java.jpg");//获取图像
JButton JB1 = new JButton("按钮",icon);
frame.setLayout(new GridLayout(1,1));
frame.add(JL1);
frame.add(JB1);
frame.setSize(200,70);
frame.setLocationRelativeTo(null);
frame.setVisible(true);
}
}
```

如图 6-20 所示。

图 6-20

6.3.4 布局管理器

在容器中所有组件的布局（位置和大小）由布局管理器来控制，Java 语言提供了 FlowLayout、BorderLayout、GridLayout、CardLayout 和 GridBagLayout 等多种布局管理器。每种容器都有自己默认的布局管理器，在默认的情况下，JPanel 使用 FlowLayout，而内容窗格 ContentPane(JApplet、JDialog 和 JFrame 对象的主容器)使用 BorderLayout。如果不希望使用默认的布局管理器，则可使用所有容器的父类 Container 的 setLayout()方法来改变默认的布局管理器。例如，下面是使得 JPanel 使用 BorderLayout 的代码：

```
JPanel pane = new JPanel();
panesetLayout(new BorderLayout());
```

当向面板或内容窗格等容器添加组件时,add 方法的参数个数和类型是不同的,这依赖于正在使用的容器的布局管理器。下面介绍几种常用的布局管理器。

1. FlowLayout

FlowLayout 布局是一种最基本的布局,这种布局指的是把组件一个接一个从左至右、从上至下地依次放在容器上,每一行中的组件默认为居中对齐。当容器的尺寸改变后,组件的大小不变,但布局将会随之变化。

FlowLayout 是 Applet 和 JPanel 的默认布局管理器,FlowLayout 类的构造方法如下:

FlowLayout()创建每行组件对齐方式为居中对齐、组件间距为 5 个像素单位的对象。

FlowLayout(int align)创建指定每行组件对齐方式、组件间距为 5 个像素单位的对象,align 可取 3 个静态常量 LEFT、CENTER 和 RIGHT 之一(分别表示左、中、右对齐方式)。

FlowLayout(int align,int hgap,int vgap)创建指定每行组件对齐方式的对象,该对象也使用参数 vgap 和 hgap 指定了组件间的以像素为单位的纵横间距。

向使用 FlowLayout 布局的容器添加组件可简单地使用下面的语句:

```
add(组件名);
```

【例 6.13】

```
import java.awt.FlowLayout;
import javax.swing.JFrame;
import javax.swing.JTextField;
public class JFlowLayout1{
public static void main(String args[]){
JFrame frame = new JFrame("流式布局");
frame.setLayout(new FlowLayout());
JTextField JT = null;
for(int i = 0;i<9;i + + ){
JT = new JTextField("文本框" + i);
frame.add(JT);
}
frame.setSize(280,123);
frame.setLocationRelativeTo(null);
frame.setVisible(true);
}
}
```

如图 6-21 所示。

图 6-21

2. BorderLayout

BorderLayout 是内容窗格的默认布局管理器,它将容器的布局分为 5 个区:北区、南区、东区、西区和中区。当容器的大小改变时,容器中的各个组件相对位置不变,其中间部分组件的尺寸会发生变化,四周组件宽度固定不变。

BorderLayout 类的构造方法如下:

- BorderLayout()构造没有间距的布局器对象。
- BorderLayout(int hgap, int vgap)构造有水平和垂直间距的布局器对象。

向 BorderLayout 布局的容器添加组件时,每添加一个组件都应指明该组件加在哪个区域中,add()方法的第二个参数指明加入的区域,区域"东、南、西、北、中"可用 5 个静态常量表示:BorderLayout.EAST、BorderLayout.SOUTH、BorderLayout.WEST、BorderLayout.NORTH 和 BorderLayout.CENTER。

【例 6.14】创建边框布局。

```java
import java.awt.BorderLayout;
import javax.swing.JFrame;
import javax.swing.JButton;
public class BorderLayout1{
    public static void main(String args[]){
        JFrame frame = new JFrame("边框布局");
        frame.setLayout(new BorderLayout());
        frame.add(new JButton("按钮 1"),BorderLayout.EAST);
        frame.add(new JButton("按钮 2"),BorderLayout.WEST);
        frame.add(new JButton("按钮 3"),BorderLayout.SOUTH);
        frame.add(new JButton("按钮 4"),BorderLayout.NORTH);
        frame.add(new JButton("按钮 5"),BorderLayout.CENTER);
        frame.setSize(280,123);
        frame.setLocationRelativeTo(null);
        frame.setVisible(true);
    }
}
```

如图 6-22 所示。

图 6-22

3. GridLayout

GridLayout 布局是将容器的空间分成若干行和列的一个个网格,可以给出网格的行数和列数,组件添加到这些网格中。当改变容器的大小后,其中的组件相对位置不变,但大小改变。容器中各个组件同高度、同宽度,各个组件默认的排列方式为:从上到下、从左到右。

GridLayout 类的构造方法如下:

- public GridLayout()创建单行每个组件一列的 GridLayout 对象。
- public GridLayout(int rows,int cols)创建指定行列数的 GridLayout 对象。
- public GridLayout(int rows, int cols, int hgap, int vgap)创建指定行列数的 GridLayout 对象。

因为没有容器默认使用 GridLayout,所以在使用 GridLayout 前,要用 setLayout()方法将容器的布局管理器设置为 GridLayout。

在向 GridLayout 添加组件时,组件加入容器要按序进行,每个网格中都必须加入组件。若希望某个网格为空,可以为该网格加入一个空的标签:add(new JLabel())。

【例 6.15】创建网格布局。

```
import java.awt.GridLayout;
import javax.swing.JFrame;
import javax.swing.JButton;
public class GridLayout1{
public static void main(String args[]){
JFrame frame = new JFrame("网格布局");
frame.setLayout(new GridLayout(3,3));
JButton but = null;
for(int i = 0;i<9;i + + ){
but = new JButton("按钮" + i);
frame.add(but);
}
frame.setSize(280,123);
frame.setLocationRelativeTo(null);
frame.setVisible(true);
}
}
```

如图 6-23 所示。

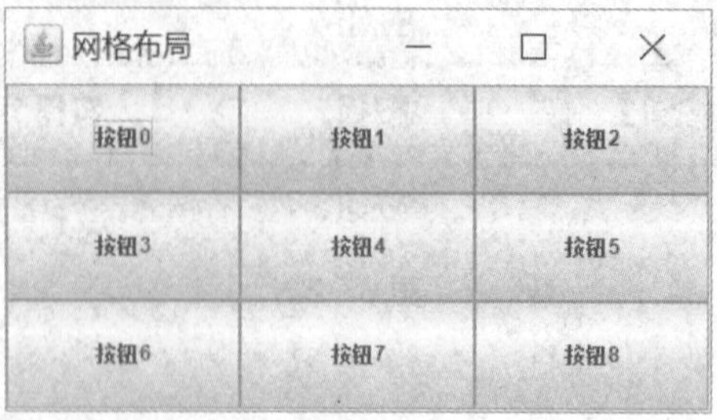

图 6-23

4. CardLayout

CardLayout 布局管理器能够使得多个组件共享同一显示空间,这些组件之间的关系像一叠重叠的扑克牌,只有最上面的组件是可见的。注意,在一个显示空间(卡片)中只能显示一个组件,因此,可使用容器嵌套的方法来显示多个组件。

CardLayout 类的构造方法如下:
- CardLayout()创建间距为零的对象。
- CardLayout(int hgap, int vgap)创建带有水平 hgap 和垂直 vgap 间距的对象。

CardLayout 类的常用方法如下:
- public void next(Container parent)翻转到下一张卡片。
- public void previous(Container parent)翻转到上一张卡片。
- public void first(Container parent)翻转到第一张卡片。
- public void last(Container parent)翻转到最后一张卡片。
- public void show(Container parent, String name)显示具有指定组件名称的卡片。

为了使用叠在下面的组件,可以为每个组件取一名字,名字在用 add()方法向容器添加组件时指定,需要某个组件时通过 show()方法指定该组件的名字来选取它。此外,还可以顺序使用这些组件(用 next()和 previous()方法),或者直接指明选取第一个组件(用 first()方法)或最后一个组件(用 last()方法)。

【例 6.16】创建卡片布局。

```
import java.awt.CardLayout;
import java.awt.Container;
import javax.swing.JFrame;
import javax.swing.JButton;
public class CardLayout1{
    public static void main(String args[]){
        JFrame frame = new JFrame("卡片布局");
        frame.setLayout(new CardLayout());
        Container con = frame.getContentPane();
        JButton b1 = new JButton("卡片一");
        JButton b2 = new JButton("卡片二");
```

```
JButton b3 = new JButton("卡片三");
con.add("card1",b1);
con.add("card2",b2);
con.add("card3",b3);
frame.setSize(280,123);
frame.setLocationRelativeTo(null);
frame.setVisible(true);
    }
}
```

如图 6-24 所示。

图 6-24

5. 绝对定位

以上的布局管理器都是依靠专门的工具完成的,在 Java 中也可以通过绝对定位的方式完成布局。如果不想在窗体中指定布局管理器,也可以通过设置绝对坐标的方式完成。Component 中提供了 setBounds()方法,可以定位一个组件的坐标,使用 X、Y 的坐标表示方式,此方法定义如下:

public void setBounds(int x, int y, int width, int height)

每个组件都可以使用 setBounds(),设置其在窗体中的位置和大小。setBounds()方法在窗体中对组件进行布局的原理如图 6-25 所示:

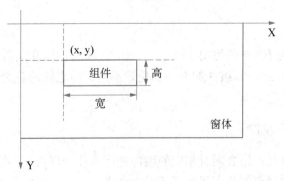

图 6-25

【例6.17】使用绝对定位进行布局。

```java
import javax.swing.JFrame;
import javax.swing.JButton;
public class AbsoluteLayout1{
public static void main(String args[]){
JFrame frame = new JFrame("绝对定位");
frame.setLayout(null);
JButton JB1 = new JButton("按钮1");
JButton JB2 = new JButton("按钮2");
frame.setSize(280,123);
JB1.setBounds(55,30,80,20);
JB2.setBounds(140,30,80,20);
frame.add(JB1);
frame.add(JB2);
frame.setSize(280,123);
frame.setLocationRelativeTo(null);
frame.setVisible(true);
}
}
```

如图 6-26 所示。

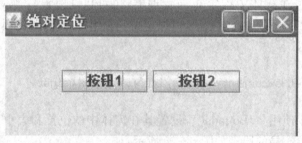

图 6-26

注意：

如果要使用绝对定位，必须使用 setLayout(null) 取消布局管理器。使用绝对定位的优点是，不管窗体如何改变大小，组件的大小和位置不变。而之前的各种布局，当窗体改变时，组件大小也随之改变。

6.3.5 事件处理机制

图形界面设计完成后，还必须对相应的组件进行事件处理，程序才能实现相应功能。所谓事件，是指一个组件对象的状态发生了变化。例如，当一个按钮按下时，该按钮的状态就发生了改变，此时就会产生一个事件。而如果要处理此事件，就需要事件的监听者不断地监听事件的变化，并根据这些事件进行相应的处理。

在 Swing 编程中，依然使用了最早 AWT 的事件处理方式，所有的事件类（基本上任意的一个组件都有对应的事件）都是 EventObject 类的子类，如图 6-27 所示：

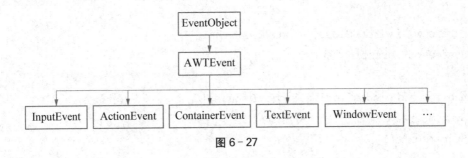

图 6-27

EventObject 类中定义的 getSource()方法，用于获取发生此事件的源对象，即哪个组件对象发生了该事件。在明确了事件源后（即哪个组件产生事件了），还必须有能够接收和处理事件的对象，即事件监听器。所有的事件监听器都是以事件监听接口的形式出现的，处理时只需要实现该接口即可，大部分图形界面的事件处理类或接口都保存在 java.awt.event 包中。

Java 语言的事件处理流程如图 6-28 所示：

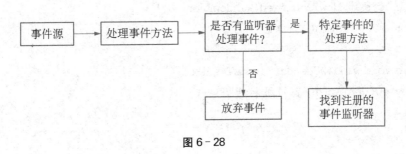

图 6-28

1. 窗体事件

（1）WindowListener 接口。

WindowListener 是专门处理窗体事件的事件监听接口，用来监听一个窗体的所有变化。例如：窗口打开、关闭等都可以使用这个接口进行监听。WindowListener 接口定义的常用方法如下：

- void windowActivated(WindowEvent e)将窗口变为活动窗口时触发。
- void windowDeactivated(WindowEvent e)将窗口变为不活动窗口时触发。
- void windowClosed(WindowEvent e)当窗口关闭时触发。
- void windowClosing(WindowEvent e)当窗口正在关闭时触发。
- void windowIconified(WindowEvent e)当窗口最小化时触发。
- void windowDeiconified(WindowEvent e)当窗口从最小化恢复到正常状态时触发。
- void windowOpened(WindowEvent e)当窗口打开时触发。

【例 6.18】实现 WindowListener 接口。

```java
class WindowEvent implements WindowListener{
public void windowActivated(WindowEvent e){
System.out.println("窗口选中");
}
public void windowClosed(WindowEvent e){
System.out.println("窗口已关闭");
}
public void windowClosing(WindowEvent e){
System.out.println("窗口关闭");
System.exit(1);
}
public void windowDeactivated(WindowEvent e){
System.out.println("取消窗口选中");
}
public void windowDeiconified(WindowEvent e){
System.out.println("窗口从最小化恢复");
}
public void windowIconified(WindowEvent e){
System.out.println("窗口最小化");
}
public void windowOpened(WindowEvent e){
System.out.println("窗口被打开");
}
}
```

WindowListener 事件监听器完成后，还必须在相应的窗体组件使用时注册监听，才能处理相应的事件。使用窗体的 addWindowListener()方法可注册 WindowListener 事件监听。

【例 6.19】实现窗体事件监听。

```java
import java.awt.event.WindowListener;
import java.awt.event.WindowEvent;
import javax.swing.JFrame;
class WindowListener1 implements WindowListener{
public void windowActivated(WindowEvent e){
System.out.println("窗口选中");
}
public void windowClosed(WindowEvent e){
System.out.println("窗口已关闭");
```

```
}
public void windowClosing(WindowEvent e){
System.out.println("窗口关闭");
System.exit(1);
}
public void windowDeactivated(WindowEvent e){
System.out.println("取消窗口选中");
}
public void windowDeiconified(WindowEvent e){
System.out.println("窗口从最小化恢复");
}
public void windowIconified(WindowEvent e){
System.out.println("窗口最小化");
}
public void windowOpened(WindowEvent e){
System.out.println("窗口被打开");
}
}
public class JFrameEvent1{
public static void main(String args[]){
JFrame frame = new JFrame("Window事件处理");
frame.addWindowListener(new WindowListener1());//加入事件
frame.setSize(300,200);
frame.setLocation(300,200);
frame.setVisible(true);
}
}
```

如图 6-29、图 6-30 所示。

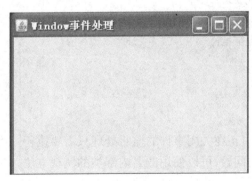

图 6-29

```
E:\>java JFrameEvent1
窗口选中
窗口被打开
窗口最小化
取消窗口选中
窗口从最小化恢复
窗口选中
窗口关闭
```

图 6-30

(2) 监听适配器。

如果在上例中,只想对窗口的正在关闭事件进行处理,而其他窗口事件发生时不必处理。此时,如果使用 WindowListener 接口处理,不管其他事件是否需要处理,所有的方法还是必须要全部实现,这样非常烦琐。为此,Java 语言在事件处理中提供了很多的适配器(Adapter)类,方便用户根据自己的需要进行事件处理的实现,即用户设计的子类继承了相应的适配器类后,就可以根据用户的需要进行相关方法的重写,而不必重写所有的方法。以 WindowAdapter 为例,如果用户只想对窗口的正在关闭事件进行处理,子类继承了 WindowAdapter 类后,只需要重写 windowClosing()方法即可。

【例 6.20】用 WindowAdapter 实现窗口关闭监听。

```java
import java.awt.event.WindowAdapter;
import java.awt.event.WindowEvent;
import javax.swing.JFrame;
class WinClose extends WindowAdapter{
public void windowClosing(WindowEvent e){
System.out.println("窗口关闭");
System.exit(1);
}
}
public class WindowClose{
public static void main(String args[]){
JFrame frame = new JFrame("Windows 适配器");
frame.addWindowListener(new WinClose());//加入事件
frame.setSize(300,150);
frame.setLocationRelativeTo(null);
frame.setVisible(true);
}
}
```

如图 6-31、图 6-32 所示。

上例中,我们用于窗口正在关闭事件的监听处理只需要执行一次,将它设置成一个类太浪费了。为了解决这样的问题,可以使用匿名内部类的实现方法。

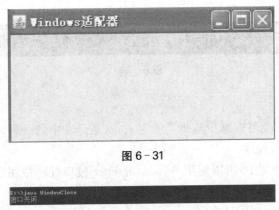

图 6-31

图 6-32

【例 6.21】用匿名内部类实现窗口关闭监听。

```
import java.awt.event.WindowAdapter;
import java.awt.event.WindowEvent;
import javax.swing.JFrame;
public class WindowClose2{
public static void main(String args[]){
JFrame frame = new JFrame("Windows 适配器");
frame.addWindowListener(new WindowAdapter(){
public void windowClosing(WindowEvent e){
System.out.println("窗口关闭");
System.exit(1);
}
});//加入事件
frame.setSize(300,150);
frame.setLocationRelativeTo(null);
frame.setVisible(true);
}
}
```

如图 6-33、图 6-34 所示。

图 6-33

图 6-34

注意：

直接编写匿名内部类可以减少监听类的定义，这是开发中的一种常用做法。

2. 动作事件及监听处理

在 Swing 的事件处理中，可以使用 ActionListener 接口处理按钮点击、文本框回车等组件动作事件(ActionEvent)。ActionListener 接口只定义了一个方法：

- void actionPerformed(ActionEvent e) 发生操作时调用。

(1) 按钮事件。

JButton 组件引发的事件为 ActionEvent，可实现 ActionListener 监听器接口的 actionPerformed()方法，用 addActionListener()方法注册，用 getActionCommand()或 getSource()方法确定事件源。此外，在 JButton 按钮的使用中，还常用到继承来的 setActionCommand()方法设置动作命令。

【例 6.22】单个按钮的监听事件(匿名内部类实现、getSource()方法确定事件源)。

```java
import java.awt.GridLayout;
import java.awt.event.ActionEvent;
import java.awt.event.ActionListener;
import javax.swing.JFrame;
import javax.swing.JButton;
import javax.swing.JTextField;
class JBL_implement{
private JFrame frame = new JFrame("按钮事件监听1");
private JButton JB1 = new JButton("测试");
private JTextField JT1 = new JTextField();
public JBL_implement(){//构造函数
frame.setLayout(new GridLayout(1,1));
frame.add(JB1);
JB1.addActionListener(new ActionListener(){
public void actionPerformed(ActionEvent e){
if(e.getSource() == JB1){
JT1.setText("点击了测试按钮");
}
}
});
frame.add(JT1);
```

```
        frame.setSize(300,150);
        frame.setLocationRelativeTo(null);
        frame.setDefaultCloseOperation(JFrame.EXIT_ON_CLOSE);
        frame.setVisible(true);
    }
}
public class JButtonListener1{
    public static void main(String args[]){
        new JBL_implement();
    }
}
```

如图 6-35 所示。

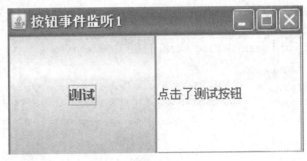

图 6-35

【例 6.23】单个按钮的监听事件(匿名内部类实现、getActionCommand( )方法确定事件源)。

```
import java.awt.GridLayout;
import java.awt.event.ActionEvent;
import java.awt.event.ActionListener;
import javax.swing.JFrame;
import javax.swing.JButton;
import javax.swing.JTextField;
class JBL_implement{
    private JFrame frame = new JFrame("按钮事件监听2");
    private JButton JB1 = new JButton("测试");
    private JTextField JT1 = new JTextField();
    public JBL_implement(){//构造函数
        frame.setLayout(new GridLayout(1,1));
        frame.add(JB1);
```

```
JB1.addActionListener(new ActionListener(){
public void actionPerformed(ActionEvent e){
if(e.getActionCommand().equals("测试")){
JT1.setText("点击了测试按钮");
}
}
});
frame.add(JT1);
frame.setSize(300,150);
frame.setLocationRelativeTo(null);
frame.setDefaultCloseOperation(JFrame.EXIT_ON_CLOSE);
frame.setVisible(true);
}
}
public class JButtonListener2{
public static void main(String args[]){
new JBL_implement();
}
}
```

如图 6-36 所示。

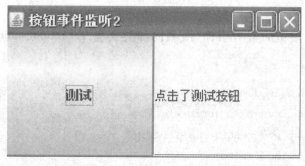

图 6-36

【例 6.24】多个按钮的监听事件(单独类)。

```
import java.awt.GridLayout;
import java.awt.event.ActionEvent;
import java.awt.event.ActionListener;
import javax.swing.JFrame;
import javax.swing.JButton;
```

```java
import javax.swing.JTextField;

public class JButtonListener3{
    private JFrame frame = new JFrame("按钮事件监听3");
    private JButton JB1 = new JButton("测试1");
    private JButton JB2 = new JButton("测试2");
    private JButton JB3 = new JButton("测试3");
    private JTextField JT1 = new JTextField();

    class JBL_implement implements ActionListener{
        public void actionPerformed(ActionEvent e){
            if(e.getSource() == JB1){
                JT1.setText("点击了测试按钮1");
            }
            if(e.getSource() == JB2){
                JT1.setText("点击了测试按钮2");
            }
            if(e.getSource() == JB3){
                JT1.setText("点击了测试按钮3");
            }
        }
    }

    public JButtonListener3(){//构造方法
        frame.setLayout(new GridLayout(2,2));
        frame.add(JB1);
        JB1.addActionListener(new JBL_implement());
        frame.add(JB2);
        JB2.addActionListener(new JBL_implement());
        frame.add(JB3);
        JB3.addActionListener(new JBL_implement());
        frame.add(JT1);
        frame.setSize(300,150);
        frame.setLocationRelativeTo(null);
        frame.setDefaultCloseOperation(JFrame.EXIT_ON_CLOSE);
        frame.setVisible(true);
    }
```

```
public static void main(String args[]){
new JButtonListener3();
}
}
```

如图 6-37 所示。

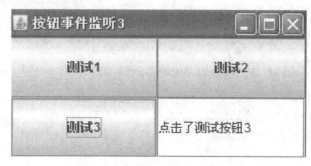

图 6-37

(2) 文本框事件。

当用户在文本框里按回车键时,就产生了一个 ActionEvent 事件。当用户在文本框中移动文本光标时,就产生 CaretEvent 事件。可注册 addCaretListener 监听器,实现 CaretListener 的 caretUpdate()进行事件处理。

【例 6.25】完成文本框的事件监听程序。

```
import java.awt.GridLayout;
import java.awt.event.ActionEvent;
import java.awt.event.ActionListener;
import javax.swing.event.CaretEvent;
import javax.swing.event.CaretListener;
import javax.swing.JFrame;
import javax.swing.JTextField;
import javax.swing.JLabel;

public class JTextFieldListener{
private JFrame frame = new JFrame("按钮事件监听3");
private JTextField JT1 = new JTextField();
private JLabel JL1 = new JLabel();

class JTL_implement implements ActionListener{
public void actionPerformed(ActionEvent e){
if(e.getSource() = = JT1){
```

```java
JL1.setText("在文本框里回车");
    }
  }
}

class JTL_implement2 implements CaretListener{
  public void caretUpdate(CaretEvent e){
    if(e.getSource() = = JT1){
      JL1.setText("选中文本框");
    }
  }
}

public JTextFieldListener(){//构造方法
  frame.setLayout(new GridLayout(1,1));
  frame.add(JT1);
  JT1.addActionListener(new JTL_implement());
  JT1.addCaretListener(new JTL_implement2());
  frame.add(JL1);
  frame.setSize(300,150);
  frame.setLocationRelativeTo(null);
  frame.setDefaultCloseOperation(JFrame.EXIT_ON_CLOSE);
  frame.setVisible(true);
}
public static void main(String args[]){
  new JTextFieldListener();
}
}
```

如图 6-38、图 6-39 所示。

图 6-38

图 6-39

3. 键盘事件

Java 程序为了输入数据、操作命令等，经常用到键盘，而键盘的按下、释放也会引发键盘事件（KeyEvent）。在 Swing 的事件处理中也可以对键盘的操作进行监视，直接使用 KeyListener 接口完成即可，此接口定义了如下的方法：

- void keyTyped(KeyEvent e)键入某个键时调用。
- void keyPressed(KeyEvent e)键盘按下时调用。
- void keyReleased(KeyEvent e)键盘松开时调用。

为识别引发键盘事件的按键，获取键盘输入的内容，常用到 KeyEvent 类的如下方法：

- public char getKeyChar()返回键入的字符，只针对 keyTyped 有意义。例如，shift+<b>按键事件返回值为字符"B"。
- public int getKeyCode()返回键入字符的键码。
- public static String getKeyText(int keyCode)返回此键的信息，即返回描述键代码的字符串，如"F1""HOME""A"等。

【例 6.26】通过 KeyListener 实现键盘监听。

```
import java.awt.event.KeyListener;
import java.awt.event.KeyEvent;
import javax.swing.JFrame;
import javax.swing.JTextArea;
import javax.swing.JScrollPane;
public class KeyListener1{
private JFrame frame = new JFrame("键盘事件监听 1");
private JTextArea JTA1 = new JTextArea();
private JScrollPane JSC1 = new JScrollPane(JTA1);
class KL_implement implements KeyListener{
public void keyPressed(KeyEvent e){
JTA1.append("键盘""+ KeyEvent.getKeyText(e.getKeyCode()) +""键按下\n");
}
public void keyReleased(KeyEvent e){
```

```
JTA1.append("键盘""+KeyEvent.getKeyText(e.getKeyCode())+"""键松开\n");
}
public void keyTyped(KeyEvent e){
JTA1.append("输入的内容是:"+e.getKeyChar()+"\n");
}
}

public KeyListener1(){//构造方法
frame.add(JSC1);
JTA1.addKeyListener(new KL_implement());
frame.setSize(300,150);
frame.setLocationRelativeTo(null);
frame.setDefaultCloseOperation(JFrame.EXIT_ON_CLOSE);
frame.setVisible(true);
}

public static void main(String args[]){
new KeyListener1();
}
}
```

如图 6-40 所示。

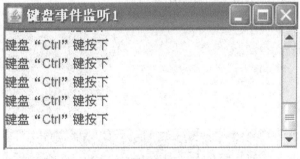

图 6-40

上例中,使用 KeyListener 接口处理键盘事件,不管处理全部键盘事件,还是个别键盘事件,都必须重写 KeyListener 接口中的全部方法。为此,Swing 也提供了键盘处理的适配器类(KeyAdapter),直接使用此类即可根据需要完成特定键盘事件的处理。

【例 6.27】通过 KeyAdapter 实现键盘松开监听。

```
import java.awt.event.KeyAdapter;
import java.awt.event.KeyEvent;
```

```java
import javax.swing.JFrame;
import javax.swing.JTextArea;
import javax.swing.JScrollPane;
public class KeyListener2{
private JFrame frame = new JFrame("键盘事件监听1");
private JTextArea JTA1 = new JTextArea();
private JScrollPane JSC1 = new JScrollPane(JTA1);
class KL_implement extends KeyAdapter{
public void keyReleased(KeyEvent e){
JTA1.append("键盘"" + KeyEvent.getKeyText(e.getKeyCode()) + ""键松开\n");
}
}

public KeyListener2(){//构造方法
frame.add(JSC1);
JTA1.addKeyListener(new KL_implement());
frame.setSize(300,150);
frame.setLocationRelativeTo(null);
frame.setDefaultCloseOperation(JFrame.EXIT_ON_CLOSE);
frame.setVisible(true);
}

public static void main(String args[]){
new KeyListener2();
}
}
```

如图6-41所示。

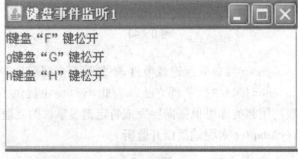

图6-41

4. 鼠标事件

在 Java 图形界面应用程序中,通常使用鼠标来进行人机交互操作,例如鼠标的移动、单击、双击等都会引发鼠标事件,MouseListener 和 MouseMotionListener 接口用于处理鼠标事件。MouseListener 接口用于处理鼠标按下、松开、单击、进入组件、离开组件,而 MouseMotionListener 主要用于处理鼠标在组件上按下并拖动和鼠标移动到组件。

(1) MouseListener 接口。

MouseListener 接口中的方法如下:
- void mouseClicked(MouseEvent e)鼠标单击时调用(按下并释放)。
- void mousePressed(MouseEvent e)鼠标按下时调用。
- void mouseReleased(MouseEvent e)鼠标松开时调用。
- void mouseEntered(MouseEvent e)鼠标进入组件时调用。
- void mouseExited(MouseEvent e)鼠标离开组件时调用。

每一种鼠标操作都会产生 MouseEvent 事件,通过 MouseEvent 事件可以获知鼠标的相关操作信息。MouseEvent 类的常用方法和常量如下:
- public static final int BUTTON1 表示鼠标左键的常量。
- public static final int BUTTON2 表示鼠标滚轴的常量。
- public static final int BUTTON3 表示鼠标右键的常量。
- public int getButton()以数字形式返回按下的鼠标键。
- public int getClickCount()返回鼠标的单击次数。
- public static String getMouseModifiersText(int modifiers)以字符串形式返回鼠标按下的键信息。
- public int getX()返回鼠标操作的 X 坐标。
- public int getY()返回鼠标操作的 Y 坐标。

【例 6.28】通过 MouseListener 实现鼠标监听。

```java
import java.awt.event.MouseListener;
import java.awt.event.MouseEvent;
import javax.swing.JFrame;
import javax.swing.JTextArea;
import javax.swing.JScrollPane;
public class MouseListener1{
private JFrame frame = new JFrame("键盘事件监听1");
private JTextArea JTA1 = new JTextArea();
private JScrollPane JSC1 = new JScrollPane(JTA1);
class ML_implement implements MouseListener{
public void mouseClicked(MouseEvent e){
int c = e.getButton();
String mouseInfo = null;
if(c = = MouseEvent.BUTTON1){
```

```java
        mouseInfo = "左键";
    }
    if(c = = MouseEvent.BUTTON3){
        mouseInfo = "右键";
    }
    if(c = = MouseEvent.BUTTON2){
        mouseInfo = "滚轴";
    }
    JTA1.append("鼠标单击:" + mouseInfo + "\n");
}
public void mouseEntered(MouseEvent e){
    JTA1.append("鼠标进入组件。\n");
}
public void mouseExited(MouseEvent e){
    JTA1.append("鼠标离开组件。\n");
}
public void mousePressed(MouseEvent e){
    JTA1.append("鼠标按下。\n");
}
public void mouseReleased(MouseEvent e){
    JTA1.append("鼠标松开。\n");
}
}

public MouseListener1(){//构造方法
    frame.add(JSC1);
    JTA1.addMouseListener(new ML_implement());
    frame.setSize(300,150);
    frame.setLocationRelativeTo(null);
    frame.setDefaultCloseOperation(JFrame.EXIT_ON_CLOSE);
    frame.setVisible(true);
}

public static void main(String args[]){
    new MouseListener1();
}
}
```

如图 6-42 所示。

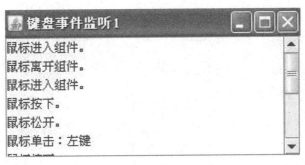

图 6-42

上例中，MouseListener 接口不管处理多少种鼠标事件，都必须重写全部方法。因此，Swing 也提供了鼠标处理的适配器类（MouseAdapter），用于特定鼠标事件的处理。

【例 6.29】通过 MouseAdapter 实现鼠标左键按下监听。

```
import java.awt.event.MouseAdapter;
import java.awt.event.MouseEvent;
import javax.swing.JFrame;
import javax.swing.JTextArea;
import javax.swing.JScrollPane;
public class MouseListener2{
private JFrame frame = new JFrame("鼠标事件监听2");
private JTextArea JTA1 = new JTextArea();
private JScrollPane JSC1 = new JScrollPane(JTA1);
class ML_implement extends MouseAdapter{
public void mouseClicked(MouseEvent e){
int c = e.getButton();
String mouseInfo = null;
if(c = = MouseEvent.BUTTON1){
mouseInfo = "左键";
}
JTA1.append("鼠标单击:" + mouseInfo + "\n");
}
}
public MouseListener2(){//构造方法
frame.add(JSC1);
JTA1.addMouseListener(new ML_implement());
frame.setSize(300,150);
frame.setLocationRelativeTo(null);
```

```
        frame.setDefaultCloseOperation(JFrame.EXIT_ON_CLOSE);
        frame.setVisible(true);
    }

    public static void main(String args[]){
        new MouseListener2();
    }
}
```

如图 6-43 所示。

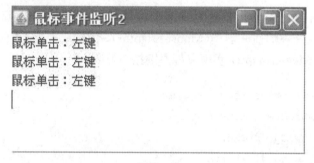

图 6-43

(2) MouseMotionListener 接口。

在一般的图形界面中经常可以看到鼠标拖拽操作的情况,在 Swing 的事件处理中可以使用 MouseMotionListener 接口完成鼠标的拖拽操作。MouseMotionListener 接口的方法如下:

- void mouseDragged(MouseEvent e)在组件上按下并拖动时调用。
- void mouseMoved(MouseEvent e)鼠标移动到组件时调用。

【例 6.30】通过 MouseMotionListener 实现鼠标拖拽监听。

```
import java.awt.event.MouseMotionListener;
import java.awt.event.MouseEvent;
import javax.swing.JFrame;
import javax.swing.JTextArea;
import javax.swing.JScrollPane;
public class MouseMotionListener1{
    private JFrame frame = new JFrame("鼠标事件监听1");
    private JTextArea JTA1 = new JTextArea();
    private JScrollPane JSC1 = new JScrollPane(JTA1);
    class MML_implement implements MouseMotionListener{
        public void mouseDragged(MouseEvent e){
```

```
JTA1.append("鼠标拖拽到:X = " + e.getX() + ",Y = " + e.getY() + "\n");

}
public void mouseMoved(MouseEvent e){
JTA1.append("鼠标移动到窗体。" + "\n");
}
}

public MouseMotionListener1(){//构造方法
frame.add(JSC1);
JTA1.addMouseMotionListener(new MML_implement());
frame.setSize(300,150);
frame.setLocationRelativeTo(null);
frame.setDefaultCloseOperation(JFrame.EXIT_ON_CLOSE);
frame.setVisible(true);
}

public static void main(String args[]){
new MouseMotionListener1();
}
}
```

如图 6-44 所示。

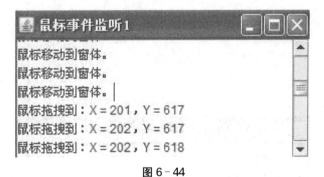

图 6-44

同理，鼠标拖拽操作也有对应的 MouseMotionAdapter 类，用于指定鼠标拖拽操作。
【例 6.31】通过 MouseMotionAdapter 实现鼠标移动到组件监听。

```
import java.awt.event.MouseMotionAdapter;
import java.awt.event.MouseEvent;
import javax.swing.JFrame;
```

```java
import javax.swing.JTextArea;
import javax.swing.JScrollPane;
public class MouseMotionListener2{
    private JFrame frame = new JFrame("鼠标事件监听1");
    private JTextArea JTA1 = new JTextArea();
    private JScrollPane JSC1 = new JScrollPane(JTA1);
    class MML_implement extends MouseMotionAdapter{

        public void mouseMoved(MouseEvent e){
            JTA1.append("鼠标移动到窗体。" + "\n");
        }
    }

    public MouseMotionListener2(){//构造方法
        frame.add(JSC1);
        JTA1.addMouseMotionListener(new MML_implement());
        frame.setSize(300,150);
        frame.setLocationRelativeTo(null);
        frame.setDefaultCloseOperation(JFrame.EXIT_ON_CLOSE);
        frame.setVisible(true);
    }

    public static void main(String args[]){
        new MouseMotionListener2();
    }
}
```

如图6-45所示。

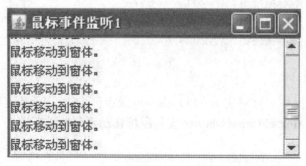

图6-45

任务 实施

6.3.6 学生成绩管理系统登录界面设计实现

```java
//用户名为fan、密码为123456时成功登录,跳转到MainJFrame
import java.awt.event.ActionEvent;
import java.awt.event.ActionListener;
import javax.swing.JButton;
import javax.swing.JFrame;
import javax.swing.JLabel;
import javax.swing.JPasswordField;
import javax.swing.JTextField;
/**
*用户登录界面
*
**/
public class LoginJFrame extends JFrame implements ActionListener{
    private JTextField uNameText = new JTextField();//用户名文本组件
    private JPasswordField uPswPwd = new JPasswordField();//密码文本组件
    private JLabel uNameLabel = new JLabel();//用户名标签组件
    private JLabel uPswLabel = new JLabel();//密码标签组件
    private JButton uLoginButton = new JButton("登录");//定义按钮对象
    private JButton uResetButton = new JButton("重置");//定义按钮对象
    public static MainJFrame mainFrame;//定义主窗体对象

    /**
    *构造方法
    **/
    public LoginJFrame(){

        uNameLabel.setText("用户名:");
        uNameLabel.setBounds(16,14,54,24);
        uNameText.setBounds(72,14,150,24);
        uPswLabel.setText("密    码:");
        uPswLabel.setBounds(16,50,54,24);
        uPswPwd.setEchoChar('*');
        uPswPwd.setBounds(72,50,150,24);
```

```java
        uLoginButton.setBounds(65,90,60,24);
        uResetButton.setBounds(140,90,60,24);
        uLoginButton.addActionListener(this);
        uResetButton.addActionListener(this);
        this.add(uNameLabel);
        this.add(uNameText);
        this.add(uPswLabel);
        this.add(uPswPwd);
        this.add(uLoginButton);
        this.add(uResetButton);
        this.setTitle("登录窗口");
        this.setLayout(null);
        this.setDefaultCloseOperation(JFrame.EXIT_ON_CLOSE);
        this.setSize(250,170);
        this.setLocationRelativeTo(null);
        this.setResizable(false);
        this.setVisible(true);
    }

    /**
     * 在文本框上回车、单击按钮激发的事件
     *
     * @param curr
     */
    public void actionPerformed(ActionEvent e){
        if(e.getSource() == uLoginButton){//点击登录按钮
            //判断登录名和密码不能为空
            String name = uNameText.getText().trim();
            String password = new String(uPswPwd.getPassword());
            if(name.length() == 0){
                uNameText.setText("登录名不能为空!");
                return;
            }

            if(password.length() == 0){
                uNameText.setText("密码不能为空!");
                return;
            }
```

```java
    //判断登录名和密码是否正确
    if(name.equals("fan")&&password.equals("123456")){
        mainFrame = new MainJFrame();
        this.dispose();
    }else{
        uNameText.setText("用户名和密码不存在,请重新输入!");
        return;
        }
    }
    if(e.getSource() = = uResetButton){//重置按钮
        uNameText.setText("");
        uPswPwd.setText("");
    }
}
/* *
* 主函数
*
* @param args
*/
public static void main(String[ ]args){
    new LoginJFrame();
    }
}
```

如图 6-46 至图 6-49 所示。

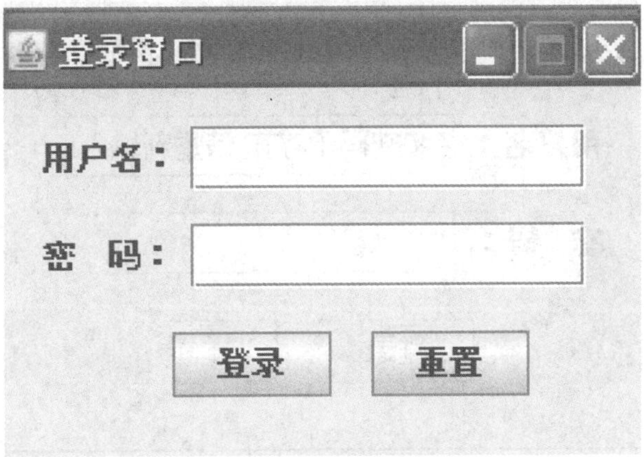

图 6-46

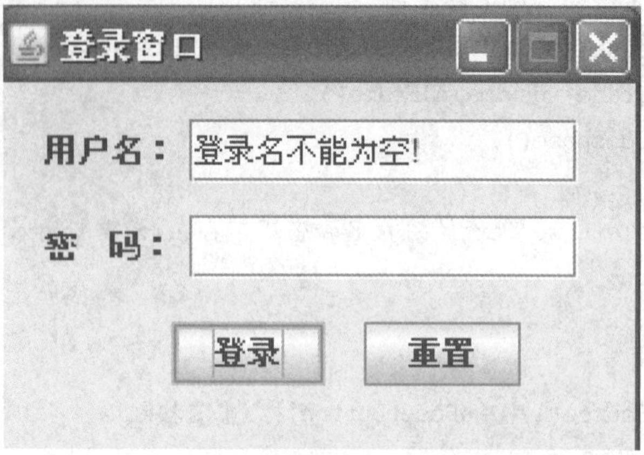

图 6 - 47

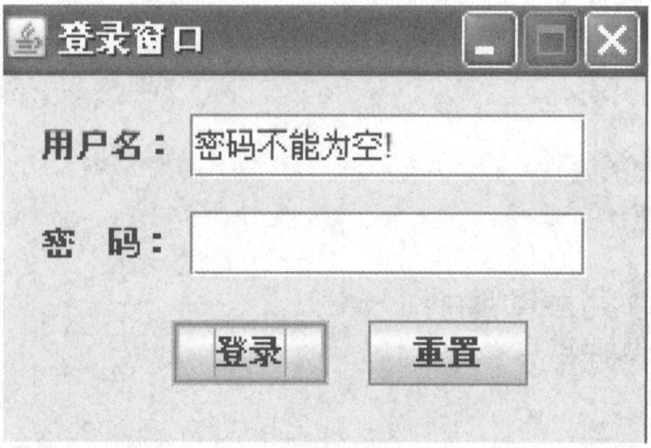

图 6 - 48

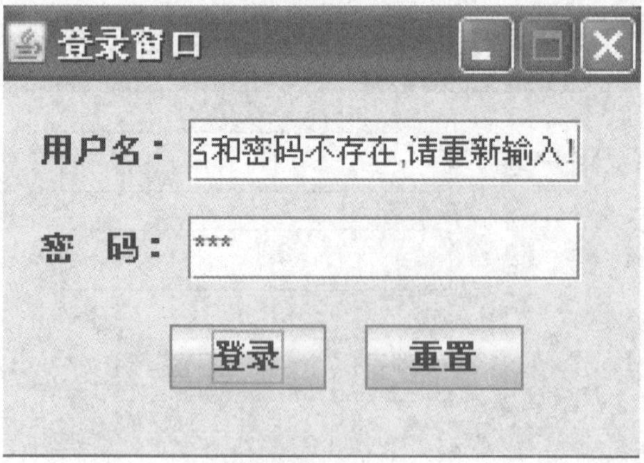

图 6 - 49

任务 拓展

设计完成自助银行服务系统的用户登录界面,要求如下:
(1) 当输入的用户名为 admin、密码为 admin 时,点击【登录】按钮进入 6.2 的拓展任务中实现的主界面。
(2) 当输入的用户名或密码有错误时,显示相应的提示信息。
(3) 当点击登录界面的【重置】按钮时,所有填写的用户信息(用户名、密码)清空。

6.4 任务 4 学生成绩管理系统信息管理窗口设计

任务描述 及分析

小明要设计实现学生成绩管理系统信息管理窗口,信息管理窗口要求如下:
(1) 在程序主界面点击"信息添加"菜单项,直接调出对应窗口,可以完成学生的学号、姓名、性别、课程名称、平时成绩、过程考核成绩、终结考核成绩的添加,并能根据平时成绩、过程考核成绩、终结考核成绩自动计算出综合成绩及其等级,然后保存到记事本文档 StudentScore.txt;当点击【返回】按钮时,可返回主界面。
(2) 在程序主界面点击"信息修改"菜单项,先弹出对话框(用于输入要修改成绩信息的学生的学号),再判断输入的学号是否存在(从记事本文档 StudentScore.txt 中读取学生学号信息进行比较判断),若存在则调出对应窗口,完成学生成绩相关信息的修改,保存到记事本文档 StudentScore.txt;当点击【返回】按钮时,可返回主界面。反之,如输入的学号不存在,提示学号不存在,信息修改窗口不打开。
(3) 在程序主界面点击"信息删除"菜单项,先弹出对话框(用于输入要删除成绩信息的学生的学号),再判断输入的学号是否存在(从记事本文档 StudentScore.txt 中读取学生学号信息进行比较判断)。若存在,则直接从记事本文档 StudentScore.txt 中删除该学号对应的学生信息,并提示信息删除成功;反之,若输入的学号不存在,提示学号不存在。

要完成这个工作任务,第一步,我们需要掌握 Swing 的复选框、单选按钮、下拉列表框、面板、对话框等组件的常用属性和方法;第二步,利用第一步和工作任务 6.2 及工作任务 6.3 所学组件知识,创建信息管理窗口及对应的组件,并合理布局;第三步,利用工作任务 6.3 所学事件处理相关知识,完成"主窗口"和"信息管理窗口"的事件处理程序设计。

相关 知识

6.4.1 复选框 JCheckBox

1. 创建复选框

在 Java 图形界面应用程序中经常会选择用户的兴趣等,而一个用户会有多个兴趣,此时，就需要使用复选框。JCheckBox 类提供复选框按钮的支持。

JCheckBox 类的层次关系如下：

Javax. swing. AbstractButton
　　-javax. swing. JToggleButton
　　　　-javax. swing. JCheckBox

JCheckBox 类的构造方法如下：

* JCheckBox()创建无文本的初始未选复选框按钮。
* JCheckBox(Icon icon)创建有图像无文本的初始未选复选框按钮。
* JCheckBox(Icon icon, boolean selected)创建带图像和选择状态但无文本的复选框按钮。
* JCheckBox(String text)创建带文本的初始未选复选框按钮。
* JCheckBox(String text, boolean selected)创建具有指定文本和状态的复选框按钮。
* JCheckBox(String text, Icon icon)创建具有指定文本和图标图像的初始未选复选框按钮。
* JCheckBox(String text, Icon icon, boolean selected)创建具有指定文本、图标图像、选择状态的复选框按钮。

其中，构造方法的参数 selected 若为真，则表示按钮初始状态为选中。

JCheckBox 类常用的方法如下：

* public boolean isSelected()返回是否被选中，当复选框按钮选中时返回 true，否则返回 false。
* public void setSelected(boolean Selected)设置是否选中，参数 selected 若为 true，则表示设置按钮状态为选中；反之，就表示设置按钮状态为未选中。
* public void setText(String text)设置显示文本。
* public void setIcon(Icon defaultIcon)设置图片。

【例 6.32】 创建 3 个复选框，分别显示选项一、选项二、选项三。

```
import java.awt.GridLayout;
import javax.swing.JFrame;
import javax.swing.JCheckBox;

class JCheckBox1{
private JFrame frame = new JFrame("单选按钮");//定义窗体
private JCheckBox jcb1 = new JCheckBox("选项一");//定义第一个复选框
private JCheckBox jcb2 = new JCheckBox("选项二");//定义第二个复选框
private JCheckBox jcb3 = new JCheckBox("选项三");//定义第三个复选框

public JCheckBox1(){
frame.setLayout(new GridLayout(1,3));//设置组件的排版为 1 行 3 列
frame.add(this.jcb1);//增加组件
```

```
frame.add(this.jcb2);//增加组件
frame.add(this.jcb3);//增加组件
frame.setSize(300,200);
frame.setLocationRelativeTo(null);
this.frame.setVisible(true);//设置可显示
frame.setDefaultCloseOperation(JFrame.EXIT_ON_CLOSE);
}

public static void main(String args[]){
    new JCheckBox1();
}
}
```

如图 6-50 所示。

图 6-50

2. 复选框事件处理

JCheckBox 类的选择事件是 ItemEvent，可实现 ItemListener 监听器接口的 itemStateChanged()方法来处理事件，用 addItemListener()方法注册。

ItemListener 接口只定义了 itemStateChanged()一个方法，其格式如下：
- void itemStateChanged(ItemEvent e)当用户取消或选定某个选项时调用。

itemStateChanged()方法中用到了 ItemEvent 事件，该事件的常用方法及常量如下所示：
- public Object getItem()返回受事件影响的选项。
- public int getStateChange()返回选定状态的类型(已选择或已取消)。
- public static final int SELECTED 表示选项被选中的常量。
- public static final int DESELECTED 表示选项未被选中的常量。

【例 6.33】创建 3 个复选框和一个标签,三个复选框分别显示选项一、选项二、选项三,标签动态显示选中了哪个复选框。如果取消选中,则动态显示取消了哪个选项的选中状态。

```java
import java.awt.GridLayout;
import javax.swing.JFrame;
import javax.swing.JCheckBox;
import javax.swing.JLabel;
import java.awt.event.ItemListener;
import java.awt.event.ItemEvent;
class JCheckBox2{
private JFrame frame = new JFrame("复选按钮");//定义窗体
private JCheckBox JCB1 = new JCheckBox("选项一");//定义第一个复选框
private JCheckBox JCB2 = new JCheckBox("选项二");//定义第二个复选框
private JCheckBox JCB3 = new JCheckBox("选项三");//定义第三个复选框
private JLabel JL1 = new JLabel("您选中了:");
class ItemListener1 implements ItemListener{
public void itemStateChanged(ItemEvent e){
JCheckBox JCB = (JCheckBox)e.getItem();//得到产生的事件
String YuanZiFu = JL1.getText();//获取标签 JL1 原来显示的内容
if(e.getStateChange() = = ItemEvent.SELECTED){//如果被选中了
JL1.setText(YuanZiFu + JCB.getText());
}
else{//取消选中状态
JL1.setText("您取消了" + JCB.getText() + "的选中状态");
}
}
}
public JCheckBox2(){//构造函数
frame.setLayout(new GridLayout(2,2));//设置组件的排版为 2 行 2 列
frame.add(JCB1);//增加组件
frame.add(JCB2);//增加组件
frame.add(JCB3);//增加组件
frame.add(JL1);//增加组件
JCB1.addItemListener(new ItemListener1());
JCB2.addItemListener(new ItemListener1());
JCB3.addItemListener(new ItemListener1());
frame.setSize(300,200);
frame.setLocationRelativeTo(null);
```

```
frame.setVisible(true);//设置可显示
frame.setDefaultCloseOperation(JFrame.EXIT_ON_CLOSE);
}
public static void main(String args[]){
new JCheckBox2();
}
}
```

如图 6-51、图 6-52 所示。

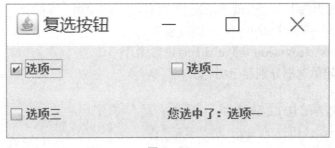

图 6-51

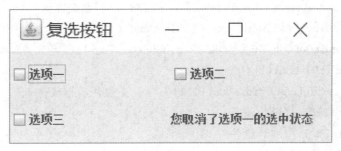

图 6-52

6.4.2 单选按钮 JRadioButton

1. 创建单选按钮

单选按钮就是在给出的多个信息中选择指定的一个,即进行多选一。例如,性别有两个选项"男"和"女",但只能二选一,此时就可以使用单选按钮实现性别选择。在 Swing 中可以使用 JRadioButton 类完成一组单选按钮的操作,JRadioButton 类的构造方法如下:

• JRadioButton(Icon icon)创建一个单选按钮,并指定图片。
• JRadioButton(Icon icon, boolean selected)创建一个单选按钮,并指定图片和其是否选中。

- JRadioButton(String text)创建一个单选按钮,并指定其文本,默认不选中。
- JRadioButton(String text, boolean selected)创建一个单选按钮,并指定文本和其是否选中。
- JRadioButton(String text, Icon icon)创建一个单选按钮,并指定其文本和图片,默认不选中。
- JRadioButton(String text, Icon icon, boolean selected)创建一个单选按钮,并指定图片、文本和其是否选中。

JRadioButton 类常用的方法如下:

- public boolean isSelected()返回是否被选中,当复选框按钮选中时返回 true,否则返回 false。
- public void setSelected(boolean Selected)设置是否选中,参数 selected 若为 true,则表示设置按钮状态为选中;反之,就表示设置按钮状态为未选中。
- public void setText(String text)设置显示文本。
- public void setIcon(Icon defaultIcon)设置图片。

【例 6.34】创建单选框分别显示男、女。

```java
import java.awt.GridLayout;
import javax.swing.JFrame;
import javax.swing.JRadioButton;
class JRadioButton1{
    private JFrame frame = new JFrame("单选按钮");//定义窗体
    private JRadioButton JRB1 = new JRadioButton("男");//定义第一个单选框
    private JRadioButton JRB2 = new JRadioButton("女");//定义第二个单选框
    public JRadioButton1(){
        frame.setLayout(new GridLayout(1,2));//设置组件的排版为1行3列
        frame.add(JRB1);//增加组件
        frame.add(JRB2);//增加组件
        frame.setSize(300,200);
        frame.setLocationRelativeTo(null);
        frame.setVisible(true);//设置可显示
        frame.setDefaultCloseOperation(JFrame.EXIT_ON_CLOSE);
    }
    public static void main(String args[]){
        new JRadioButton1();
    }
}
```

如图 6-53 所示。

图 6-53

在上例中,单选按钮虽然创建成功,但是并不能实现"单选"功能,而是形成了"多选"。这是因为单选按钮是在一组按钮中选择一个,因此必须将单选按钮分组,即指明在一个组中包含哪些单选按钮。可用 ButtonGroup 创建按钮组对象,应用按钮组对象的 add( )方法顺序加入各个单选按钮。

【例 6.35】创建单选按钮,显示男、女,并实现性别的单项选择。

```
import java.awt.GridLayout;
import javax.swing.JFrame;
import javax.swing.JRadioButton;
import javax.swing.ButtonGroup;
class JRadioButton2{
private JFrame frame = new JFrame("单选按钮");//定义窗体
private JRadioButton JRB1 = new JRadioButton("男");//定义第一个单选框
private JRadioButton JRB2 = new JRadioButton("女");//定义第二个单选框
private ButtonGroup BG1 = new ButtonGroup();//创建分组 BG1
public JRadioButton2(){
frame.setLayout(new GridLayout(1,2));//设置组件的排版为1行3列
BG1.add(JRB1);
BG1.add(JRB2);
frame.add(JRB1);//增加组件
frame.add(JRB2);//增加组件
frame.setSize(300,200);
frame.setLocationRelativeTo(null);
frame.setVisible(true);//设置可显示
frame.setDefaultCloseOperation(JFrame.EXIT_ON_CLOSE);
}
```

```
public static void main(String args[]){
    new JRadioButton2();
}
}
```

如图 6-54 所示。

图 6-54

2. 单选按钮事件处理

单选按钮的事件包括 ActionEvent 类事件和 ItemEvent 类事件，因此单选按钮的事件可以通过实现之前讲过的 ActionListener 接口或 ItemListener 接口处理。

（1）实现 ActionListener 接口。

JRadioButton 组件引发的 ActionEvent 事件，可实现 ActionListener 监听器接口的 actionPerformed()方法，用 addActionListener()方法注册，用 getActionCommand()或 getSource()方法确定事件源。

【例 6.36】创建两个单选按钮和一个标签，单选按钮用于显示男、女，标签用于动态显示选中的内容（用 ActionListener 监听器接口实现）。

```
import java.awt.GridLayout;
import javax.swing.JFrame;
import javax.swing.JRadioButton;
import javax.swing.JLabel;
import javax.swing.ButtonGroup;
import java.awt.event.ActionEvent;
import java.awt.event.ActionListener;
class JRadioButton3{
    private JFrame frame = new JFrame("单选按钮");//定义窗体
    private JRadioButton JRB1 = new JRadioButton("男");//定义第一个单选框
```

```java
private JRadioButton JRB2 = new JRadioButton("女");//定义第二个单选框
private ButtonGroup BG1 = new ButtonGroup();//创建分组 BG1
private JLabel JL1 = new JLabel("您选择的是:");

class RadioListener implements ActionListener{
public void actionPerformed(ActionEvent e){
if(e.getSource() = = JRB1)
JL1.setText("您选中了:男");
else
JL1.setText("您选中了:女");
}
}

public JRadioButton3(){//构造函数
frame.setLayout(new GridLayout(1,3));//设置组件的排版为 1 行 3 列
BG1.add(JRB1);
BG1.add(JRB2);
frame.add(JRB1);//增加组件
frame.add(JRB2);//增加组件
JRB1.addActionListener(new RadioListener());
JRB2.addActionListener(new RadioListener());
frame.add(JL1);//增加组件
frame.setSize(300,200);
frame.setLocationRelativeTo(null);
frame.setVisible(true);//设置可显示
frame.setDefaultCloseOperation(JFrame.EXIT_ON_CLOSE);
}
public static void main(String args[]){
new JRadioButton3();
}
}
```

如图 6-55 所示。

(2) 实现 ItemListener 接口。

JRadioButton 组件引发的 ItemEvent 事件,可实现 ItemListener 监听器接口的 itemStateChanged()方法,用 addItemListener()方法注册,用 getActionCommand()或 getSource()方法确定事件源。

图 6-55

【例 6.37】创建两个单选按钮和一个标签,单选按钮用于显示男、女,标签用于动态显示选中的内容(用 ItemListener 监听器接口实现)。

```
import java.awt.GridLayout;
import javax.swing.JFrame;
import javax.swing.JRadioButton;
import javax.swing.JLabel;
import javax.swing.ButtonGroup;
import java.awt.event.ItemListener;
import java.awt.event.ItemEvent;
class JRadioButton4{
private JFrame frame = new JFrame("单选按钮");//定义窗体
private JRadioButton JRB1 = new JRadioButton("男");//定义第一个单选框
private JRadioButton JRB2 = new JRadioButton("女");//定义第二个单选框
private ButtonGroup BG1 = new ButtonGroup();//创建分组 BG1
private JLabel JL1 = new JLabel("您选择的是:");

class RadioListener implements ItemListener{
public void itemStateChanged(ItemEvent e){
if(e.getSource() = = JRB1)
JL1.setText("您选中了:男");
else
JL1.setText("您选中了:女");
}
}
```

```
public JRadioButton4(){//构造函数
frame.setLayout(new GridLayout(1,3));//设置组件的排版为1行3列
BG1.add(JRB1);
BG1.add(JRB2);
frame.add(JRB1);//增加组件
frame.add(JRB2);//增加组件
JRB1.addItemListener(new RadioListener());
JRB2.addItemListener(new RadioListener());
frame.add(JL1);//增加组件
frame.setSize(300,200);
frame.setLocationRelativeTo(null);
frame.setVisible(true);//设置可显示
frame.setDefaultCloseOperation(JFrame.EXIT_ON_CLOSE);
}
public static void main(String args[]){
new JRadioButton4();
}
}
```

如图 6-56 所示。

图 6-56

6.4.3 下拉列表框 JComboBox

在 Java 图形界面应用程序中,用户经常会遇到在多个选项中选择一项的操作,在 Swing 中可以使用下拉列表框 JComboBox 实现该功能。在未选择下拉列表框时,下拉列表框显示为带按钮的一个选项的形式;当对下拉列表框单击时,下拉列表框会打开并列出多个选项的一个列表提供给用户选择。

1. 创建下拉列表框

JComboBox 类的构造方法有 4 种,常用的构造方法有如下 2 种:
- public JComboBox(Object[]items)用对象数组创建下拉列表框。
- public JComboBox(Vector<?>items)用 Vector 创建下拉列表框。

JComboBox 类的常用方法如下:
- public void addItem(Object anObject)为列表添加选项。
- public Object getItemAt(int index)返回指定索引处的列表项。
- public int getItemCount()返回列表中的项数。
- public Object getSelectedItem()返回当前所选项。
- public void setSelectedIndex(int anIndex)设置默认选项的索引号。
- public void setMaximumRowCount(int count)设置下拉列表显示的最大行数。

【例6.38】使用对象数组创建下拉列表框,显示北京、上海、天津、重庆。

```java
import javax.swing.JFrame;
import javax.swing.JComboBox;
class JComboBox1{
private JFrame frame = new JFrame("下拉列表框");//定义窗体
String[]Str_JCB = {"北京","上海","天津","重庆"};//创建字符串对象数组
private JComboBox JCB1 = new JComboBox(Str_JCB);//创建下拉列表框

public JComboBox1(){//构造函数
frame.setLayout(null);
frame.add(JCB1);//增加组件
JCB1.setBounds(10,10,80,30);
frame.setSize(300,200);
frame.setLocationRelativeTo(null);
frame.setVisible(true);//设置可显示
frame.setDefaultCloseOperation(JFrame.EXIT_ON_CLOSE);
}
public static void main(String args[]){
new JComboBox1();
}
}
```

如图 6-57 所示。

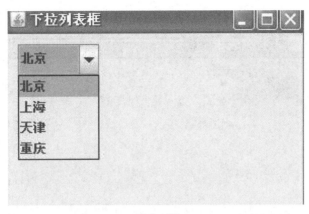

图 6-57

【例 6.39】使用 Vector 创建下拉列表框,显示北京、上海、天津、重庆。

```java
import javax.swing.JFrame;
import javax.swing.JComboBox;
import java.util.Vector;
class JComboBox2{
    private JFrame frame = new JFrame("下拉列表框");//定义窗体
    private Vector<String>v = new Vector<String>();
    private JComboBox JCB1 = new JComboBox(v);//创建下拉列表框
    public JComboBox2(){//构造函数
        v.add("北京");
        v.add("上海");
        v.add("天津");
        v.add("重庆");
        frame.setLayout(null);
        frame.add(JCB1);//增加组件
        JCB1.setBounds(10,10,80,30);
        frame.setSize(300,200);
        frame.setLocationRelativeTo(null);
        frame.setVisible(true);//设置可显示
        frame.setDefaultCloseOperation(JFrame.EXIT_ON_CLOSE);
    }
    public static void main(String args[]){
        new JComboBox2();
    }
}
```

如图 6-58 所示。

图 6-58

2. 下拉列表框事件处理

下拉列表框的事件包括 ActionEvent 类事件和 ItemEvent 类事件,因此下拉列表框的事件也可以通过实现之前讲过的 ActionListener 接口或 ItemListener 接口处理。

(1) 实现 ActionListener 接口。

【例 6.40】使用对象数组创建下拉列表框,显示北京、上海、天津、重庆,再创建一个文本框用于显示选中的选项。

```java
import javax.swing.JFrame;
import javax.swing.JComboBox;
import javax.swing.JTextField;
import java.awt.event.ActionEvent;
import java.awt.event.ActionListener;
class JComboBox3{
private JFrame frame = new JFrame("下拉列表框");//定义窗体
String[]Str_JCB = {"北京","上海","天津","重庆"};//创建字符串对象数组
private JComboBox JCB1 = new JComboBox(Str_JCB);//创建下拉列表框
private JTextField JT1 = new JTextField();

class JComboBoxListener implements ActionListener{
public void actionPerformed(ActionEvent e){
if(e.getSource() = = JCB1)
JT1.setText("您选择的直辖市是:" + JCB1.getSelectedItem().toString());
}
}
```

```
public JComboBox3(){//构造函数
frame.setLayout(null);
frame.add(JCB1);//增加组件
frame.add(JT1);//增加组件
JCB1.setBounds(10,10,70,30);
JT1.setBounds(80,10,150,30);
JCB1.addActionListener(new JComboBoxListener());
frame.setSize(300,200);
frame.setLocationRelativeTo(null);
frame.setVisible(true);//设置可显示
frame.setDefaultCloseOperation(JFrame.EXIT_ON_CLOSE);
}
public static void main(String args[]){
new JComboBox3();
}
}
```

如图 6-59 所示。

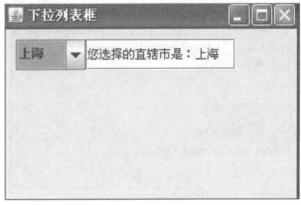

图 6-59

（2）实现 ItemListener 接口。

【例 6.41】使用 Vector 创建下拉列表框，显示北京、上海、天津、重庆，再创建一个文本框用于显示选中的选项。

```
import javax.swing.JFrame;
import javax.swing.JComboBox;
import javax.swing.JTextField;
import java.awt.event.ItemListener;
```

```java
import java.awt.event.ItemEvent;
import java.util.Vector;
class JComboBox4{
private JFrame frame = new JFrame("下拉列表框");//定义窗体
private Vector<String>v = new Vector<String>();
private JComboBox JCB1 = new JComboBox(v);//创建下拉列表框
private JTextField JT1 = new JTextField();

class JComboBoxListener implements ItemListener{
public void itemStateChanged(ItemEvent e){
if(e.getSource() == JCB1)
JT1.setText("您选择的直辖市是:"+JCB1.getSelectedItem().toString());
}
}

public JComboBox4(){//构造函数
v.add("北京");
v.add("上海");
v.add("天津");
v.add("重庆");
frame.setLayout(null);
frame.add(JCB1);//增加组件
frame.add(JT1);//增加组件
JCB1.setBounds(10,10,70,30);
JT1.setBounds(80,10,150,30);
JCB1.addItemListener(new JComboBoxListener());
frame.setSize(300,200);
frame.setLocationRelativeTo(null);
frame.setVisible(true);//设置可显示
frame.setDefaultCloseOperation(JFrame.EXIT_ON_CLOSE);
}
public static void main(String args[]){
new JComboBox4();
}
}
```

如图 6-60 所示。

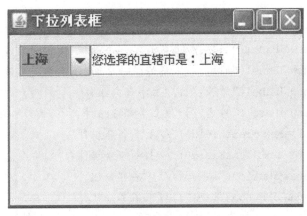

图 6-60

6.4.4 面板 JPanel

面板 JPanel 是一种添加到其他容器使用的容器组件,可将组件添加到 JPanel,在 JPanel 中完成各个组件的排列,然后再将所有独立的 JPanel 添加到某个容器(如 JFrame),使用 JPanel 可以实现各种复杂的界面设计。

JPanel 类继承于 javax.swing.JComponent 类,其构造方法有:
- public JPanel() 创建具有默认 FlowLayout 布局的 JPanel 对象。
- public JPanel(LayoutManager layout) 创建具有指定布局管理器的 JPanel 对象。

JPanel 可指定边界,可用的边界有 titled、etched、beveled、line、matte、compound 和 empty 等,也可以创建自己的边界。可用 JComponent 类的 setBorder()方法设置边界,其用法如下:

```
public void setBorder(Border border)
```

其中,Border 类的参数可用 javax.swing.BorderFactory 类中的方法获得,获取各种相应边界的方法为:createTitledBorder()、createEtchedBorder()、createBevelBorder()、createRaisedBevelBorder()、createLoweredBevelBorder()、createLineBorder()、createMatteBorder()、createCompoundBorder() 和 createEmptyBorder()。

【例 6.42】 创建两个 JPanel 面板并用面板将窗体分为上下两部分,上半部分的面板显示标签和文本框,下半部分的面板显示按钮。

```
import javax.swing.JPanel;
import javax.swing.BorderFactory;
import javax.swing.JButton;
import javax.swing.JLabel;
import javax.swing.JTextField;
import javax.swing.JFrame;
```

```java
class JPanel1{
    JFrame frame = new JFrame("面板");//实例化窗体对象
    JPanel JP1 = new JPanel(null);//准备好了一个面板,面板是空布局
    JPanel JP2 = new JPanel(null);//准备好了一个面板,面板是空布局
    JLabel JL1 = new JLabel("学号:");//上半部分组件
    JLabel JL2 = new JLabel("姓名:");//上半部分组件
    JTextField JT1 = new JTextField();//上半部分组件
    JTextField JT2 = new JTextField();//上半部分组件
    JButton JB1 = new JButton("添加");//下半部分组件
    JButton JB2 = new JButton("删除");//下半部分组件
    JButton JB3 = new JButton("修改");//下半部分组件
    JButton JB4 = new JButton("查询");//下半部分组件
    public JPanel1(){
        frame.setLayout(null);
        //将上半部分组件都加入了JP1中并布局位置和大小
        JP1.add(JL1);
        JL1.setBounds(10,20,40,20);
        JP1.add(JT1);
        JT1.setBounds(70,20,70,20);
        JP1.add(JL2);
        JL2.setBounds(150,20,40,20);
        JP1.add(JT2);
        JT2.setBounds(200,20,70,20);
        frame.add(JP1);//将面板加到窗体之上
        JP1.setBounds(10,10,290,50);//设置面板在窗体中的位置和大小
        JP1.setBorder(BorderFactory.createTitledBorder("上半部分"));//设置标题
        //将下半部分组件都加入了JP2中并布局位置和大小
        JP2.add(JB1);
        JB1.setBounds(10,30,60,30);
        JP2.add(JB2);
        JB2.setBounds(80,30,60,30);
        JP2.add(JB3);
        JB3.setBounds(150,30,60,30);
        JP2.add(JB4);
        JB4.setBounds(220,30,60,30);
        frame.add(JP2);//将面板加到窗体之上
        JP2.setBounds(10,60,290,80);//设置面板在窗体中的位置和大小
        JP2.setBorder(BorderFactory.createTitledBorder("下半部分"));//设置标题
```

```
        frame.setSize(320,200);
        frame.setLocationRelativeTo(null);
        frame.setVisible(true);//设置可显示
        frame.setDefaultCloseOperation(JFrame.EXIT_ON_CLOSE);
    }

    public static void main(String args[]){

        new JPanel1();
    }
}
```

如图 6-61 所示。

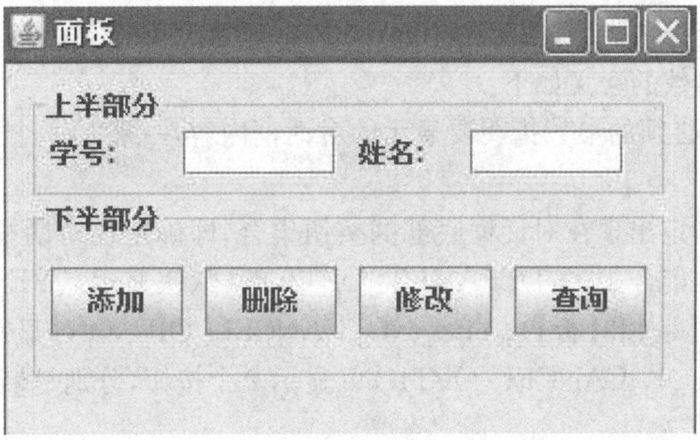

图 6-61

6.4.5 对话框

对话框是为人机对话过程提供交互模式的工具,应用程序通过对话框,或给用户提供信息,或从用户获得信息。对话框是一种大小不能变化、不能有菜单的容器窗口,可以在其中放置用于得到用户输入的控件。但对话框只是一个临时窗口,它不能作为一个应用程序的主框架,必须包含在其他的容器中(如 JFrame 窗体)。由于对话框依赖其他窗口,所以,当它所依赖的窗口消失或最小化时,对话框也将消失;窗口还原时,对话框又会自动恢复。

对话框分为强制和非强制两种。强制型对话框不能中断对话过程,直至对话框结束,才让程序响应对话框以外的事件。非强制型对话框可以中断对话过程,去响应对话框以外的事件。强制型对话框也称有模式对话框,非强制型对话框也称非模式对话框。

Java 语言提供了多种对话框类来支持多种形式的对话框:

(1) JOptionPane 类支持简单标准的对话框。

(2) JDialog 类支持定制用户自己的对话框。

(3) JFileChooser 类支持文件打开保存对话框、JColorChooser 类支持颜色选择对话框等。

1. JOptionPane 对话框

JOptionPane 提供的对话框是模式对话框,使用 JOptionPane 可以创建和自定义问题、信息、警告以及错误等几种类型的对话框,JOptionPane 提供标准对话框的布局支持、图标、指定对话框的标题和文本、自定义按钮文本、允许自定义组件的对话框显示、指定对话框在屏幕上的显示位置等特性。

(1) JOptionPane 类有 7 种构造方法,常用的有 5 种:

• JOptionPane()创建具有测试信息的 JOptionPane 对象。

• JOptionPane(Object message)创建显示 message 和默认选项的 JOptionPane 对象。

• JOptionPane(Object message, int messageType)创建以 messageType 类型显示 message 和使用默认选项的 JOptionPane 对象。

• JOptionPane(Object message, int messageType, int optionType)创建显示指定类型信息和指定选项类型的 JOptionPane 对象。

• JOptionPane(Object message, int messageType, int optionType, Icon icon)创建显示指定类型信息和指定选项类型、图标的 JOptionPane 对象。

构造函数中各参数的含义如下:

• message 这个必须的参数指明要显示的对话框内容,一般是一个字符串,显示在对话框的一个标签中。

• optionType 指定出现在对话框底部的按钮集合(即确定在对话框出现几个按钮),可以选择下面 4 个标准集合之一:DEFAULT_OPTION(默认显示)、YES_NO_OPTION(显示 2 个按钮,分别是【是】和【否】)、YES_NO_CANCEL_OPTION(显示 3 个按钮,分别是【是】【否】和【取消】)、OK_CANCEL_OPTION(显示 2 个按钮,分别是【确认】和【取消】)。

注意:

如果点击【YES】(【是】)按钮,其返回值为 YES_OPTION;如果点击【NO】(【否】)按钮,其返回值为 NO_OPTION;如果点击【CANEL】(【取消】)按钮,其返回值为 CANCEL_OPTION;如果点击【OK】(【确定】)按钮,其返回值为 OK_OPTION;如果没有点击按钮,直接关闭对话框,其返回值为 CLOSED_OPTION。

• messageType 确定显示在对话框中的图标,从下列值中选择一个:PLAIN_MESSAGE(无 icon 图标)、ERROR_MESSAGE(带错误图标)、INFORMATION_MESSAGE(带普通信息图标)、WARNING_MESSAGE(带警示图标)、QUESTION_MESSAGE(带询问信息图标)。

• icon 指明了在对话框中显示用户定义图标。

(2) JOptionPane 类主要用 4 种消息提示框方法显示不同类型的对话框:

• showConfirmDialog():确认对话框。

• showInputDialog():输入对话框。

• showMessageDialog():消息对话框。

• showOptionDialog():选择对话框。

① showConfirmDialog()。

showConfirmDialog()方法有 4 种参数设置类型:

showConfirmDialog(Component parentComponent, Object message)

showConfirmDialog(Component parentComponent, Object message, String title, int optionType)

showConfirmDialog(Component parentComponent, Object message, String title, int optionType, int messageType)

showConfirmDialog(Component parentComponent, Object message, String title, int optionType, int messageType, Icon icon)

showConfirmDialog()方法中各参数的含义如下：

- parentComponent 每个 showConfirmDialog()方法的第一个参数总是父组件，必须是一个框架、一个框架中的组件或 null 值，JOptionPane 构造方法不包含这个参数。
- message 这个必须的参数指明要显示的对话框，一般是一个字符串，显示在对话框的一个标签中。
- title 对话框的标题。
- optionType 指定出现在对话框底部的按钮集合（即确定在对话框出现几个按钮），可以选择下面 4 个标准集合之一：DEFAULT_OPTION（默认显示）、YES_NO_OPTION（显示 2 个按钮，分别是【是】和【否】）、YES_NO_CANCEL_OPTION（显示 3 个按钮，分别是【是】【否】和【取消】）、OK_CANCEL_OPTION（显示 2 个按钮，分别是【确定】和【取消】）。

注意：

点击对话框中的按钮后，对话框返回一个整数。如果点击【YES】(【是】)按钮，其返回值为 YES_OPTION；如果点击【NO】(【否】)按钮，其返回值为 NO_OPTION；如果点击【CANEL】(【取消】)按钮，其返回值为 CANCEL_OPTION；如果点击【OK】(【确定】)按钮，其返回值为 OK_OPTION；如果没有点击按钮，直接关闭对话框，其返回值为 CLOSED_OPTION。

- messageType 确定显示在对话框中的图标，从下列值中选择一个：PLAIN_MESSAGE（无 icon 图标）、ERROR_MESSAGE（带错误图标）、INFORMATION_MESSAGE（带普通信息图标）、WARNING_MESSAGE（带警示图标）、QUESTION_MESSAGE（带询问信息图标）。
- icon 指明了在对话框中显示用户定义图标。

【例 6.43】创建一个按钮（按钮显示文本"显示确认对话框"）和一个标签，当点击按钮时，弹出确认对话框并显示"您喜欢 Java 吗？"。如果点击对话框的【是】按钮，标签显示"您选择是的您喜欢 Java!"；如果点击对话框的【否】按钮，标签显示"您选择的是您不喜欢 Java!"；如果点击对话框的【取消】按钮，标签显示"很遗憾,您没有做出选择!"。

```
import javax.swing.JFrame;
import javax.swing.JOptionPane;
import javax.swing.JButton;
import javax.swing.JLabel;
import java.awt.event.ActionListener;
import java.awt.event.ActionEvent;
```

```java
public class showConfirmDialog1{
public JFrame frame = new JFrame("确认对话框");
public JButton JB1 = new JButton("显示确认对话框");
public JOptionPane JOP1 = new JOptionPane();
public JLabel JL1 = new JLabel("请做出您的选择:");
public showConfirmDialog1(){
frame.setLayout(null);
frame.add(JB1);
JB1.setBounds(10,10,130,30);
JB1.addActionListener(new OptionListener());
frame.add(JL1);
JL1.setBounds(160,10,160,30);
frame.setSize(360,100);
frame.setLocationRelativeTo(null);
frame.setVisible(true);//设置可显示
frame.setDefaultCloseOperation(JFrame.EXIT_ON_CLOSE);
}
public class OptionListener implements ActionListener{
public void actionPerformed(ActionEvent e){
int n = JOP1.showConfirmDialog(frame,"您喜欢 Java 吗?","确认对话框",
JOptionPane.YES_NO_CANCEL_OPTION,JOptionPane.QUESTION_MESSAGE);
if(n = = JOptionPane.YES_OPTION)
JL1.setText("您选择是的您喜欢 Java!");
if(n = = JOptionPane.NO_OPTION)
JL1.setText("您选择的是您不喜欢 Java!");
if(n = = JOptionPane.CANCEL_OPTION)
JL1.setText("很遗憾,您没有做出选择!");
}
}
public static void main(String[]args){
new showConfirmDialog1();
}
}
```

如图 6-62 至图 6-64 所示。

② showInputDialog()。

showInputDialog()方法有 6 种参数设置类型:

- showInputDialog(Object message)
- showInputDialog(Component parentComponent, Object message)

图 6-62

图 6-63

图 6-64

- showInputDialog(Object message, Object initialSelectionValue)
- showInputDialog (Component parentComponent, Object message, Object initialSelectionValue)
- showInputDialog(Component parentComponent, Object message, String title, int messageType)
- showInputDialog(Component parentComponent, Object message, String title, int messageType, Icon icon, Object[]selectionValues, Object initialSelectionValue)

showInputDialog()方法中各参数的含义如下：

- parentComponent 表示父组件，必须是一个框架、一个框架中的组件或 null 值，JOptionPane 构造方法不包含这个参数。
- message 这个必须的参数指明要显示的对话框内容，一般是一个字符串，显示在对话框的一个标签中。
- title 对话框标题。
- icon 指明了在对话框中显示用户定义图标。
- messageType 确定显示在对话框中的图标，从下列值中选择一个：PLAIN_MESSAGE（无 icon 图标）、ERROR_MESSAGE（带错误图标）、INFORMATION_MESSAGE（带普通信息图标）、WARNING_MESSAGE（带警示图标）、QUESTION_MESSAGE（带询问信息图标）。
- selectionValues 表示用户可能的选择值，以数组方式赋值，显示形式为 ComboBox。

- initialSelectionValue 表示对话框初始化时输入框中的显示值。

注意：

showInputDialog()方法的返回值有两种：String 类型和 Object 类型，当用户按下确定按钮时会返回用户输入（或选择）的信息，若按下取消按钮则会返回 null。

【例6.44】创建一个按钮（按钮显示文本"显示输入对话框"）和一个标签，当点击按钮时，弹出输入对话框并显示"请输入您的学号："。在输入对话框里输入学号后，如果点击对话框的【确定】按钮，标签显示"您输入的学号是 XXX！"（XXX 代表刚才输入的学号）；如果点击对话框的【取消】按钮，标签显示"您没有输入您的学号！"，在对话框显示询问信息图标。

```java
import javax.swing.JFrame;
import javax.swing.JOptionPane;
import javax.swing.JButton;
import javax.swing.JLabel;
import java.awt.event.ActionListener;
import java.awt.event.ActionEvent;
public class showInputDialog1{
public JFrame frame = new JFrame("输入对话框");
public JButton JB1 = new JButton("显示输入对话框");
public JOptionPane JOP1 = new JOptionPane();
public JLabel JL1 = new JLabel("请输入您的学号：");
public showInputDialog1(){
frame.setLayout(null);
frame.add(JB1);
JB1.setBounds(10,10,130,30);
JB1.addActionListener(new OptionListener());
frame.add(JL1);
JL1.setBounds(160,10,160,30);
frame.setSize(360,100);
frame.setLocationRelativeTo(null);
frame.setVisible(true);//设置可显示
frame.setDefaultCloseOperation(JFrame.EXIT_ON_CLOSE);
}
public class OptionListener implements ActionListener{
public void actionPerformed(ActionEvent e){
String fanhui = JOP1.showInputDialog(frame,"请输入您的学号！");
JL1.setText("您输入的学号是：" + fanhui);
}
}
```

```
public static void main(String[ ]args){
new showInputDialog1();
}
}
```

如图 6-65 至图 6-67 所示。

图 6-65

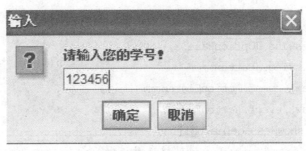

图 6-66

图 6-67

③ showMessageDialog()。

showMessageDialog()方法有 3 种参数设置：

• showMessageDialog(Component parentComponent, Object message)

• showMessageDialog(Component parentComponent, Object message, String title, int messageType)

• showMessageDialog(Component parentComponent, Object message, String title, int messageType, Icon icon)

showMessageDialog()方法的各个参数含义如下：

• parentComponent 表示父组件，必须是一个框架、一个框架中的组件或 null 值，JOptionPane 构造方法不包含这个参数。

- message 这个必须的参数指明要显示的对话框内容,一般是一个字符串,显示在对话框的一个标签中。
- title 对话框标题。
- messageType 确定显示在对话框中的图标,从下列值中选择一个:PLAIN_MESSAGE(无 icon 图标)、ERROR_MESSAGE(带错误图标)、INFORMATION_MESSAGE(带普通信息图标)、WARNING_MESSAGE(带警示图标)、QUESTION_MESSAGE(带询问信息图标)。
- icon 指明了在对话框中显示用户定义图标。

注意:

showMessageDialog()方法无返回值,Message Dialog 只是告知用户某些信息,用户除了点击确定按钮外不能与其进行交互。

【例6.45】创建一个按钮(按钮显示文本"显示信息对话框"),当点击按钮时,弹出信息对话框并显示"这是用于提示的信息对话框!",在对话框显示警示图标。

```java
import javax.swing.JFrame;
import javax.swing.JOptionPane;
import javax.swing.JButton;
import java.awt.event.ActionListener;
import java.awt.event.ActionEvent;
public class showMessageDialog1{
public JFrame frame = new JFrame("信息对话框");
public JButton JB1 = new JButton("显示信息对话框");
public JOptionPane JOP1 = new JOptionPane();
public showMessageDialog1(){
frame.setLayout(null);
frame.add(JB1);
JB1.setBounds(10,10,130,30);
JB1.addActionListener(new OptionListener());
frame.setSize(150,80);
frame.setLocationRelativeTo(null);
frame.setVisible(true);//设置可显示
frame.setDefaultCloseOperation(JFrame.EXIT_ON_CLOSE);
}
public class OptionListener implements ActionListener{
public void actionPerformed(ActionEvent e){
JOP1.showMessageDialog(frame,"这是用于提示的信息对话框!","信息对话框",JOptionPane.WARNING_MESSAGE);
}
```

```
}
public static void main(String[ ]args){
new showMessageDialog1();
}
}
```

如图 6-68、图 6-69 所示。

图 6-68

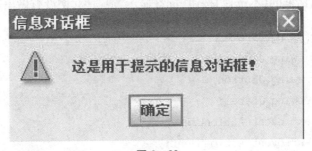

图 6-69

④ showOptionDialog()。

showOptionDialog()方法只有一种参数设置：

• showOptionDialog(Component parentComponent, Object message, String title, int optionType, int messageType, Icon icon, Object[]options, Object initialValue)

showOptionDialog()方法中各个参数的含义如下：

• parentComponent 表示父组件，必须是一个框架、一个框架中的组件或 null 值，JOptionPane 构造方法不包含这个参数。

• message 这个必须的参数指明要显示的对话框内容，一般是一个字符串，显示在对话框的一个标签中。

• title 对话框标题。

• optionType 指定出现在对话框底部的按钮集合(即确定在对话框出现几个按钮)，可以选择下面 4 个标准集合之一：DEFAULT_OPTION(默认显示)、YES_NO_OPTION(显示 2 个按钮，分别是【是】和【否】)、YES_NO_CANCEL_OPTION(显示 3 个按钮，分别是

【是】【否】和【取消】)、OK_CANCEL_OPTION(显示 2 个按钮,分别是【确认】和【取消】)。

注意:

如果点击【YES】(【是】)按钮,其返回值为 YES_OPTION;如果点击【NO】(【否】)按钮,其返回值为 NO_OPTION;如果点击【CANEL】(【取消】)按钮,其返回值为 CANCEL_OPTION;如果点击【OK】(【确定】)按钮,其返回值为 OK_OPTION;如果没有点击按钮,直接关闭对话框,其返回值为 CLOSED_OPTION。

- messageType 确定显示在对话框中的图标,从下列值中选择一个:PLAIN_MESSAGE(无 icon 图标)、ERROR_MESSAGE(带错误图标)、INFORMATION_MESSAGE(带普通信息图标)、WARNING_MESSAGE(带警示图标)、QUESTION_MESSAGE(带询问信息图标)。
- icon 指明了在对话框中显示用户定义图标。
- options 进一步指明任选对话框底部的按钮显示的文本。一般地,对按钮指定一个字符串数组。
- initialValue 指明选择的初始值。

【例6.46】创建一个按钮(按钮显示文本"显示选择对话框")和一个标签,当点击按钮时,弹出选择对话框并显示"您喜欢Java吗?"。如果点击对话框的【喜欢】按钮,标签显示"您选择是的您喜欢 Java!";如果点击对话框的【不喜欢】按钮,标签显示"您选择的是您不喜欢Java!";如果点击对话框的【不选择】按钮,标签显示"很遗憾,您没有做出选择!"。

```java
import javax.swing.JFrame;
import javax.swing.JOptionPane;
import javax.swing.JButton;
import javax.swing.JLabel;
import java.awt.event.ActionListener;
import java.awt.event.ActionEvent;
public class showOptionDialog1{
public JFrame frame = new JFrame("选择对话框");
public JButton JB1 = new JButton("显示选择对话框");
public JOptionPane JOP1 = new JOptionPane();
public JLabel JL1 = new JLabel("请做出您的选择:");
public showOptionDialog1(){
frame.setLayout(null);
frame.add(JB1);
JB1.setBounds(10,10,130,30);
JB1.addActionListener(new OptionListener());
frame.add(JL1);
JL1.setBounds(160,10,160,30);
frame.setSize(360,100);
frame.setLocationRelativeTo(null);
```

```
frame.setVisible(true);//设置可显示
frame.setDefaultCloseOperation(JFrame.EXIT_ON_CLOSE);
}
public class OptionListener implements ActionListener{
public void actionPerformed(ActionEvent e){
Object[]options = {"喜欢","不喜欢","不选择"};
int n = JOP1.showOptionDialog(frame,"您喜欢Java吗?","选择对话框",
JOptionPane.YES_NO_CANCEL_OPTION,
JOptionPane.QUESTION_MESSAGE,null,options,options[0]);
if(n = = JOptionPane.YES_OPTION)
JL1.setText("您选择是的您喜欢Java!");
if(n = = JOptionPane.NO_OPTION)
JL1.setText("您选择的是您不喜欢Java!");
if(n = = JOptionPane.CANCEL_OPTION)
JL1.setText("很遗憾,您没有做出选择!");
}
}
public static void main(String[]args){
new showOptionDialog1();
}
}
```

如图6-70至图6-72所示。

图6-70

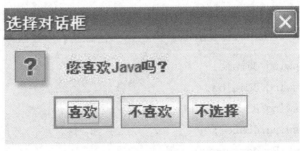

图6-71

图 6-72

2. JDialog 类

JDialog 类是对话框的基类，JDialog 对象也是一种容器，因此也可以给 JDialog 对话框指派布局管理器，JDialog 对话框的默认布局为 BorderLayout 布局。但组件不能直接加到对话框中，对话框也包含一个内容面板，应当把组件加到 JDialog 对象的内容面板中。由于对话框依赖窗口，因此要建立对话框，必须先要创建一个窗口。

JDialog 类常用的构造方法如下：

- JDialog()创建一个初始化不可见的且没有标题、没有指定 Frame 所有者的非强制型对话框。
- JDialog(JFrame f,String s)构造一个初始化不可见的非强制型对话框，参数 f 设置对话框所依赖的窗口，参数 s 用于设置标题。通常先声明一个 JDialog 类的子类，然后创建这个子类的一个对象，就建立了一个对话框。
- JDialog(JFrame f,String s,boolean b)构造一个标题为 s、初始化不可见的对话框。参数 f 设置对话框所依赖的窗口，参数 b 决定对话框是否强制或非强制型。

JDialog 类的其他常用方法如下：

- getTitle()获取对话框的标题。
- setTitle(String s)设置对话框的标题。
- setModal(boolean b)设置对话框的模式。
- setSize()设置框的大小。
- setResizable(boolean b)不可更改对话框的大小。
- setVisible(boolean b)显示或隐藏对话框。
- setLocationRelativeTo(parentComponent)设置对话框的相对位置。

【例 6.47】在窗体中创建一个按钮（按钮显示文本"显示自定义对话框"），当点击按钮时，弹出一个自定义对话框。自定义对话框的大小为(180,80)且大小不可更改，标题为"自定义对话框"，模式为强制模式，对话框可见，对话框在窗口中央显示。此外在对话框中添加一个标签（标签显示文本"如果想隐藏对话框请点击下边的隐藏按钮。"）和一个按钮（按钮显示文本"隐藏"），当点击【隐藏】按钮后，对话框隐藏。

```
import javax.swing.JFrame;
import javax.swing.JDialog;
import javax.swing.JButton;
import javax.swing.JLabel;
import java.awt.BorderLayout;
import java.awt.event.ActionListener;
```

```java
import java.awt.event.ActionEvent;
public class JDialog1{
public JFrame frame = new JFrame("自定义对话框");
public JButton JB1 = new JButton("显示自定义对话框");
public JButton JB2 = new JButton("隐藏");
public JLabel JL1 = new JLabel("如果想隐藏对话框请点击下边的隐藏按钮。");
public JDialog JD1 = new JDialog(frame,"自定义对话框",true);
public JDialog1(){
frame.setLayout(null);
frame.add(JB1);
JB1.setBounds(10,10,160,30);
JB1.addActionListener(new OptionListener());
frame.setSize(200,100);
frame.setLocationRelativeTo(null);
frame.setVisible(true);//设置可显示
frame.setDefaultCloseOperation(JFrame.EXIT_ON_CLOSE);
}
public class OptionListener implements ActionListener{
public void actionPerformed(ActionEvent e){
JD1.setSize(300,80);
JD1.setResizable(false);
JD1.setLocationRelativeTo(frame);
JD1.add("North",JL1);
JD1.add("Center",JB2);
JB2.addActionListener(new OptionListener1());
JD1.setVisible(true);
}
}
public class OptionListener1 implements ActionListener{
public void actionPerformed(ActionEvent e){
JD1.setVisible(false);
}
}
public static void main(String[ ]args){
new JDialog1();
}
}
```

如图 6-73、图 6-74 所示。

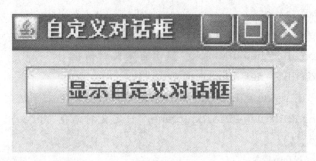

图 6-73

图 6-74

任务 实施

6.4.6　学生成绩管理系统信息管理窗口设计实现

1. 实现信息添加功能

(1) 修改主界面 MainJFrame 的代码如下：

```
import javax.swing.JFrame;
import javax.swing.JMenu;
import javax.swing.JMenuBar;
import javax.swing.JMenuItem;

import java.util.Scanner;
import java.io.File;
import java.io.FileInputStream;
import java.io.FileWriter;
import java.util.Scanner;
import java.util.ArrayList;
import java.awt.event.ActionEvent;
import java.awt.event.ActionListener;
```

```java
public class MainJFrame extends JFrame implements ActionListener   {
    //创建菜单栏
    JMenuBar jMenuBar = new JMenuBar();
    //创建菜单
    JMenu jMenu01 = new JMenu("信息管理");
    //创建菜单项
    JMenuItem jMenuItem01_00 = new JMenuItem("信息添加");
    JMenuItem jMenuItem01_01 = new JMenuItem("信息修改");
    JMenuItem jMenuItem01_02 = new JMenuItem("信息删除");
    //创建菜单
    JMenu jMenu02 = new JMenu("信息查询");
    //创建菜单项
    JMenuItem jMenuItem02_00 = new JMenuItem("成绩查询");
    //创建菜单
    JMenu jMenu03 = new JMenu("系统退出");
    //创建菜单项
    JMenuItem jMenuItem03_00 = new JMenuItem("退出");

public MainJFrame(){
        super("学生成绩管理系统");
        //菜单栏
        jMenu01.add(jMenuItem01_00);
        jMenu01.addSeparator();
        jMenu01.add(jMenuItem01_01);
        jMenu01.addSeparator();
        jMenu01.add(jMenuItem01_02);
        jMenu02.add(jMenuItem02_00);
        jMenu03.add(jMenuItem03_00);
        jMenuBar.add(jMenu01);
        jMenuBar.add(jMenu02);
        jMenuBar.add(jMenu03);
        //在窗口上设置菜单栏
        this.setJMenuBar(jMenuBar);
        jMenuItem01_00.addActionListener(this);
        jMenuItem01_01.addActionListener(this);
        jMenuItem01_02.addActionListener(this);
        jMenuItem02_00.addActionListener(this);
        jMenuItem03_00.addActionListener(this);
```

```java
        //设置窗口
        this.setDefaultCloseOperation(JFrame.EXIT_ON_CLOSE);
        this.setSize(700,580);
        this.setLocationRelativeTo(null);
        this.setVisible(true);
    }
    public void actionPerformed(ActionEvent e){
        if(e.getSource() = = jMenuItem01_00){
        new AddJFrame();
        }
        if(e.getSource() = = jMenuItem01_01){
        new SnoInputJFrame();
        }
        if(e.getSource() = = jMenuItem01_02){
        new DeleteJFrame();
        }
        if(e.getSource() = = jMenuItem02_00){
        new SelectJFrame();
        }
        if(e.getSource() = = jMenuItem03_00){
        System.exit(0);
        }
    }

    public static void main(String args[]){
        new MainJFrame();
    }
}
```

(2) 信息添加窗口代码如下：

```java
import javax.swing.JButton;
import javax.swing.JCheckBox;
import javax.swing.JComboBox;
import javax.swing.JFrame;
import javax.swing.JLabel;
import javax.swing.JOptionPane;
import javax.swing.JPanel;
```

```java
import javax.swing.JRadioButton;
import javax.swing.ButtonGroup;
import javax.swing.JTextField;
import javax.swing.BorderFactory;
import java.awt.event.ActionEvent;
import java.awt.event.ActionListener;
import java.io.File;
import java.io.FileInputStream;
import java.io.FileWriter;
import java.util.Scanner;
import java.util.ArrayList;

public class AddJFrame extends JFrame implements ActionListener{
    private JPanel jp1;
    private JPanel jp2;
    private JLabel snoLabel;
    private JTextField snoText;
    private JLabel snameLabel;
    private JTextField snameText;
    private JLabel sexLabel;
    private JRadioButton sexJRB[];
    private JLabel cnameLabel;
    private JComboBox cnameJCB;
    String strCname = "==请选择==,Java程序设计,网络数据库技术与应用,JSP动态网页设计,网页设计与制作,图形图像处理";
    private JLabel pingShiLabel;
    private JTextField pingShiText;
    private JLabel guoChengLabel;
    private JTextField guoChengText;
    private JLabel zhongJieLabel;
    private JTextField zhongJieText;
    private JLabel zongHeLabel;
    private JTextField zongHeText;
    private JLabel zongHeJiBieLabel;
    private JTextField zongHeJiBieText;
    private JLabel loveLabel;
    private JCheckBox loveJCK[];
    private JButton addButton;
```

```java
        private JButton resetButton;
        private JButton returnButton;

        public AddJFrame(){
            this.setTitle("添加学生成绩信息");
            this.setLayout(null);
            this.setSize(520,450);
            jp1 = new JPanel(null);

            //第一行
            snoLabel = new JLabel("学号:");
            jp1.add(snoLabel);
            snoLabel.setBounds(10,20,40,30);

            snoText = new JTextField();
            jp1.add(snoText);
            snoText.setBounds(60,20,100,30);

            snameLabel = new JLabel("姓名:");
            jp1.add(snameLabel);
            snameLabel.setBounds(170,20,40,30);

            snameText = new JTextField();
            snameText.setBounds(220,20,100,30);
            jp1.add(snameText);

            sexLabel = new JLabel("性别:");
            sexLabel.setBounds(330,20,40,30);
            jp1.add(sexLabel);

            sexJRB = new JRadioButton[2];
            ButtonGroup bg = new ButtonGroup();
            sexJRB[0] = new JRadioButton("男");
            sexJRB[0].setSelected(true);
            jp1.add(sexJRB[0]);
            sexJRB[0].setBounds(380,20,40,30);
            sexJRB[1] = new JRadioButton("女");
            jp1.add(sexJRB[1]);
```

```java
sexJRB[1].setBounds(440,20,40,30);
bg.add(sexJRB[0]);
bg.add(sexJRB[1]);

//第二行
cnameLabel = new JLabel("课程名称：");
cnameLabel.setBounds(10,80,80,30);
jp1.add(cnameLabel);

cnameJCB = new JComboBox();
//将字符串分割成字符串数组 split
String[]cnameArray = strCname.split(",");
for(int i = 0;i<cnameArray.length;i++){
    cnameJCB.addItem(cnameArray[i]);
}
cnameJCB.setBounds(80,80,85,30);
jp1.add(cnameJCB);

pingShiLabel = new JLabel("平时成绩：");
pingShiLabel.setBounds(165,80,70,30);
jp1.add(pingShiLabel);

pingShiText = new JTextField();
pingShiText.setBounds(235,80,80,30);
jp1.add(pingShiText);

guoChengLabel = new JLabel("过程考核成绩：");
guoChengLabel.setBounds(315,80,100,30);
jp1.add(guoChengLabel);

guoChengText = new JTextField();
guoChengText.setBounds(410,80,80,30);
jp1.add(guoChengText);

//第三行
zhongJieLabel = new JLabel("终结考核成绩：");
zhongJieLabel.setBounds(10,140,100,30);
jp1.add(zhongJieLabel);
```

```java
zhongJieText = new JTextField();
zhongJieText.setBounds(105,140,80,30);
jp1.add(zhongJieText);

zongHeLabel = new JLabel("综合成绩:");
zongHeLabel.setBounds(185,140,80,30);
jp1.add(zongHeLabel);

zongHeText = new JTextField();
zongHeText.setBounds(255,140,80,30);
zongHeText.setEnabled(false);
jp1.add(zongHeText);

zongHeJiBieLabel = new JLabel("综合成绩级别:");
zongHeJiBieLabel.setBounds(335,140,95,30);
jp1.add(zongHeJiBieLabel);

zongHeJiBieText = new JTextField();
zongHeJiBieText.setBounds(430,140,60,30);
zongHeJiBieText.setEnabled(false);
jp1.add(zongHeJiBieText);

//第四行
loveLabel = new JLabel("爱好:");
loveLabel.setBounds(10,200,40,30);
jp1.add(loveLabel);

loveJCK = new JCheckBox[5];
loveJCK[0] = new JCheckBox("篮球");
loveJCK[0].setBounds(60,200,80,30);
jp1.add(loveJCK[0]);

loveJCK[1] = new JCheckBox("足球");
loveJCK[1].setBounds(150,200,80,30);
jp1.add(loveJCK[1]);

loveJCK[2] = new JCheckBox("乒乓球");
loveJCK[2].setBounds(240,200,80,30);
```

```java
jp1.add(loveJCK[2]);

loveJCK[3] = new JCheckBox("羽毛球");
loveJCK[3].setBounds(330,200,80,30);
jp1.add(loveJCK[3]);

loveJCK[4] = new JCheckBox("排球");
loveJCK[4].setBounds(420,200,60,30);
jp1.add(loveJCK[4]);

this.add(jp1);
jp1.setBounds(5,5,500,260);
jp1.setBorder(BorderFactory.createTitledBorder("学生成绩信息"));//设置标题

jp2 = new JPanel(null);
//将相关操作的按钮组件添加到 jp2 上
//设置窗口的大小

addButton = new JButton("添　加");
addButton.setBounds(40,40,100,40);
jp2.add(addButton);
addButton.addActionListener(this);

resetButton = new JButton("重　置");
resetButton.setBounds(200,40,100,40);
resetButton.addActionListener(this);
jp2.add(resetButton);

returnButton = new JButton("返　回");
returnButton.setBounds(360,40,100,40);
returnButton.addActionListener(this);
jp2.add(returnButton);

this.add(jp2);
jp2.setBounds(5,280,500,120);
 jp2.setBorder(BorderFactory.createTitledBorder("添加界面相关操作"));//设置标题
```

```java
        this.setLocationRelativeTo(null);
        this.setResizable(false);
        this.setDefaultCloseOperation(JFrame.DO_NOTHING_ON_CLOSE);
        this.setVisible(true);
    }

    public void actionPerformed(ActionEvent e){
        if(e.getSource() == addButton){
            if(find()){
                if(isNullForm()){
                    addButtonExe();
                }
            }
        }
        if(e.getSource() == resetButton){
            snoText.setText("");
            snameText.setText("");
            pingShiText.setText("");
            guoChengText.setText("");
            zhongJieText.setText("");
            zongHeText.setText("");
            zongHeJiBieText.setText("");
            cnameJCB.setSelectedIndex(0);
            for(int i = 0;i<5;i++){
                loveJCK[i].setSelected(false);
            }
        }
        if(e.getSource() == returnButton){
            this.dispose();
        }
    }

    public boolean find(){
        String input = snoText.getText().trim();
        if(input.equals("")){
            JOptionPane.showMessageDialog(this,"请输入学号!");
            return false;
        }
```

```java
try{
int number = Integer.parseInt(snoText.getText());
File file = new File("studentScore.txt");
if(file.exists()){
FileInputStream is = new FileInputStream(file);
Scanner sc = new Scanner(is);
while(sc.hasNext()){
int sno = sc.nextInt();
if(sno = = number)  {
snoText.setText("");
JOptionPane.showMessageDialog(this,"该学号已经存在,请重新输入!");
return false;
              }
String sname = sc.next();
char sex = sc.next().charAt(0);
String cname = sc.next();
byte pingShiScore = sc.nextByte();
byte guoChengScore = sc.nextByte();
byte zhongJieScore = sc.nextByte();
double zhongHeScore = sc.nextDouble();
String zongHeJiBie = sc.next();
String love = sc.next();
}
is.close();
return true;
}else{
JOptionPane.showMessageDialog(this,"文件不存在!");
return false;
}
}
catch(Exception e){
JOptionPane.showMessageDialog(this,e.toString());
return false;
}
}

public void addButtonExe(){
try{
```

```java
int sno = Integer.parseInt(snoText.getText());
String sname = snameText.getText();
char sex;
if(sexJRB[0].isSelected()){
sex = '男';
}else{
sex = '女';}
String cname = cnameJCB.getSelectedItem().toString();
byte pingShiScore = Byte.parseByte(pingShiText.getText());
byte guoChengScore = Byte.parseByte(guoChengText.getText());
byte zhongJieScore = Byte.parseByte(zhongJieText.getText());
double
zongHeScore = pingShiScore * 0.1 + guoChengScore * 0.4 + zhongJieScore * 0.5;
zongHeText.setText(String.valueOf(zongHeScore));
String  zongHeScoreJiBie = null;
if(zongHeScore<60){
zongHeScoreJiBie = "不及格";
}else if(zongHeScore<70){
zongHeScoreJiBie = "及格";
}else if(zongHeScore<80){
zongHeScoreJiBie = "中";
}
else if(zongHeScore<90){
zongHeScoreJiBie = "良";
}
else if(zongHeScore< = 100){
zongHeScoreJiBie = "优";
}else{
zongHeScoreJiBie = "初始成绩有误!";
}
zongHeJiBieText.setText(zongHeScoreJiBie);
String love = "";
if(loveJCK[0].isSelected())
love = love + "篮球" + "   ";
if(loveJCK[1].isSelected())
love = love + "足球" + "   ";
if(loveJCK[2].isSelected())
love = love + "乒乓球" + "   ";
```

```java
        if(loveJCK[3].isSelected())
        love = love + "羽毛球" + "   ";
        if(loveJCK[4].isSelected())
        love = love + "排球" + "   ";
        if(love.trim().length() = = 0)
        love = "未选爱好";

        StudentScore student = new StudentScore(sno,
sname, sex, cname, pingShiScore, guoChengScore, zhongJieScore, zongHeScore,
zongHeScoreJiBie);
        File file = new File("studentScore.txt");
        if(!file.exists()){//判断输出文件是否存在
        file.createNewFile();//创建文件
        }
        FileWriter wr = new FileWriter(file,true);

    wr.append(student.sno + "\t" + student.sname + "\t" + student.sex + "\t" +
student.cname + "\t" + student.pingShiScore + "\t" + student.guoChengScore + "\t"
+ student.zhongJieScore + " \t" + student.zongHeScore + " \t" + student.
zongHeScoreJiBie + "\t" + love + "\n");
        wr.flush();
        wr.close();
        JOptionPane.showMessageDialog(this,"保存成功!");
        }
        catch(Exception e){
        JOptionPane.showMessageDialog(this,e.toString());
        }
        }
        /**
        *判断组件的内容是否为空
        *
        *@return
        */
        private boolean isNullForm(){
        if(snoText.getText().trim().length() = = 0){
          JOptionPane.showMessageDialog(this,"【学号】不能为空!");
          snoText.requestFocus(true);
          return false;
```

```java
        }
        if(snameText.getText().trim().length() = = 0){
            JOptionPane.showMessageDialog(this,"【姓名】不能为空!");
            snameText.requestFocus(true);
            return false;
        }
        if(pingShiText.getText().trim().length() = = 0){
            JOptionPane.showMessageDialog(this,"【平时成绩】不能为空!");
            pingShiText.requestFocus(true);
            return false;
        }
        if(guoChengText.getText().trim().length() = = 0){
            JOptionPane.showMessageDialog(this,"【过程考核成绩】不能为空!");
            guoChengText.requestFocus(true);
            return false;
        }
        if(zhongJieText.getText().trim().length() = = 0){
            JOptionPane.showMessageDialog(this,"【终结考核成绩】不能为空!");
            zhongJieText.requestFocus(true);
            return false;
        }
        return true;
    }
    public static void main(String[ ]args){
        new AddJFrame();
    }
}


class StudentScore{
    int sno;//定义变量sno,用于表示学生学号
    String sname;//定义变量sname,用于表示学生姓名
    char sex;//定义变量sex,用于表示学生性别
    String cname;//定义变量Cname,用于表示课程名称
    byte pingShiScore;//定义变量pingShiScore,用于表示学生平时成绩
    byte guoChengScore;//定义变量guoChengScore,用于表示学生过程考核成绩
    byte zhongJieScore;//定义变量zhongJieScore,用于表示学生终结考核成绩
    double zongHeScore;//定义变量zongHeScore,用于表示学生综合成绩
    //定义变量zongHeScoreJiBie,用于表示学生综合成绩的等级
```

```java
        String zongHeScoreJiBie;

    public StudentScore(){//空构造函数
    }

     public StudentScore(int sno, String sname, char sex, String cname, byte pingShiScore, byte guoChengScore, byte zhongJieScore, double zongHeScore, String zongHeScoreJiBie){//带参数的构造函数
        this.sno = sno;
        this.sname = sname;
        this.sex = sex;
        this.cname = cname;
        this.pingShiScore = pingShiScore;
        this.guoChengScore = guoChengScore;
        this.zhongJieScore = zhongJieScore;
        this.zongHeScore = zongHeScore;
        this.zongHeScoreJiBie = zongHeScoreJiBie;
    }

    public int getSno(){
        return sno;
    }
    public void setSno(int sno){
        this.sno = sno;
    }
    public String getSname(){
        return sname;
    }
    public void setSname(String sname){
        this.sname = sname;
    }
    public char getSex(){
        return sex;
    }
    public void setSex(char sex){
        this.sex = sex;
    }
    public String getCname(){
```

```java
            return cname;
        }
        public void setCname(String cname){
            this.cname = cname;
        }
        public byte getPingShiScore(){
            return pingShiScore;
        }
        public void setPingShiScore(byte pingShiScore){
            this.pingShiScore = pingShiScore;
        }
        public byte getGuoChengScore(){
            return guoChengScore;
        }
        public void setGuoChengScore(byte guoChengScore){
            this.guoChengScore = guoChengScore;
        }
        public byte getZhongJieScore(){
            return zhongJieScore;
        }
        public void setZhongJieScore(byte zhongJieScore){
            this.zhongJieScore = zhongJieScore;
        }
        public double getZongHeScore(){
            return zongHeScore;
        }
        public void setZongHeScore(double zongHeScore){
            this.zongHeScore = zongHeScore;
        }
        public String getZongHeScoreJiBie(){
            return zongHeScoreJiBie;
        }
        public void setZongHeScoreJiBie(String zongHeScoreJiBie){
            this.zongHeScoreJiBie = zongHeScoreJiBie;
        }
}
```

如图 6-75 至图 6-77 所示。

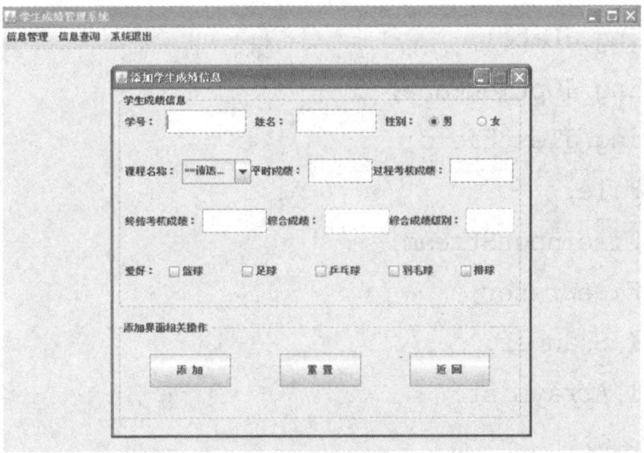

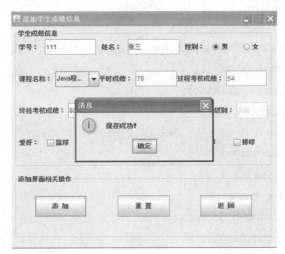

图 6-75

图 6-76

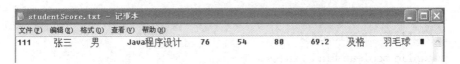

图 6-77

2. 实现信息修改功能

(1) 输入学号对话框窗口代码如下：

```
import java.awt.event.ActionEvent;
import java.awt.event.ActionListener;
import java.awt.event.WindowAdapter;
import java.awt.event.WindowEvent;
import javax.swing.JButton;
import javax.swing.JDialog;
import javax.swing.JFrame;
```

```java
import javax.swing.JLabel;
import javax.swing.JOptionPane;
import javax.swing.JTextField;
import java.io.File;
import java.io.FileInputStream;
import java.io.FileWriter;
import java.util.Scanner;
import java.util.ArrayList;

/**
 *学号输入界面
 */
public class SnoInputJFrame extends JDialog implements ActionListener{
    private JLabel lbLTitle;
    private JTextField txtNo;
    private JButton btnNext;
    public int sno;
    public String sname;
    public char sex;
    public String cname;
    public byte pingShiScore;
    public byte guoChengScore;
    public byte zhongJieScore;
    public double zhongHeScore;
    public String zongHeJiBie;
    public String love;
    public boolean find = false;
    public static int fileindex;
    public static int xuehao;
    public SnoInputJFrame(){
        this.setTitle("请输入学生学号!");
        this.setLayout(null);
        lbLTitle = new JLabel("请输入学号:");
        lbLTitle.setBounds(10,10,200,30);
        this.add(lbLTitle);
        txtNo = new JTextField();
        txtNo.setBounds(5,50,200,20);
        this.add(txtNo);
```

```java
        btnNext = new JButton("下一步");
        btnNext.setBounds(60,90,80,20);
        btnNext.addActionListener(this);
        this.add(btnNext);
        this.setDefaultCloseOperation(JFrame.DO_NOTHING_ON_CLOSE);
        //匿名内部类
        this.addWindowListener(new WindowAdapter(){
           public void windowClosing(WindowEvent arg0){
               dispose();
           }
        });
        this.setSize(220,150);
        this.setLocationRelativeTo(null);
        this.setVisible(true);
    }

    public void actionPerformed(ActionEvent e){
    try{
        fileindex = 0;
        String input = txtNo.getText().trim();
        if(input.equals("")){
        JOptionPane.showMessageDialog(this,"请输入学号!");
        return;
        }
        int number = Integer.parseInt(txtNo.getText());
        File file = new File("studentScore.txt");
        if(file.exists()){
        FileInputStream is = new FileInputStream(file);
        Scanner sc = new Scanner(is);
        while(sc.hasNext()){
        sno = sc.nextInt();
        if(sno = = number)   {
        xuehao = sno;
        find = true;
        this.dispose();
        new UpdateJFrame();
        break;
        }
```

```
                sname = sc.next();
                sex = sc.next().charAt(0);
                cname = sc.next();
                pingShiScore = sc.nextByte();
                guoChengScore = sc.nextByte();
                zhongJieScore = sc.nextByte();
                zhongHeScore = sc.nextDouble();
                zongHeJiBie = sc.next();
                love = sc.next();
                fileindex++;
                }
                is.close();
                if(!find){
                JOptionPane.showMessageDialog(this,"该学号未找到!");
                txtNo.setText("");
                }
                }else{
                JOptionPane.showMessageDialog(this,"记录文件 studentScore.txt 不存在!");
                }
                }
                catch(Exception e1){
                    JOptionPane.showMessageDialog(this,e1.toString());
        }
        }
    }
```

如图 6-78、图 6-79 所示。

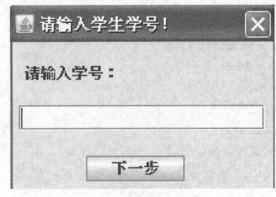

图 6-78

图 6-79

（2）信息修改窗口代码如下：

```
import java.awt.event.ActionEvent;
import java.awt.event.ActionListener;
import javax.swing.JButton;
import javax.swing.JCheckBox;
import javax.swing.JComboBox;
import javax.swing.JFrame;
import javax.swing.JLabel;
import javax.swing.JOptionPane;
import javax.swing.JPanel;
import javax.swing.JRadioButton;
import javax.swing.ButtonGroup;
import javax.swing.JTextField;
import javax.swing.BorderFactory;
import java.io.File;
import java.io.FileInputStream;
import java.io.FileWriter;
import java.util.Scanner;
import java.util.ArrayList;

public class UpdateJFrame extends JFrame implements ActionListener{
    private JPanel jp1;
    private JPanel jp2;
    private JLabel snoLabel;
    private JTextField snoText;
    private JLabel snameLabel;
    private JTextField snameText;
    private JLabel sexLabel;
```

```java
        private JRadioButton sexJRB[];
        private JLabel cnameLabel;
        private JComboBox cnameJCB;
        String strCname = "===请选择==,Java程序设计,网络数据库技术与应用,
JSP动态网页设计,网页设计与制作,图形图像处理";
        private JLabel pingShiLabel;
        private JTextField pingShiText;
        private JLabel guoChengLabel;
        private JTextField guoChengText;
        private JLabel zhongJieLabel;
        private JTextField zhongJieText;
        private JLabel zongHeLabel;
        private JTextField zongHeText;
        private JLabel zongHeJiBieLabel;
        private JTextField zongHeJiBieText;
        private JLabel loveLabel;
        private JCheckBox loveJCK[];
        private JButton updateButton;
        private JButton resetButton;
        private JButton returnButton;
        private ArrayList<StudentScore> arr = new ArrayList<StudentScore>();
        public ArrayList<String> happy = new ArrayList<String>();
        int xuehao1;
        SnoInputJFrame s;
        public UpdateJFrame(){
            this.setTitle("修改学生成绩信息");
            this.setLayout(null);
            this.setSize(520,450);
            jp1 = new JPanel(null);
            s = new SnoInputJFrame();
            xuehao1 = s.xuehao;
            s.dispose();
            snoLabel = new JLabel("学号:");
            jp1.add(snoLabel);
            snoLabel.setBounds(10,20,40,30);

            snoText = new JTextField();
            jp1.add(snoText);
```

```java
snoText.setBounds(60,20,100,30);
snoText.setText(String.valueOf(xuehao1));

snameLabel = new JLabel("姓名:");
jp1.add(snameLabel);
snameLabel.setBounds(170,20,40,30);

snameText = new JTextField();
snameText.setBounds(220,20,100,30);
jp1.add(snameText);

sexLabel = new JLabel("性别:");
sexLabel.setBounds(330,20,40,30);
jp1.add(sexLabel);

sexJRB = new JRadioButton[2];
ButtonGroup bg = new ButtonGroup();
sexJRB[0] = new JRadioButton("男");
sexJRB[0].setSelected(true);
jp1.add(sexJRB[0]);
sexJRB[0].setBounds(380,20,40,30);
sexJRB[1] = new JRadioButton("女");
jp1.add(sexJRB[1]);
sexJRB[1].setBounds(440,20,40,30);
bg.add(sexJRB[0]);
bg.add(sexJRB[1]);

//第二行
cnameLabel = new JLabel("课程名称:");
cnameLabel.setBounds(10,80,80,30);
jp1.add(cnameLabel);

cnameJCB = new JComboBox();
//将字符串分割成字符串数组split
String[]cnameArray = strCname.split(",");
for(int i = 0;i<cnameArray.length;i++){
    cnameJCB.addItem(cnameArray[i]);
}
```

```java
cnameJCB.setBounds(80,80,85,30);
jp1.add(cnameJCB);

pingShiLabel = new JLabel("平时成绩:");
pingShiLabel.setBounds(165,80,70,30);
jp1.add(pingShiLabel);

pingShiText = new JTextField();
pingShiText.setBounds(235,80,80,30);
jp1.add(pingShiText);

guoChengLabel = new JLabel("过程考核成绩:");
guoChengLabel.setBounds(315,80,100,30);
jp1.add(guoChengLabel);

guoChengText = new JTextField();
guoChengText.setBounds(410,80,80,30);
jp1.add(guoChengText);

//第三行
zhongJieLabel = new JLabel("终结考核成绩:");
zhongJieLabel.setBounds(10,140,100,30);
jp1.add(zhongJieLabel);

zhongJieText = new JTextField();
zhongJieText.setBounds(105,140,80,30);
jp1.add(zhongJieText);

zongHeLabel = new JLabel("综合成绩:");
zongHeLabel.setBounds(185,140,80,30);
jp1.add(zongHeLabel);

zongHeText = new JTextField();
zongHeText.setBounds(255,140,80,30);
zongHeText.setEnabled(false);
jp1.add(zongHeText);

zongHeJiBieLabel = new JLabel("综合成绩级别:");
```

```java
zongHeJiBieLabel.setBounds(335,140,95,30);
jp1.add(zongHeJiBieLabel);

zongHeJiBieText = new JTextField();
zongHeJiBieText.setBounds(430,140,60,30);
zongHeJiBieText.setEnabled(false);
jp1.add(zongHeJiBieText);

//第四行
loveLabel = new JLabel("爱好:");
loveLabel.setBounds(10,200,40,30);
jp1.add(loveLabel);

loveJCK = new JCheckBox[5];
loveJCK[0] = new JCheckBox("篮球");
loveJCK[0].setBounds(60,200,80,30);
jp1.add(loveJCK[0]);

loveJCK[1] = new JCheckBox("足球");
loveJCK[1].setBounds(150,200,80,30);
jp1.add(loveJCK[1]);

loveJCK[2] = new JCheckBox("乒乓球");
loveJCK[2].setBounds(240,200,80,30);
jp1.add(loveJCK[2]);

loveJCK[3] = new JCheckBox("羽毛球");
loveJCK[3].setBounds(330,200,80,30);
jp1.add(loveJCK[3]);

loveJCK[4] = new JCheckBox("排球");
loveJCK[4].setBounds(420,200,60,30);
jp1.add(loveJCK[4]);

this.add(jp1);
jp1.setBounds(5,5,500,260);
jp1.setBorder(BorderFactory.createTitledBorder("学生成绩信息"));//设
置标题
```

```java
            jp2 = new JPanel(null);
            //将相关操作的按钮组件添加到 jp2 上
            //设置窗口的大小
            updateButton = new JButton("修　改");
            updateButton.setBounds(40,40,100,40);
            jp2.add(updateButton);
            updateButton.addActionListener(this);


            resetButton = new JButton("重　置");
            resetButton.setBounds(200,40,100,40);
            resetButton.addActionListener(this);
            jp2.add(resetButton);


            returnButton = new JButton("返　回");
            returnButton.setBounds(360,40,100,40);
            returnButton.addActionListener(this);
            jp2.add(returnButton);


            this.add(jp2);
            jp2.setBounds(5,280,500,120);
            jp2.setBorder(BorderFactory.createTitledBorder("添加界面相关操作"));//
设置标题


            this.setLocationRelativeTo(null);
            this.setResizable(false);
            this.setDefaultCloseOperation(JFrame.EXIT_ON_CLOSE);
            this.setVisible(true);
        }

        public void actionPerformed(ActionEvent e){
            if(e.getSource() = = updateButton){
                if(isNullForm()){
                    read();
                    update();
                    save();
                }
            }
            if(e.getSource() = = resetButton){
```

```java
                    snoText.setText("");
                    snameText.setText("");
                    pingShiText.setText("");
                    guoChengText.setText("");
                    zhongJieText.setText("");
                    zongHeText.setText("");
                    zongHeJiBieText.setText("");
                    cnameJCB.setSelectedIndex(0);
                    for(int i = 0;i<5;i++){
                    loveJCK[i].setSelected(false);
                    }
                }
                if(e.getSource() == returnButton){
                    this.dispose();
                }
            }

            public void read()  {
                try{
                    File file = new File("studentScore.txt");
                    if(file.exists()){
                        FileInputStream is = new FileInputStream(file);
                        arr.clear();
                        happy.clear();
                        Scanner sc = new Scanner(is);
                        while(sc.hasNext()){
                            int sno = sc.nextInt();
                            String sname = sc.next();
                            char sex = sc.next().charAt(0);
                            String cname = sc.next();
                            byte pingShiScore = sc.nextByte();
                            byte guoChengScore = sc.nextByte();
                            byte zhongJieScore = sc.nextByte();
                            double zongHeScore = sc.nextDouble();
                            String zongHeScoreJiBie = sc.next();
                            String love = sc.next();
                            StudentScore student = new StudentScore(sno, sname, sex, cname, pingShiScore,guoChengScore,zhongJieScore,zongHeScore,zongHeScoreJiBie);
```

```java
            arr.add(student);
            happy.add(love);
            }
            is.close();
        }else{
            JOptionPane.showMessageDialog(this,"文件不存在!");
        }
    }
    catch(Exception e){
        JOptionPane.showMessageDialog(this,e.toString());
    }
}

public void update(){
    try{
        s = new SnoInputJFrame();
        int index = s.fileindex;
        s.dispose();
        int sno = Integer.parseInt(snoText.getText());
        String sname = snameText.getText();
        char sex;
        if(sexJRB[0].isSelected()){
            sex = '男';
        }else{
            sex = '女';}
        String cname = cnameJCB.getSelectedItem().toString();
        byte pingShiScore = Byte.parseByte(pingShiText.getText());
        byte guoChengScore = Byte.parseByte(guoChengText.getText());
        byte zhongJieScore = Byte.parseByte(zhongJieText.getText());
        double zongHeScore = pingShiScore * 0.1 + guoChengScore * 0.4 + zhongJieScore * 0.5;
        zongHeText.setText(String.valueOf(zongHeScore));
        String zongHeScoreJiBie = null;
        if(zongHeScore<60){
            zongHeScoreJiBie = "不及格";
        }else if(zongHeScore<70){
            zongHeScoreJiBie = "及格";
        }else if(zongHeScore<80){
```

```java
zongHeScoreJiBie = "中";
}
else if(zongHeScore<90){
zongHeScoreJiBie = "良";
}
else if(zongHeScore<=100){
zongHeScoreJiBie = "优";
}else{
zongHeScoreJiBie = "初始成绩有误!";
}
zongHeJiBieText.setText(zongHeScoreJiBie);
String love = "";
if(loveJCK[0].isSelected())
love = love + "篮球" + "  ";
if(loveJCK[1].isSelected())
love = love + "足球" + "  ";
if(loveJCK[2].isSelected())
love = love + "乒乓球" + "  ";
if(loveJCK[3].isSelected())
love = love + "羽毛球" + "  ";
if(loveJCK[4].isSelected())
love = love + "排球" + "  ";
if(love.trim().length() == 0)
love = "未选爱好";
    StudentScore student = new StudentScore(sno, sname, sex, cname, pingShiScore, guoChengScore, zhongJieScore, zongHeScore, zongHeScoreJiBie);
    arr.set(index, student);
    happy.set(index, love);
}
catch(Exception e){
JOptionPane.showMessageDialog(this, e.toString());
}
}

public void save()  {
try{
FileWriter wr = new FileWriter("studentScore.txt");
for(int i = 0; i<arr.size(); i++){
```

```java
        wr.write(arr.get(i).sno + "\t" + arr.get(i).sname + "\t" + arr.get(i).sex + "\
t" + arr.get(i).cname + "\t" + arr.get(i).pingShiScore + "\t" + arr.get(i).
guoChengScore + "\t" + arr.get(i).zhongJieScore + "\t" + arr.get(i).zongHeScore
 + "\t" + arr.get(i).zongHeScoreJiBie + "\t" + happy.get(i) + "\n");
        }
        wr.close();
        s.dispose();
        JOptionPane.showMessageDialog(this,"学生成绩信息修改成功!");
    }
    catch(Exception e){
        JOptionPane.showMessageDialog(this,e.toString());
    }
}
/**
 * 判断组件的内容是否为空
 *
 * @return
 */
private boolean isNullForm(){
    if(snoText.getText().trim().length() == 0){
        JOptionPane.showMessageDialog(this,"【学号】不能为空!");
        snoText.requestFocus(true);
        return false;
    }
    if(snameText.getText().trim().length() == 0){
        JOptionPane.showMessageDialog(this,"【姓名】不能为空!");
        snameText.requestFocus(true);
        return false;
    }
    if(pingShiText.getText().trim().length() == 0){
        JOptionPane.showMessageDialog(this,"【平时成绩】不能为空!");
        pingShiText.requestFocus(true);
        return false;
    }
    if(guoChengText.getText().trim().length() == 0){
        JOptionPane.showMessageDialog(this,"【过程考核成绩】不能为空!");
        guoChengText.requestFocus(true);
        return false;
```

```
            }
            if(zhongJieText.getText().trim().length() = = 0){
            JOptionPane.showMessageDialog(this,"【终结考核成绩】不能为空!");
                zhongJieText.requestFocus(true);
                return false;
            }
            return true;
        }

        public static void main(String[ ]args){
            new UpdateJFrame();
        }
    }
```

如图 6-80、图 6-81 所示。

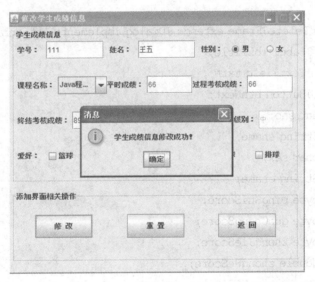

图 6-80

图 6-81

3. 实现信息删除功能

信息删除窗口代码如下:

```java
import java.awt.event.ActionEvent;
import java.awt.event.ActionListener;
import java.awt.event.WindowAdapter;
import java.awt.event.WindowEvent;
import javax.swing.JButton;
import javax.swing.JDialog;
import javax.swing.JFrame;
import javax.swing.JLabel;
import javax.swing.JOptionPane;
import javax.swing.JTextField;
import java.io.File;
import java.io.FileInputStream;
import java.io.FileWriter;
import java.util.Scanner;
import java.util.ArrayList;

public class DeleteJFrame extends JDialog implements ActionListener{
    private JLabel lbLTitle;
    private JTextField txtNo;
    private JButton btnNext;
    public int sno;
    public String sname;
    public char sex;
    public String cname;
    public byte pingShiScore;
    public byte guoChengScore;
    public byte zhongJieScore;
    public double zhongHeScore;
    public String zongHeJiBie;
    public String love;
    public boolean find = false;
    public static int fileindex;
    private ArrayList<StudentScore> arr = new ArrayList<StudentScore>();
    public ArrayList<String> happy = new ArrayList<String>();

    public DeleteJFrame(){
        this.setTitle("请输入学生学号!");
        this.setLayout(null);
```

```java
lbLTitle = new JLabel("请输入学号:");
lbLTitle.setBounds(10,10,200,30);
this.add(lbLTitle);
txtNo = new JTextField();
txtNo.setBounds(5,50,200,20);
this.add(txtNo);
btnNext = new JButton("下一步");
btnNext.setBounds(60,90,80,20);
btnNext.addActionListener(this);
this.add(btnNext);
this.setDefaultCloseOperation(JFrame.DO_NOTHING_ON_CLOSE);
//匿名内部类
this.addWindowListener(new WindowAdapter(){
    public void windowClosing(WindowEvent arg0){
        dispose();
    }
});
this.setSize(220,150);
this.setLocationRelativeTo(null);
this.setVisible(true);
}

public void actionPerformed(ActionEvent e){
    try{
        String input = txtNo.getText().trim();
        if(input.equals("")){
            JOptionPane.showMessageDialog(this,"请输入学号!");
            return;
        }
        fileindex = 0;
        int number = Integer.parseInt(txtNo.getText());
        File file = new File("studentScore.txt");
        if(file.exists()){
            FileInputStream is = new FileInputStream(file);
            Scanner sc = new Scanner(is);
            while(sc.hasNext()){
                sno = sc.nextInt();
                if(sno = = number)    {
```

```java
                find = true;
                int n = JOptionPane.showConfirmDialog(this,"您确认删除该学生的成绩信息吗?","确认对话框", JOptionPane.YES_NO_OPTION, JOptionPane.QUESTION_MESSAGE);
                if(n == JOptionPane.YES_OPTION){
                read();
                arr.remove(fileindex);
                happy.remove(fileindex);
                save();
                }
                if(n == JOptionPane.NO_OPTION){
                this.dispose();
                }
                break;
                }
                sname = sc.next();
                sex = sc.next().charAt(0);
                cname = sc.next();
                pingShiScore = sc.nextByte();
                guoChengScore = sc.nextByte();
                zhongJieScore = sc.nextByte();
                zhongHeScore = sc.nextDouble();
                zongHeJiBie = sc.next();
                love = sc.next();
                fileindex++;
                }
                is.close();
                if(!find){
                JOptionPane.showMessageDialog(this,"该学号未找到!");
                txtNo.setText("");
                return;
                }
                }else{
                JOptionPane.showMessageDialog(this,"记录文件 studentScore.txt 不存在!");
                }
                }
                catch(Exception e1){
```

```java
            JOptionPane.showMessageDialog(this,e1.toString());
        }
    }

            public void read()  {
            try{
            File file = new File("studentScore.txt");
            if(file.exists()){
            FileInputStream is = new FileInputStream(file);
            arr.clear();
            happy.clear();
            Scanner sc = new Scanner(is);
            while(sc.hasNext()){
            int sno = sc.nextInt();
            String sname = sc.next();
            char sex = sc.next().charAt(0);
            String cname = sc.next();
            byte pingShiScore = sc.nextByte();
            byte guoChengScore = sc.nextByte();
            byte zhongJieScore = sc.nextByte();
            double zongHeScore = sc.nextDouble();
            String zongHeScoreJiBie = sc.next();
            String love = sc.next();
    StudentScore student = new StudentScore(sno,
sname, sex, cname, pingShiScore, guoChengScore, zhongJieScore, zongHeScore,
zongHeScoreJiBie);
            arr.add(student);
            happy.add(love);
            }
            is.close();
            }else{
            JOptionPane.showMessageDialog(this,"文件不存在!");
            }
            }
            catch(Exception e){
            JOptionPane.showMessageDialog(this,e.toString());
            }
        }
```

```java
public void save()  {
try{
FileWriter wr = new FileWriter("studentScore.txt");
for(int i = 0;i<arr.size();i++){
    wr.write(arr.get(i).sno + "\t" + arr.get(i).sname + "\t" + arr.get(i).sex + "\t" + arr.get(i).cname + "\t" + arr.get(i).pingShiScore + "\t" + arr.get(i).guoChengScore + "\t" + arr.get(i).zhongJieScore + "\t" + arr.get(i).zongHeScore + "\t" + arr.get(i).zongHeScoreJiBie + "\t" + happy.get(i) + "\n");
}
wr.close();
txtNo.setText("");
JOptionPane.showMessageDialog(this,"成功删除学生成绩信息!");
this.dispose();
}
catch(Exception e){
JOptionPane.showMessageDialog(this,e.toString());
}
}
```

如图 6-82 至图 6-84 所示。

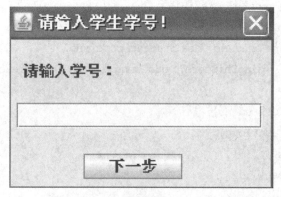

图 6-82

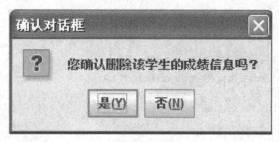

图 6-83

图 6-84

任务拓展

设计完成自助银行服务系统的开户、存/取款、销户图形界面,并实现相应功能,要求如下:

(1) 用户的账户相关信息包括用户名、账号、密码、身份证号、余额,将上述信息保存到记事本文件 bank.txt。

(2) 当执行开户操作时,在记事本文件 bank.txt 中新增一条用户信息的记录。

(3) 当执行存/取款操作时,应根据用户的账号找到用户的余额信息,然后对余额进行相应的加/减操作。

(4) 当执行销户操作时,根据用户的账号从记事本文件 bank.txt 中删除相应的用户信息。

任务5 学生成绩管理系统信息查询窗口设计

任务描述及分析

小明要设计实现学生成绩管理系统信息查询窗口,信息查询窗口要求如下:

(1) 用户可以按照学号、姓名、班级等不同的关键字进行成绩信息的查询,并且选择了相应的关键字后,可以在对应的文本框输入要查询关键字对应的值。例如,选择按照学号关键字进行查询,可以在对应的文本框中输入要查询的学生的学号。

(2) 如果用户选择了关键字类型,但没有在对应的文本框输入关键字对应的值,则查询所有的成绩信息,并按照该关键字分组显示。例如,选择按照班级关键字进行查询,且没有在文本框输入对应的班级,就查询所有的学生成绩,并按班级分组显示。

(3) 查询到的信息显示在表格中。

要完成这个工作任务,第一步,我们需要掌握表格的常用属性和方法;第二步,利用所学组件知识,创建信息查询窗口及对应的组件,并合理布局;第三步,利用所学事件处理相关知识,完成"信息查询窗口"的事件处理程序设计。此外,为了扩展知识面,我们在该工作任务中还将补充讲解常用的其他容器(JTabbedPane、JSplitPane)、列表框、文件选择框、滚动条。

相关知识

6.5.1 滚动条 JScrollPane

在一般的图形界面中如果显示的区域不够大了,往往会出现滚动条以方便用户浏览,在 Swing 中 JScrollPane 的主要功能就是为显示的内容加入水平滚动条。JScrollPane 主要由 JViewPort 和 JScrollBar 两部分组成,前者主要是显示一个巨型的区域让用户浏览,而后者主要是形成水平或垂直的滚动条。JScrollPane 类的主要方法及常量如下所示:

(1) 常用方法。

- public JScrollPane(Component view)构造方法,将指定的组件加入滚动条,根据大小显示水平或垂直滚动条。
- public JScrollPane(Component view, int vsbPolicy, int hsbPolicy)构造方法,将指定的组件加入滚动条,根据需要设置是否显示水平或垂直滚动条。
- public void setHorizontalScrollBarPolicy(int policy)该方法用于设置水平滚动条的显示策略。
- public void setVerticalScrollBarPolicy(int policy)该方法用于设置垂直滚动条的显示策略。

(2) 常量。

- static final int HORIZONTAL_SCROLLABR_ALWAYS 表示始终显示水平滚动条的常量。
- static final int HORIZONTAL_SCROLLABR_NEVER 表示任何情况下都不显示水平滚动条的常量。
- static final int HORIZONTAL_SCROLLABR_AS_NEEDED 表示根据自身需要显示水平滚动条的常量。
- static final int VERTICAL_SCROLLABR_ALWAYS 表示始终显示垂直滚动条的常量。
- static final int VERTICAL_SCROLLABR_NEVER 表示任何情况下都不显示垂直滚动条的常量。
- static final int VERTICAL_SCROLLABR_AS_NEEDED 表示根据自身需要显示垂直滚动条的常量。

【例 6.48】利用 JTextArea 和 JScrollPane 组件设计一个带滚动条的编辑器(要求始终显示水平和垂直滚动条)。

```
import javax.swing.JFrame;
import javax.swing.JScrollPane;
import javax.swing.JTextArea;
public class JScrollPane1 extends JFrame{
```

```java
    public JScrollPane1(){//定义一个构造方法
        //创建文本区域组件
        JTextArea JTA1 = new JTextArea(20,50);
        //由于滚动条必须在水平和垂直方向始终显示,所以选择构造方法
JScrollPane(Component,vsbPolicy,hsbPolicy)
        //创建JScrollPane()面板对象,并将文本域对象添加到面板中
         JScrollPane JSP1 = new JScrollPane(JTA1,JScrollPane.VERTICAL_SCROLLBAR_ALWAYS,JScrollPane.HORIZONTAL_SCROLLBAR_ALWAYS);
        //将该面板添加到该容器中
        add(JSP1);
        //设置窗体的外部特性
        setTitle("带滚动条的文字编辑器");//设置窗口的标题文字
        setSize(400,400);//设置窗口的大小
        setLocationRelativeTo(null);//窗口居中显示
        setVisible(true);//设置可视化
        //设置窗口的关闭方式
      setDefaultCloseOperation(JFrame.EXIT_ON_CLOSE);
    }
    public static void main(String[]args){
        //TODO Auto-generated method stub
        JScrollPane1 jp = new JScrollPane1();
    }
  }
```

如图6-85所示。

图6-85

6.5.2 表格 JTable

1. 使用 JTable 构造函数创建表格

表格组件提供了以行和列的形式显示数据的视图,在程序开发中是一个非常重要的组件,尤其在需要将一堆数据有条理地展现给用户的时候,表格设计更能显示出它的重要性。在 Swing 中可以通过 JTable 组件非常轻松地构造出所需要的表格,JTable 类的常用构造方法如下:

- JTable(Object[][]rowData,Object[]columnNames)构造一个 JTable 表格来显示二维数组 rowData 中的值,其列标题(列名)为一维数组 columnNames 中的值。
- JTable(Vector rowData, Vector columnNames)构造一个 JTable 表格,通过 Vector 设置显示数据和列标题,rowData 中是要显示的数据,columnNames 中是列标题。
- JTable(TableModel dm)使用 TableModel 构造一个 JTable 表格。

【例6.49】使用带有二维数组和一维数组参数的构造方法创建表格。

```java
import javax.swing.JFrame;
import javax.swing.JTable;
public class JTable1 extends JFrame{
    public JTable1(){//定义一个构造方法
    //表格标题
    String[]titles = {"学号","姓名","年龄","性别","Java 成绩","是否及格"};
    //表格显示内容
    Object[][]cj = {
    {20180101,"小明",12,"男",72,true},
    {20180102,"小红",13,"女",79,true},
    {20180103,"小刚",12,"男",80,true}
    };//定义数据
    //建立表格
    JTable JT = new JTable(cj,titles);
        add(JT);
        //设置窗体的外部特性
        setTitle("表格");//设置窗口的标题文字
        setSize(300,200);//设置窗口的大小
        setLocationRelativeTo(null);//窗口居中显示
        setVisible(true);//设置可视化
        //设置窗口的关闭方式
        setDefaultCloseOperation(JFrame.EXIT_ON_CLOSE);
    }
    public static void main(String[]args){
        //TODO Auto-generated method stub
```

```
            JTable1 jt = new JTable1();
    }
}
```

如图 6-86 所示。

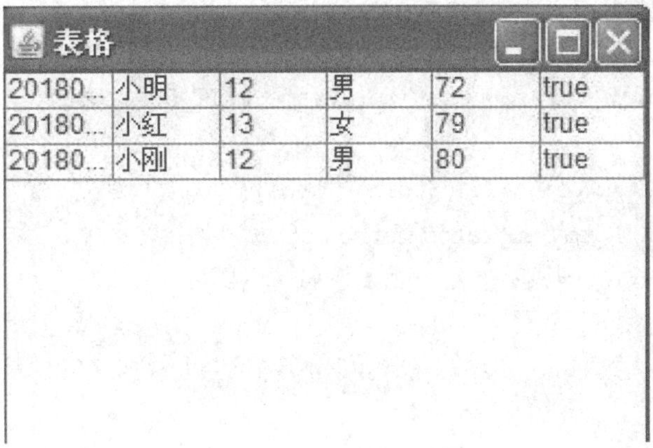

图 6-86

在上例中,我们会发现表格的标题行显示不出来。如果想要完整显示,需要将表格加入滚动条,如下例所示。

【例 6.50】使用带有二维数组和一维数组参数的构造方法创建表格,并将表格标题显示出来(将表格加入滚动条)。

```
import javax.swing.JFrame;
import javax.swing.JTable;
import javax.swing.JScrollPane;
public class JTable2 extends JFrame{
    public JTable2(){//定义一个构造方法
    //表格标题
    String[]titles = {"学号","姓名","年龄","性别","Java 成绩","是否及格"};
    //表格显示内容
    Object[][]cj = {
    {20180101,"小明",12,"男",72,true},
    {20180102,"小红",13,"女",79,true},
    {20180103,"小刚",12,"男",80,true}
    };//定义数据
    //建立表格
    JTable JT = new JTable(cj,titles);
    JScrollPane JSP1 = new JScrollPane(JT);
```

```
        add(JSP1);
        //设置窗体的外部特性
        setTitle("表格");//设置窗口的标题文字
        setSize(300,200);//设置窗口的大小
        setLocationRelativeTo(null);//窗口居中显示
        setVisible(true);//设置可视化
        //设置窗口的关闭方式
        setDefaultCloseOperation(JFrame.EXIT_ON_CLOSE);
    }
    public static void main(String[]args){
        //TODO Auto-generated method stub
        JTable2 jt = new JTable2();
    }
}
```

如图 6-87 所示。

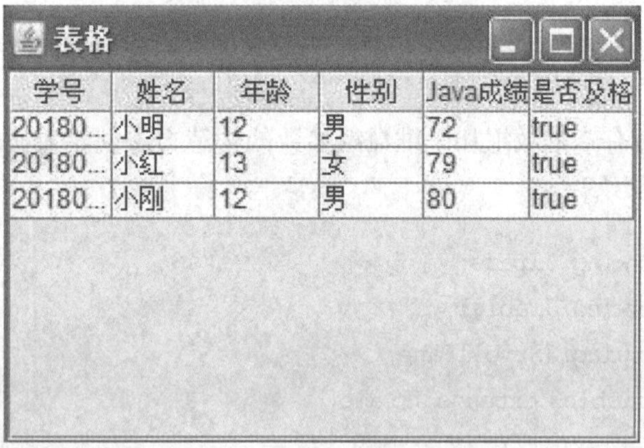

图 6-87

【例 6.51】 使用带有 Vector 参数的构造方法创建表格。

```
import javax.swing.JFrame;
import javax.swing.JTable;
import javax.swing.JScrollPane;
import java.util.Vector;
public class JTable3 extends JFrame{
    public JTable3(){//定义一个构造方法
        //表格标题
        Vector<String>title1 = new Vector<String>();
```

```java
title1.add("学号");
title1.add("姓名");
title1.add("年龄");
title1.add("性别");
title1.add("Java 成绩");
title1.add("是否及格");
//表格显示内容
Vector<Vector<Object>> hang = new Vector<Vector<Object>>();
//第一行数据
Vector<Object> hang1 = new Vector<Object>();
hang1.add(20180101);
hang1.add("小明");
hang1.add(12);
hang1.add("男");
hang1.add(72);
hang1.add(true);
//第二行数据
Vector<Object> hang2 = new Vector<Object>();
hang2.add(20180102);
hang2.add("小红");
hang2.add(13);
hang2.add("女");
hang2.add(79);
hang2.add(true);
//第三行数据
Vector<Object> hang3 = new Vector<Object>();
hang3.add(20180103);
hang3.add("小刚");
hang3.add(12);
hang3.add("男");
hang3.add(80);
hang3.add(true);
//将各行加入 Vector 对象 hang 中
hang.add(hang1);
hang.add(hang2);
hang.add(hang3);
//建立表格
JTable JT = new JTable(hang,title1);
```

```
        JScrollPane JSP1 = new JScrollPane(JT);
        add(JSP1);
        //设置窗体的外部特性
        setTitle("表格");//设置窗口的标题文字
        setSize(300,200);//设置窗口的大小
        setLocationRelativeTo(null);//窗口居中显示
        setVisible(true);//设置可视化
        //设置窗口的关闭方式
        setDefaultCloseOperation(JFrame.EXIT_ON_CLOSE);
    }
    public static void main(String[]args){
        //TODO Auto-generated method stub
        JTable3 jt = new JTable3();
    }
}
```

如图 6-88 所示。

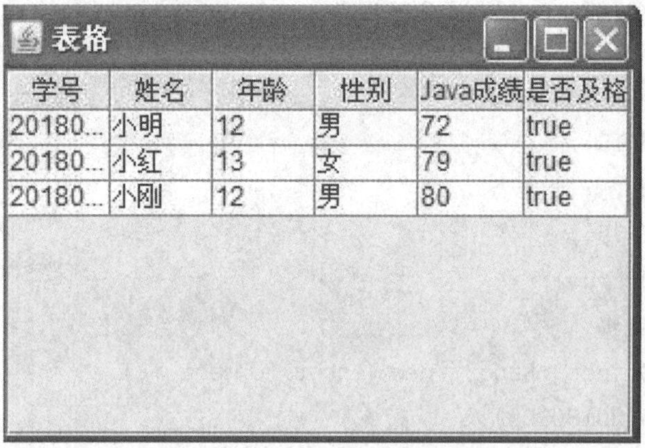

图 6-88

注意：
JTable 使用时要加入 JScrollPane 中，否则表格的标题无法显示。

2. 使用 AbstractTableModel 和 DefaultTableModel 创建表格

使用 JTable 创建的表格较为单一，如果想制作界面更加友好的表格，就要借助 TableModel 接口。TableModel 接口的常用方法如下：
- public Class getColumnClass(int columnIndex)返回字段数据类型的类名称。
- public int getColumnCount()返回字段(行)数量。
- public String getColumnName(int columnIndex)返回字段名称。
- public int getRowCount()返回数据列数量。

- public Object getValueAt(int rowIndex, int columnIndex)返回数据某个 cell 中的值。
- public boolean isCellEditable(int rowIndex, int columnIndex)返回 cell 是否可编辑，true 的话为可编辑。
- public void removeTableModelListener(TableModelListener l)从 TableModelListener 中移除一个 listener。
- public void setValueAt(Object aValue, int rowIndex, int columnIndex)设置某个 cell (rowIndex, columnIndex)的值。

在实际开发中很少直接实现 TableModel 接口，一般是使用该接口的子类 AbstractTableModel 或 DefaultTableModel。在学习 AbstractTableModel 或 DefaultTableModel 之前，我们必须先了解 TableColumnModel 接口，该接口定义了许多与表格的行、列有关的方法，如增加行、删除行等操作。但在实际应用中，我们一般也不直接实现 TableColumnModel 接口，而是通过 JTable 类的 getColumnModel()方法获得 TableColumnModel 的实例，再通过该实例就可以对表格进行相应的操作。TableColumnModel 接口的常用方法如下：

- public void addColumn(TableColumn aColumn)增加列。
- public void removeColumn(TableColumn aColumn)删除指定列。
- public TableColumn getColumn(int columnIndex)根据索引取得指定列。
- public int getColumnCount(int columnIndex)取得全部列数。
- public Enumeration<TableColumn>getColumns()返回全部列。

通过上述方法可以发现，增加或删除等表格操作都是以 TableColumn 类的形式进行的，TableColumn 类表示一列数据。

(1) 使用 AbstractTableModel 创建表格。

Java 提供的 AbstractTableModel 是一个抽象类，利用这个抽象类可以设计出不同格式的表格。AbstractTableModel 类已经实现了大部分的 TableModel 方法，但 getRowCount()、getColumnCount()、getValueAt() 这 3 个方法没有实现，因此我们使用 AbstractTableModel 类创建表格时，除了要根据程序设计需要重写相应的方法外，必须还要实现 getRowCount()、getColumnCount()、getValueAt()方法。

【例 6.52】使用 AbstractTableModel 创建表格。

```
import javax.swing.JScrollPane;
import javax.swing.JFrame;
import javax.swing.JTable;
import javax.swing.table.AbstractTableModel;
    //定义表格模板
class ATM1 extends AbstractTableModel{
    //表格列标题
    String[]titles={"学号","姓名","年龄","性别","Java 成绩","是否及格"};
    //表格显示内容
```

```java
        Object[][]cj = {
        {20180101,"小明",12,"男",72,true},
        {20180102,"小红",13,"女",79,true},
        {20180103,"小刚",12,"男",80,false}
        };//定义数据

        //getRowCount()、getColumnCount()、getValueAt()这3个方法必须要实现
        public int getColumnCount(){//取得列的数量
        return titles.length;
        }
        public int getRowCount(){//取得行的数量
        return cj.length;
        }
         public Object getValueAt( int rowIndex, int columnIndex) {//返回索引rowIndex、columnIndex指定的数据
        return cj[rowIndex][columnIndex];
        }
        }
        public class JTable4 extends JFrame{
        ATM1 ATM11 = new ATM1();//实例化表格模板对象ATM11
        JTable JT1 = new JTable(ATM11);//根据表格模板ATM11创建表格JT1
        JScrollPane JSP1 = new JScrollPane(JT1);
        public JTable4(){
        add(JSP1);
        setTitle("AbstractTableModel 表格 1");
        setSize(300,200);
        setLocationRelativeTo(null);
        setDefaultCloseOperation(JFrame.EXIT_ON_CLOSE);
        setVisible(true);
        }
        public static void main(String args[]){
        new JTable4();
        }
        }
```

如图 6-89 所示。

在上例中,我们发现表格的标题与我们设定的标题不一致,原因是没有重写 getColumnName()方法,即如果要按照我们的要求显示相应标题,必须重写 getColumnName()方法,如下所示。

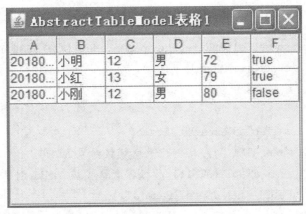

图 6-89

【例 6.53】 使用 AbstractTableModel 创建表格,并按要求显示列标题。

```
import javax.swing.JScrollPane;
import javax.swing.JFrame;
import javax.swing.JTable;
import javax.swing.table.AbstractTableModel;
    //定义表格模板
class ATM1 extends AbstractTableModel{
    //表格列标题
    String[]titles = {"学号","姓名","年龄","性别","Java 成绩","是否及格"};
    //表格显示内容
    Object[][]cj = {
    {20180101,"小明",12,"男",72,true},
    {20180102,"小红",13,"女",79,true},
    {20180103,"小刚",12,"男",80,false}
    };//定义数据
    //getRowCount()、getColumnCount()、getValueAt()这 3 个方法必须要实现
    public int getColumnCount(){//取得列的数量
    return titles.length;
    }
    public int getRowCount(){//取得行的数量
    return cj.length;
    }
     public Object getValueAt( int rowIndex, int columnIndex) {//返回索引 rowIndex、columnIndex 指定的数据
    return cj[rowIndex][columnIndex];
```

```java
}
//注意如果要正确显示列标题,别重写 getColumnName(int columnIndex)方法
public String getColumnName(int columnIndex){//返回列标题
return titles[columnIndex];
}
}
public class JTable5 extends JFrame{
ATM1 ATM11 = new ATM1();//实例化表格模板对象 ATM11
JTable JT1 = new JTable(ATM11);//根据表格模板 ATM11 创建表格 JT1
JScrollPane JSP1 = new JScrollPane(JT1);
public JTable5(){
add(JSP1);
setTitle("AbstractTableModel 表格 2");
setSize(300,200);
setLocationRelativeTo(null);
setDefaultCloseOperation(JFrame.EXIT_ON_CLOSE);
setVisible(true);
}
public static void main(String args[]){
new JTable5();
}
}
```

如图 6-90 所示。

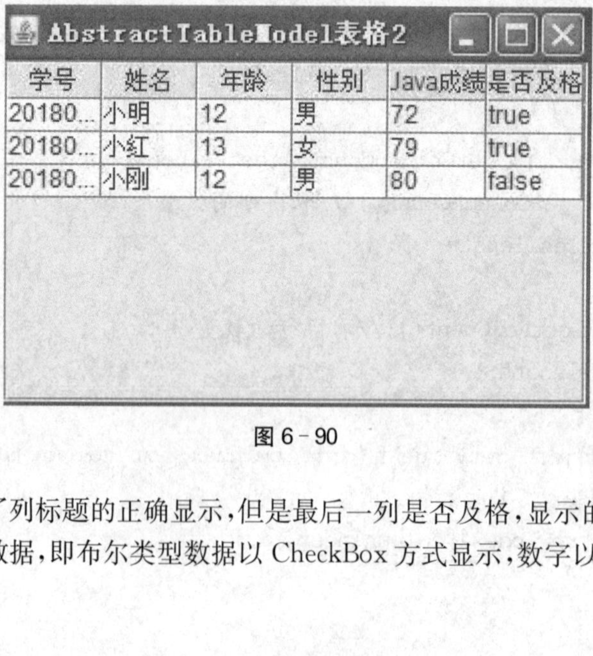

图 6-90

上例已经实现了列标题的正确显示,但是最后一列是否及格,显示的是 true、false,而我们希望按格式显示数据,即布尔类型数据以 CheckBox 方式显示,数字以 JLabel 方式显示且

文字向右排列。要达到上述目的,就需要重写 getColumnClass(int columnIndex)方法,如下例所示。

【例 6.54】使用 AbstractTableModel 创建表格,要求显示列标题且按格式显示数据。

```java
import javax.swing.JScrollPane;
import javax.swing.JFrame;
import javax.swing.JTable;
import javax.swing.table.AbstractTableModel;
    //定义表格模板
class ATM1 extends AbstractTableModel{
    //表格列标题
    String[]titles = {"学号","姓名","年龄","性别","Java 成绩","是否及格"};
    //表格显示内容
    Object[][]cj = {
    {20180101,"小明",12,"男",72,true},
    {20180102,"小红",13,"女",79,true},
    {20180103,"小刚",12,"男",80,false}
    };//定义数据

    //getRowCount()、getColumnCount()、getValueAt()这3个方法必须要实现
    public int getColumnCount(){//取得列的数量
    return titles.length;
    }
    public int getRowCount(){//取得行的数量
    return cj.length;
    }
    public Object getValueAt(int rowIndex,int columnIndex){//返回索引 rowIndex、columnIndex 指定的数据
    return cj[rowIndex][columnIndex];
    }
    //注意如果要正确显示列标题,别重写 getColumnName(int columnIndex)方法
    public String getColumnName(int columnIndex){//返回列标题
    return titles[columnIndex];
    }
    public Class<?>getColumnClass(int columnIndex){//得到指定列的类型
    return getValueAt(0,columnIndex).getClass();
    }
```

```java
}

public class JTable6 extends JFrame{
    ATM1 ATM11 = new ATM1();//实例化表格模板对象 ATM11
    JTable JT1 = new JTable(ATM11);//根据表格模板 ATM11 创建表格 JT1
    JScrollPane JSP1 = new JScrollPane(JT1);
    public JTable6(){
        add(JSP1);
        setTitle("AbstractTableModel 表格3");
        setSize(300,200);
        setLocationRelativeTo(null);
        setDefaultCloseOperation(JFrame.EXIT_ON_CLOSE);
        setVisible(true);
    }
    public static void main(String args[]){
        new JTable6();
    }
}
```

如图6-91所示。

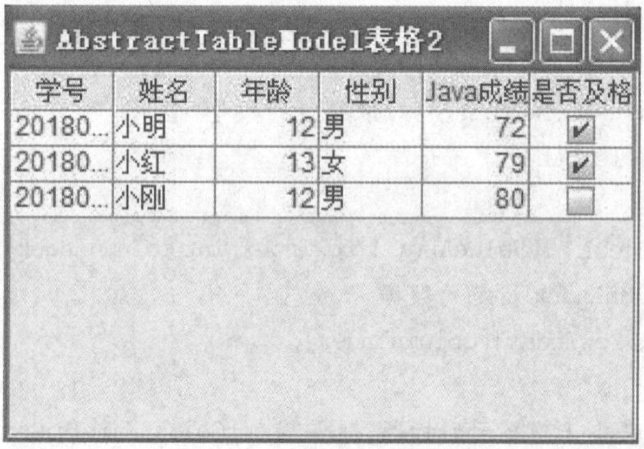

图6-91

上例已经实现了数据的按格式显示,但是突然发现小红的年龄错误(应该也是12),但是此时表格无法编辑修改,可以通过重写 isCellEditable(int rowIndex, int columnIndex)和 setValueAt(Object aValue, int rowIndex, int columnIndex)方法解决这个问题,如下例所示。

【例 6.55】 使用 AbstractTableModel 创建表格，要求显示列标题并按格式显示数据，同时还要求表格数据可编辑。

```java
import javax.swing.JScrollPane;
import javax.swing.JFrame;
import javax.swing.JTable;
import javax.swing.table.AbstractTableModel;
    //定义表格模板
class ATM1 extends AbstractTableModel{
    //表格列标题
    String[]titles={"学号","姓名","年龄","性别","Java成绩","是否及格"};
    //表格显示内容
    Object[][]cj={
    {20180101,"小明",12,"男",72,true},
    {20180102,"小红",13,"女",79,true},
    {20180103,"小刚",12,"男",80,false}
    };//定义数据
    //getRowCount()、getColumnCount()、getValueAt()这3个方法必须要实现
    public int getColumnCount(){//取得列的数量
    return titles.length;
    }
    public int getRowCount(){//取得行的数量
    return cj.length;
    }
    public Object getValueAt(int rowIndex,int columnIndex){//返回索引rowIndex、columnIndex指定的数据
    return cj[rowIndex][columnIndex];
    }
    //注意如果要正确显示列标题,别重写getColumnName(int columnIndex)方法
    public String getColumnName(int columnIndex){//返回列标题
    return titles[columnIndex];
    }
    public Class<?> getColumnClass(int columnIndex){//得到指定列的类型
    return getValueAt(0,columnIndex).getClass();
    }
    public boolean isCellEditable(int rowIndex,int columnIndex){//所有内容都可以编辑
    return true;
```

```java
}
public void setValueAt(Object aValue, int rowIndex, int columnIndex){
cj[rowIndex][columnIndex] = aValue;
}
}
public class JTable7 extends JFrame{
ATM1 ATM11 = new ATM1();//实例化表格模板对象 ATM11
JTable JT1 = new JTable(ATM11);//根据表格模板 ATM11 创建表格 JT1
JScrollPane JSP1 = new JScrollPane(JT1);
public JTable7(){
add(JSP1);
setTitle("AbstractTableModel 表格 2");
setSize(300,200);
setLocationRelativeTo(null);
setDefaultCloseOperation(JFrame.EXIT_ON_CLOSE);
setVisible(true);
}
public static void main(String args[]){
new JTable7();
}
}
```

如图 6-92、图 6-93 所示。

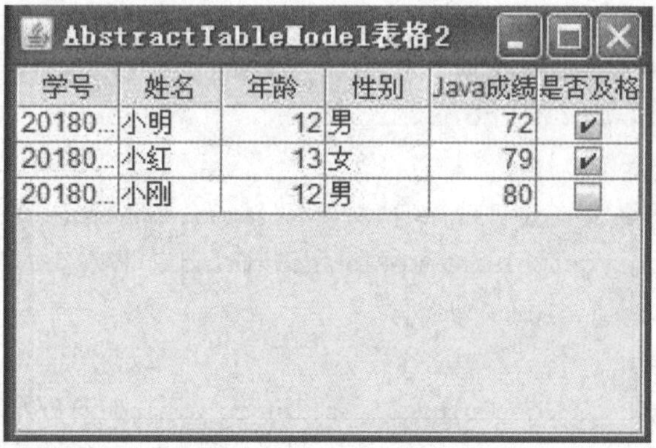

图 6-92

(2) 使用 DefaultTableModel 创建表格。

使用 AbstractTableModel 类可以方便地创建表格,而使用 DefaultTableModel 类创建表格,可以对表格进行动态的操作,如增加行(列)等。DefaultTableModel 类是继承 AbstractTableModel 抽象类而来,并且实现了 getColumnCount( )、getRowCount( ) 和

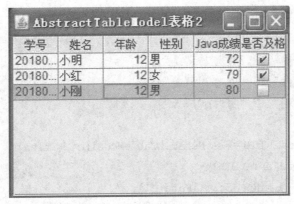

图 6-93

getValueAt()方法。此外,DefaultTableModel 类还提供了很多实用的方法,如 getColumnCount()、getRowCount()、getValueAt()、isCellEditable()、setValueAt()、addColumn()、addRow()等。因此,使用 DefaultTableModel 类创建表格比使用 AbstractTableModel 类简单,实际应用中 DefaultTableModel 类使用较多。

DefaultTableModel 类的常用方法如下:

- public void addColumn(Object columnName)增加列,并指定列名。
- public void addRow(Object[]rowData)增加行。
- public int getRowCount()返回表格行数。
- public int getColumnCount()返回表格列数。
- public void setRowCount(int rowCount)设置表格中的行数。
- public void setColumnCount(int rowCount)设置表格中的列数。
- public void removeRow(int rowCount)删除表格中的指定行。

【例 6.56】使用 **DefaultTableModel** 创建表格,然后创建 4 个按钮,通过按钮分别实现表格的增加行、删除行、增加列、删除列功能。

```
import javax.swing.JButton;
import javax.swing.JLabel;
import javax.swing.JScrollPane;
import javax.swing.JFrame;
import javax.swing.JTable;
import javax.swing.table.DefaultTableModel;
import javax.swing.table.TableColumnModel;
import javax.swing.table.TableColumn;
import java.awt.event.ActionListener;
import java.awt.event.ActionEvent;
public class JTable8 extends JFrame implements ActionListener{
//表格列标题
String[]titles={"学号","姓名","年龄","性别","Java 成绩","是否及格"};
```

```java
//表格显示内容
Object[][]cj = {
{20180101,"小明",12,"男",72,"是"},
{20180102,"小红",13,"女",79,"是"},
{20180103,"小刚",12,"男",80,"否"}
};//定义数据
DefaultTableModel DTM = new DefaultTableModel(cj,titles);
JTable JT = new JTable(DTM);
JScrollPane JSP1 = new JScrollPane(JT);
JButton addRowBtn = new JButton("增加行");//定义按钮
JButton removeRowBtn = new JButton("删除行");//定义按钮
JButton addColBtn = new JButton("增加列");//定义按钮
JButton removeColBtn = new JButton("删除列");//定义按钮
public JTable8(){
setLayout(null);
add(JSP1);
JSP1.setBounds(10,10,480,180);
add(addRowBtn);
addRowBtn.setBounds(10,200,80,50);
addRowBtn.addActionListener(this);
add(removeRowBtn);
removeRowBtn.setBounds(150,200,80,50);
removeRowBtn.addActionListener(this);
add(addColBtn);
addColBtn.setBounds(270,200,80,50);
addColBtn.addActionListener(this);
add(removeColBtn);
removeColBtn.setBounds(400,200,80,50);
removeColBtn.addActionListener(this);
setTitle("DefaultTableModel 表格");
setSize(500,300);
setLocationRelativeTo(null);
setDefaultCloseOperation(JFrame.EXIT_ON_CLOSE);
setVisible(true);
}
public void actionPerformed(ActionEvent e){
if(e.getSource() == addColBtn){//增加列
DTM.addColumn("新增列");
```

```java
}
if(e.getSource() == addRowBtn){
DTM.addRow(new Object[]{});
}
if(e.getSource() == removeColBtn){
int colCount = DTM.getColumnCount()-1;//取得列的数量
if(colCount >= 0){
//如果要想删除,则必须找到TableColumnModel
TableColumnModel columMode = JT.getColumnModel();
TableColumn taleColumn = columMode.getColumn(colCount);
columMode.removeColumn(taleColumn);//删除指定的列
DTM.setColumnCount(colCount);//设置新的列数
}
}

if(e.getSource() == this.removeRowBtn){
int rowCount = DTM.getRowCount()-1;
if(rowCount >= 0){//判断是否还有行可以删除
DTM.removeRow(rowCount);
DTM.setRowCount(rowCount);//设置新的行数
}
}
}
public static void main(String args[]){
new JTable8();
}
}
```

如图 6-94 所示。

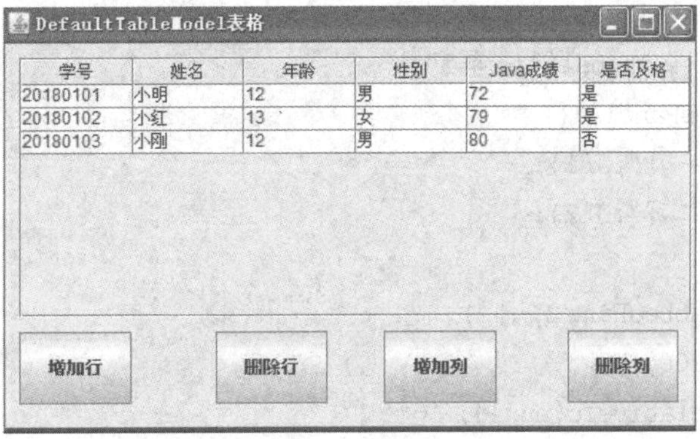

图 6-94

6.5.3 JTabbedPane

javax.swing.JTabbedPane 类继承于 javax.swing.JComponent，它的对象反映为一组带标签的面板（即在一个面板上设置多个选项卡供用户选择），每个面板都可以存放组件，因此 JTabbedPane 是一个容器组件。

JTabbedPane 类的构造方法有：

- JTabbedPane()创建空对象，该对象具有默认的标签位置 JTabbedPane.TOP 和默认的布局策略 JTabbedPane.WRAP_TAB_LAYOUT。
- JTabbedPane(int tabPlacement)创建空对象，该对象具有指定的标签位置 JTabbedPane.TOP、JTabbedPane.BOTTOM、JTabbedPane.LEFT 和 JTabbedPane.RIGHT 以及默认的布局策略 JTabbedPane.WRAP_TAB_LAYOUT。
- JTabbedPane(int tabPlacement, int tabLayoutPolicy)创建空对象，该对象具有指定的标签位置和布局策略。

JTabbedPane 类的其他常用方法如下：

- public void addTab(String title, Component component)添加一个有标题但没有图标的组件。
- public void addTab(String title, Icon icon, Component component)添加一个有标题、图标的组件。
- public void addTab(String title, Icon icon, Component component, String tip)添加一个有标题、图标和提示信息的组件。

【例 6.57】创建一个 JTabbedPane 容器，标签在顶部显示，共两页标签，第一页标题为"第一页"，将第一个标签（标签显示"标签 1"）添加到该页，第二页标题为"第二页"，将第二个标签（标签显示"标签 2"）添加到该页。

```java
import javax.swing.JFrame;
import javax.swing.JTabbedPane;
import javax.swing.JLabel;
public class JTabbedPane1 extends JFrame{
JTabbedPane JTP1 = new JTabbedPane(JTabbedPane.TOP);
JLabel JL1 = new JLabel("标签 1");
JLabel JL2 = new JLabel("标签 2");
public JTabbedPane1(){
JTP1.add("第一页",JL1);
JTP1.add("第二页",JL2);
add(JTP1);
setTitle("JTabbedPane 容器");
setSize(300,200);
setLocationRelativeTo(null);
setDefaultCloseOperation(JFrame.EXIT_ON_CLOSE);
```

```
setVisible(true);
}
public static void main(String args[]){
new JTabbedPane1();
}
}
```

如图 6-95、图 6-96 所示。

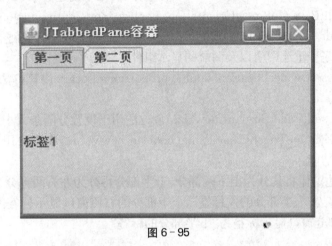

图 6-95

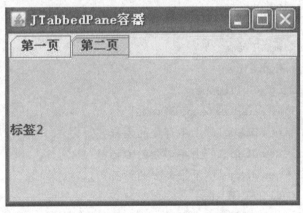

图 6-96

6.5.4 JSplitPane

JSplitPane 主要功能是分割面板,可以将一个窗体分为两个子窗体,可以是水平排列也可以是垂直排列,如图 6-97 所示。

JSplitPane 类的常用方法及常量如下所示:
- public static final int HORIZONTAL_SPLIT 表示水平分割的常量。

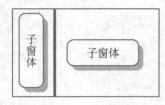

图 6-97

- public static final int VERTICAL_SPLIT 表示垂直分割的常量。
- public JSplitPane(int newOrientation)构造方法,用于创建对象,并指明分割方式。
- public JSplitPane(int newOrientation, boolean newContinuousLayout, Component newLeftComponent, Component newRightComponent)构造方法,用于创建对象、指明分割方式、分割条改变是否重绘图像以及两端的显示组件。
- public void setDividerLocation(double proportionalLocation)该方法用于设置分隔条的位置,按百分比。
- public void setDividerSize(int newSize)该方法用于设置分隔条大小。
- public void setOneTouchExpandable(boolean newValue)该方法用于设置是否提供快速展开/折叠的功能。

【例6.58】创建窗体将其分为上下两部分,上半部分再分为左右两部分,上半部分的左侧窗口显示标签内容为"上半部分的左标签",上半部分的右侧窗口显示标签内容为"上半部分的右标签",下半部分窗口显示标签为"下半部分的标签"。

```java
import javax.swing.JPanel;
import javax.swing.JButton;
import javax.swing.JLabel;
import javax.swing.JFrame;
import java.awt.Container;
import javax.swing.JSplitPane;
public class JSplitPane1 extends JFrame{
    JLabel JL1 = new JLabel("上半部分的左标签");
    JLabel JL2 = new JLabel("上半部分的右标签");
    JLabel JL3 = new JLabel("下半部分的标签");
    JSplitPane split_L_R = new JSplitPane(JSplitPane.HORIZONTAL_SPLIT,JL1,JL2);//上半部分左右分割
    JSplitPane split_T_B = new JSplitPane(JSplitPane.VERTICAL_SPLIT,split_L_R,JL3);;//上下分割
    public JSplitPane1(){
        split_L_R.setDividerSize(10);//设置上半部分左右分割条的分割线大小
        split_T_B.setDividerSize(10);//设置上下部分的分割条的分割线大小
        add(split_T_B);
        setTitle("上下左右分割窗体");
```

```
            setSize(300,200);
            setLocationRelativeTo(null);
            setVisible(true);
            setDefaultCloseOperation(JFrame.EXIT_ON_CLOSE);
        }
        public static void main(String[]args){
            //TODO Auto-generated method stub
            new JSplitPane1();
        }
    }
```

如图 6-98 所示。

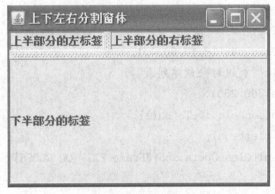

图 6-98

6.5.5 列表框

1. 使用构造函数创建列表框

列表框可以同时将多个选项信息以列表的方式展现给用户,它使用户易于操作大量的选项。使用 JList 可以构建一个列表框,JList 类的常用方法如下所示:

- public JList(ListModel dataMode)构造方法,用于根据 ListModel 构造 JList。
- public JList(Object[]listData)构造方法,用于根据对象数组构造 JList。
- public JList(Vector<?>listData)构造方法,用于根据一个 Vector 构造 JList。
- public void setSelectionMode(int selectionMode)该方法用于设置选择模式,是多选还是单选。
- public int[]getSelectedIndices()该方法用于返回所选择的全部数组。

列表框是可以设置多选或单选的,列表框是多选还是单选可以通过 ListSelectionModel 接口完成,ListSelectionModel 接口定义了如下常量:

- static final int MULTIPLE_INTERVAL_SELECTION 表示一次选择一个或多个连续的索引范围的常量。

- static final int SINGLE_INTERVAL_SELECTION 表示一次选择一个连续范围的值的常量。
- static final int SINGLE_SELECTION 表示一次选择一个值的常量。

【例6.59】使用对象数组构造列表,显示"重庆、北京、上海、天津"。

```java
import javax.swing.JFrame;
import javax.swing.JList;
import javax.swing.JScrollPane;
public class JList1 extends JFrame{
    String city[] = {"重庆","上海","天津","北京"};
    JList JL1 = new JList(city);
    JScrollPane JSP1 = new JScrollPane(JL1);
    public JList1(){
        setLayout(null);
        add(JSP1);
        JSP1.setBounds(10,10,80,40);
        setTitle("数组对象创建列表");
        setSize(300,200);
        setLocationRelativeTo(null);
        setVisible(true);
        setDefaultCloseOperation(JFrame.EXIT_ON_CLOSE);
    }
    public static void main(String[]args){
        //TODO Auto-generated method stub
        new JList1();
    }
}
```

如图6-99所示。

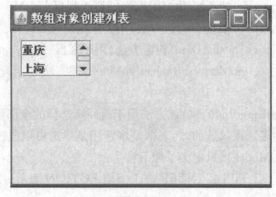

图6-99

【例 6.60】使用 Vector 构造列表，显示"重庆、北京、上海、天津"。

```java
import java.util.Vector;
import javax.swing.JFrame;
import javax.swing.JList;
import javax.swing.JScrollPane;
public class JList2 extends JFrame{
    Vector<String>v1 = new Vector<String>();
    JList JL1 = new JList(v1);
    JScrollPane JSP1 = new JScrollPane(JL1);
    public JList2(){
        v1.add("北京");
        v1.add("天津");
        v1.add("上海");
        v1.add("重庆");
        setLayout(null);
        add(JSP1);
        JSP1.setBounds(10,10,80,40);
        setTitle("Vector 创建列表");
        setSize(300,200);
        setLocationRelativeTo(null);
        setVisible(true);
        setDefaultCloseOperation(JFrame.EXIT_ON_CLOSE);
    }

    public static void main(String[ ]args){
        //TODO Auto-generated method stub
        new JList2();
    }
}
```

如图 6-100 所示。
注意：
如果 JList 中显示的内容过多，必须将其加入滚动条 JScrollPane 中，才能显示滚动条。
2. 使用 AbstractListModel 接口创建列表框
在上例中，用字符串数组和 Vector 构建列表框，但在实际应用中列表框的选项经常会发生变化，为此，我们可以用列表模式 ListModel 来构建列表框，这样更具灵活性和实用性。ListModel 是一个专门用于创建 JList 列表内容操作的接口，其定义方法如下：

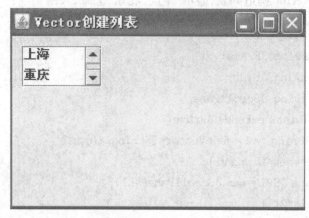

图 6-100

- void addListDataListener(ListDataListener l)该方法用于加入数据改变事件的监听。
- void removeListDataListener(ListDataListener l)该方法用于加入数据删除时的监听。
- Object getElementAt(int index)该方法用于返回指定索引处的内容。
- int getSize()该方法用于返回列表长度。

ListModel 接口虽然定义了 4 个方法,但在实际开发时常常只用到 getElementAt() 和 getSize() 方法,而使用 ListModel 接口构建列表框,必须将 4 个方法全部实现。因此,在实际应用时,我们一般通过继承 AbstractListModel 完成列表框的创建。

【例 6.61】使用 AbstractListModel 创建列表框,显示国内的直辖市。

```
import javax.swing.AbstractListModel;
import javax.swing.JFrame;
import javax.swing.JList;
import javax.swing.BorderFactory;
import javax.swing.ListSelectionModel;
import javax.swing.JScrollPane;
class ALM extends AbstractListModel{//定义列表内容
String city[] = {"北京","上海","天津","重庆"};
public Object getElementAt(int index){
if(index<city.length){
return city[index];
}else{
return null;
}
}
public int getSize(){
```

```
        return city.length;
    }
}
public class JList3 extends JFrame{
    ALM ALM1 = new ALM();
    JList JL1 = new JList(ALM1);
    JScrollPane JSP1 = new JScrollPane(JL1);
    public JList3(){
JL1.setBorder(BorderFactory.createTitledBorder("你喜欢哪个直辖市?"));
JL1.setSelectionMode(ListSelectionModel.MULTIPLE_INTERVAL_SELECTION);//多选
        setLayout(null);
        add(JSP1);
        JSP1.setBounds(10,10,200,80);
        setTitle("数组对象创建列表");
        setSize(300,200);
        setLocationRelativeTo(null);
        setVisible(true);
        setDefaultCloseOperation(JFrame.EXIT_ON_CLOSE);
    }

    public static void main(String[ ]args){
        //TODO Auto-generated method stub
        new JList3();
    }

}
```

如图 6-101 所示。

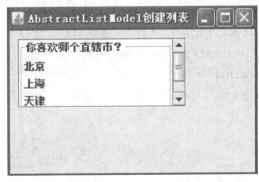

图 6-101

3. JList 事件处理

在 JList 中可以使用 ListSelectionListener 的监听接口实现对 JList 中所选项进行监听，该接口只定义了一个方法：

- void valueChanged(ListSelectionEvent e)当值发生改变时调用。

valueChanged()方法会对 ListSelectionEvent 事件处理，此事件中的常用方法如下所示：

- getSelectedIndices()返回一个整型数组，包含被选中的所有 index。
- isSelectedIndex(int index)判断指定 index 是否被选中。
- clearSelection()清除选中。
- getSelectedValues()返回一个 object 数组，包含被选中的所有元素对象。
- getSelectedValuesList()返回一个 objectList，包含被选中的所有元素对象。
- getFirstIndex()返回第一个选择状态可能发生更改的行的索引。
- getLastIndex()返回最后一个选择状态可能已发生更改的行的索引。
- getValueIsAdjusting()如果此事件是多个更改事件之一，则返回 true。例如，在一个 JList 中，单击第一行不放，然后鼠标下移到第二行不放，之后松开鼠标。其检测结果是：在第一行显示为 true，第二行还是 true，之后显示 false。

此外，ListSelectionListener 监听接口可以使用 addListSelectionListener()方法注册。

【例 6.62】 使用 AbstractListModel 创建列表框，显示国内的直辖市，并创建标签显示选中的直辖市。

```java
import javax.swing.AbstractListModel;
import javax.swing.JFrame;
import javax.swing.JList;
import javax.swing.BorderFactory;
import javax.swing.ListSelectionModel;
import javax.swing.JScrollPane;
import javax.swing.JLabel;
import javax.swing.event.ListSelectionListener;
import javax.swing.event.ListSelectionEvent;
class ALM extends AbstractListModel{//定义列表内容
String city[] = {"北京","上海","天津","重庆"};
public Object getElementAt(int index){
if(index<city.length){
return city[index];
}else{
return null;
}
}
```

```java
    public int getSize(){
    return city.length;
    }
}

public class JList4 extends JFrame implements ListSelectionListener{
    ALM ALM1 = new ALM();
    JList JL1 = new JList(ALM1);
    JScrollPane JSP1 = new JScrollPane(JL1);
        JLabel JLa1 = new JLabel();
    public JList4(){
JL1.setBorder(BorderFactory.createTitledBorder("你喜欢哪个直辖市?"));
JL1.setSelectionMode(ListSelectionModel.MULTIPLE_INTERVAL_SELECTION);//多选
        setLayout(null);
        add(JSP1);
        JSP1.setBounds(10,10,200,80);
        JL1.addListSelectionListener(this);
        add(JLa1);
        JLa1.setBounds(10,100,200,40);
        setTitle("AbstractListModel 创建列表及事件处理");
        setSize(300,200);
        setLocationRelativeTo(null);
        setVisible(true);
        setDefaultCloseOperation(JFrame.EXIT_ON_CLOSE);
    }

public void valueChanged(ListSelectionEvent e){//事件处理
int temp[] = JL1.getSelectedIndices();
String xuanze = "";
if(!e.getValueIsAdjusting()){
for(int i = 0;i<temp.length;i++){
int j = temp[i];
//String xianshi = JLa1.getText();
xuanze = xuanze + JL1.getModel().getElementAt(j).toString();
JLa1.setText(xuanze);
}
}
```

```
    }

    public static void main(String[] args){
        //TODO Auto-generated method stub
        new JList4();
    }

}
```

如图 6-102 所示。

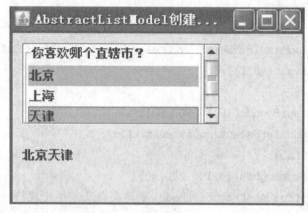

图 6-102

6.5.6 文件选择框

在一般的文本程序都具备打开和关闭的操作。例如,在使用记事本或者 Word 的时候,可以通过文件选择框选择要打开或保存的文件,在 Swing 中可以使用 JFileChooser 组件实现此功能。JFileChooser 类的常用构造方法如下:

- JFileChooser()构造一个指向用户默认目录的 JFileChooser 对象。
- JFileChooser(File currentDirectory)构造一个以给定 File 为路径的 JFileChooser 对象。

构造 JFileChooser 对象后,要利用该类的方法 showOpenDialog()或 showSaveDialog()来显示文件打开或文件关闭对话框,它们的格式为:

- public int showOpenDialog(Component parent) throws HeadlessException
- public int showSaveDialog(Component parent) throws HeadlessException

它们的参数都是包含对话框容器的对象,其返回值为下面 3 种情况:

- JFileChooser.CANCEL_OPTION 表示选择了撤销按钮。
- JFileChooser.APPROVE_OPTION 表示选择了打开或保存按钮。
- JFileChooser.ERROR_OPTION 表示出现错误。

在打开或关闭文件对话框中作出选择后,可用 JFileChooser 类的方法 getSelectedFile()

获取选中的文件。此外，还可用 setDialogTitle( ) 方法设置文件选择框的标题，用 setApproveButtonText( )方法设置文件选择框的接收按钮内容。

【例 6.63】 使用 JFileChooser 类实现将选中的记事本文件的内容显示到文本区域中。

```java
import java.io.File;
import java.io.FileInputStream;
import java.io.FileOutputStream;
import java.io.PrintStream;
import java.util.Scanner;
import java.awt.BorderLayout;
import java.awt.event.ActionEvent;
import java.awt.event.ActionListener;
import javax.swing.JFrame;
import javax.swing.JPanel;
import javax.swing.JTextArea;
import javax.swing.JLabel;
import javax.swing.JButton;
import javax.swing.JFileChooser;
import javax.swing.JScrollPane;
public class JFileChooser1 extends JFrame implements ActionListener{
    JTextArea JTA1 = new JTextArea(10,10);//定义文本区
    JScrollPane JSP1 = new JScrollPane(JTA1);
    JPanel JP1 = new JPanel();
    JButton open = new JButton("打开文件");
    JButton save = new JButton("保存文件");
    JLabel JL1 = new JLabel("现在没有打开的文件");
    public JFileChooser1(){
        setTitle("JFileChooser 打开记事本文件");
        setLayout(new BorderLayout());
        add(JL1,BorderLayout.NORTH);
        add(JSP1,BorderLayout.CENTER);
        JP1.add(open);//在面板中加入按钮
        JP1.add(save);//在面板中加入按钮
        open.addActionListener(this);
        save.addActionListener(this);
        add(JP1,BorderLayout.SOUTH);
        setSize(330,180);
        setLocationRelativeTo(null);
```

```java
setVisible(true);
setDefaultCloseOperation(JFrame.EXIT_ON_CLOSE);
}
public void actionPerformed(ActionEvent e){
File file=null;//接收文件
int result=0;//接收操作状态
JFileChooser fileChooser=new JFileChooser();//文件选择框
if(e.getSource()==open){//表示执行的是打开操作
JTA1.setText("");//打开将文字区域的内容清空
fileChooser.setApproveButtonText("确定");
fileChooser.setDialogTitle("打开文件");
result=fileChooser.showOpenDialog(this);
if(result==JFileChooser.APPROVE_OPTION){//选择的是确定按钮
file=fileChooser.getSelectedFile();//得到选择的文件
JL1.setText("打开的文件名称为:"+file.getName());
}else if(result==JFileChooser.CANCEL_OPTION){
JL1.setText("没有选择任何文件");
}else{
JL1.setText("操作出现错误");
}
if(file!=null){
try{
Scanner scan=new Scanner(new FileInputStream(file));
scan.useDelimiter("\n");
while(scan.hasNext()){
JTA1.append(scan.next());
JTA1.append("\n");
}
scan.close();
}catch(Exception e1){}
}
}
if(e.getSource()==save){//判断是否是保存操作
result=fileChooser.showSaveDialog(this);//显示保存框
if(result==JFileChooser.APPROVE_OPTION){//选择的是确定按钮
file=fileChooser.getSelectedFile();//得到选择的文件
JL1.setText("选择的存储文件名称为:"+file.getName());
```

```
}else if(result = = JFileChooser.CANCEL_OPTION){
JL1.setText("没有选择任何文件");
}else{
JL1.setText("操作出现错误");
}
if(file! = null){
try{
PrintStream out = new PrintStream(new FileOutputStream(file));
out.print(JTA1.getText());
out.close();
}catch(Exception e1){}
}
}
}
public static void main(String args[]){
new JFileChooser1();
}
}
```

如图 6-103、图 6-104 所示。

图 6-103

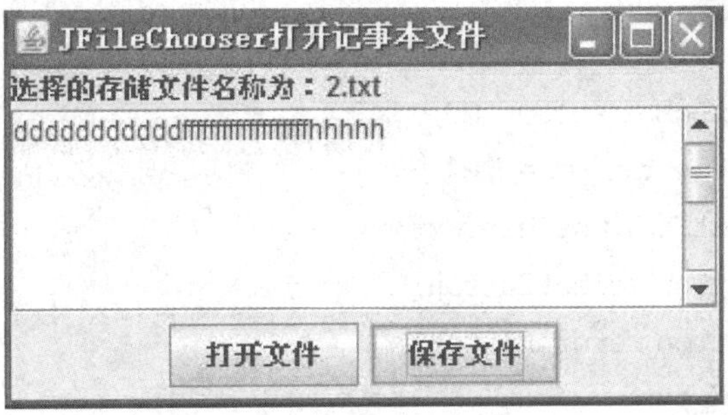

图 6-104

任务 实施

6.5.7 学生成绩管理系统信息查询窗口设计实现

成绩查询窗口代码如下：

```java
import java.awt.event.ActionEvent;
import java.awt.event.ActionListener;
import javax.swing.JButton;
import javax.swing.JComboBox;
import javax.swing.JFrame;
import javax.swing.JLabel;
import javax.swing.JOptionPane;
import javax.swing.JPanel;
import javax.swing.JTextField;
import javax.swing.BorderFactory;
import javax.swing.JScrollPane;
import javax.swing.JScrollPane;
import javax.swing.JTable;
import javax.swing.table.DefaultTableModel;
import javax.swing.table.TableColumnModel;
import javax.swing.table.TableColumn;
import java.io.File;
import java.io.FileInputStream;
import java.io.FileWriter;
import java.util.Scanner;

public class SelectJFrame extends JFrame implements ActionListener{
    private JPanel jp1;
    private JPanel jp2;
    private JLabel selectLabel;
    private JComboBox selectJCB;
    String strSelect = "= =请选择= =,学    号,姓    名,所    有";
    private JTextField selectText;
    private JButton selectButton;
    private JButton returnButton;
    //表格列标题
```

```java
String[]titles = {"学号","姓名","性别","课程名称","平时成绩","过程考核成绩","终结考核成绩","综合成绩","综合成绩级别","爱好"};
//表格显示内容
Object[][]cj;
DefaultTableModel DTM;
JTable JT;
JScrollPane JSP1;

public SelectJFrame(){
this.setTitle("查询学生成绩信息");
this.setLayout(null);
this.setSize(580,450);
jp1 = new JPanel(null);

selectLabel = new JLabel("请选择关键字并输入查询信息:");
selectLabel.setBounds(10,20,190,30);
jp1.add(selectLabel);

selectJCB = new JComboBox();
//将字符串分割成字符串数组 split
String[]selectArray = strSelect.split(",");
for(int i = 0;i<selectArray.length;i++){
    selectJCB.addItem(selectArray[i]);
}
selectJCB.setBounds(210,20,90,30);
jp1.add(selectJCB);
selectJCB.addActionListener(this);

selectText = new JTextField();
selectText.setBounds(310,20,90,30);
jp1.add(selectText);

selectButton = new JButton("查询");
selectButton.setBounds(410,20,70,30);
jp1.add(selectButton);
selectButton.addActionListener(this);

returnButton = new JButton("返回");
```

```java
        returnButton.setBounds(490,20,70,30);
        jp1.add(returnButton);
        returnButton.addActionListener(this);

        this.add(jp1);
        jp1.setBounds(5,5,565,60);
        jp1.setBorder(BorderFactory.createTitledBorder("查询方式选择"));//设置标题

        jp2 = new JPanel(null);
        //将相关操作的按钮组件添加到 jp2 上
        //设置窗口的大小

        DTM = new DefaultTableModel(cj,titles);
        JT = new JTable(DTM);
        JSP1 = new JScrollPane(JT);
        jp2.add(JSP1);
        JSP1.setBounds(10,30,550,300);

        this.add(jp2);
        jp2.setBounds(5,70,565,340);
        jp2.setBorder(BorderFactory.createTitledBorder("学生成绩信息"));//设置标题

        this.setLocationRelativeTo(null);
        this.setResizable(false);
        this.setDefaultCloseOperation(JFrame.DO_NOTHING_ON_CLOSE);
        this.setVisible(true);
    }

    private boolean isNull(){
        String input = selectText.getText().trim();
        if(input.equals("")){
            JOptionPane.showMessageDialog(this,"请输入查询信息!");
            return false;
        }else{
            return true;
        }
    }
```

```java
public void actionPerformed(ActionEvent e){
if(e.getSource() = = selectButton){
DTM.setRowCount(0);
String selected = selectJCB.getSelectedItem().toString();
if(selected.equals("学   号")){
if(isNull()){
readBySno();
}
}
if(selected.equals("姓   名")){
if(isNull()){
readBySname();
}
}
if(selected.equals("所   有")){
readAll();
}
if(selected.equals(" = =请选择= = ")){
JOptionPane.showMessageDialog(this,"请选择查询方式!");
}
}
if(e.getSource() = = selectJCB){
String selected = selectJCB.getSelectedItem().toString();
if(selected.equals("所   有")){
selectText.setText("");
selectText.setEditable(false);
}else{
selectText.setText("");
selectText.setEditable(true);
}
}
if(e.getSource() = = returnButton){
this.dispose();
}
}

public void readAll()   {
try{
```

```java
            File file = new File("studentScore.txt");
            if(file.exists()){
                FileInputStream is = new FileInputStream(file);
                Scanner sc = new Scanner(is);
                while(sc.hasNext()){
                    int sno = sc.nextInt();
                    String sname = sc.next();
                    char sex = sc.next().charAt(0);
                    String cname = sc.next();
                    byte pingShiScore = sc.nextByte();
                    byte guoChengScore = sc.nextByte();
                    byte zhongJieScore = sc.nextByte();
                    double zongHeScore = sc.nextDouble();
                    String zongHeScoreJiBie = sc.next();
                    String love = sc.next();
                    DTM.addRow(new Object[]{sno,sname,
sex,cname,pingShiScore,guoChengScore,zhongJieScore,zongHeScore,zongHeScoreJiBie,
love});
                }
                is.close();
                selectText.setText("");
                JOptionPane.showMessageDialog(this,"查询完毕!");
            }else{
                JOptionPane.showMessageDialog(this,"文件不存在!");
            }
        }
        catch(Exception e){
            JOptionPane.showMessageDialog(this,e.toString());
        }
    }

    public void readBySno()  {
        try{
            int number = Integer.parseInt(selectText.getText());
            File file = new File("studentScore.txt");
            if(file.exists()){
                FileInputStream is = new FileInputStream(file);
                Scanner sc = new Scanner(is);
```

```java
        while(sc.hasNext()){
        int sno = sc.nextInt();
        String sname = sc.next();
        char sex = sc.next().charAt(0);
        String cname = sc.next();
        byte pingShiScore = sc.nextByte();
        byte guoChengScore = sc.nextByte();
        byte zhongJieScore = sc.nextByte();
        double zongHeScore = sc.nextDouble();
        String zongHeScoreJiBie = sc.next();
        String love = sc.next();
        if(sno == number){
        DTM.addRow(new Object[]{sno,sname,
sex,cname,pingShiScore,guoChengScore,zhongJieScore,zongHeScore,zongHeScoreJiBie,
love});
        break;
        }
        }
        is.close();
        selectText.setText("");
        JOptionPane.showMessageDialog(this,"查询完毕!");
        }else{
        JOptionPane.showMessageDialog(this,"文件不存在!");
        }
        }
        catch(Exception e){
        JOptionPane.showMessageDialog(this,e.toString());
        }
        }

        public void readBySname(){
        try{
        String name = selectText.getText();
        File file = new File("studentScore.txt");
        if(file.exists()){
        FileInputStream is = new FileInputStream(file);
        Scanner sc = new Scanner(is);
        while(sc.hasNext()){
```

```java
            int sno = sc.nextInt();
            String sname = sc.next();
            char sex = sc.next().charAt(0);
            String cname = sc.next();
            byte pingShiScore = sc.nextByte();
            byte guoChengScore = sc.nextByte();
            byte zhongJieScore = sc.nextByte();
            double zongHeScore = sc.nextDouble();
            String zongHeScoreJiBie = sc.next();
            String love = sc.next();
            if(sname.equals(name)){
            DTM.addRow(new Object[]{sno,sname,
sex,cname,pingShiScore,guoChengScore,zhongJieScore,zongHeScore,zongHeScoreJiBie,
love});
            break;
            }
            }
            is.close();
            selectText.setText("");
            JOptionPane.showMessageDialog(this,"查询完毕!");
            }else{
            JOptionPane.showMessageDialog(this,"文件不存在!");
            }
            }
            catch(Exception e){
            JOptionPane.showMessageDialog(this,e.toString());
            }
            }

            public static void main(String[]args){
                new SelectJFrame();
            }
            }
```

如图 6-105、图 6-106 所示。

任务 拓展

设计自助银行服务系统的余额查询界面并实现相关查询功能,要求如下:

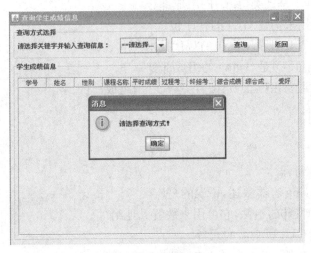

图 6‑105

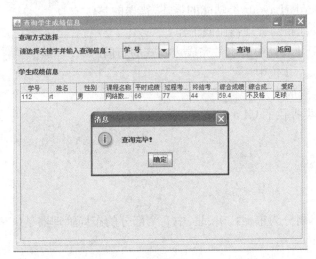

图 6‑106

(1) 用户的账户相关信息包括用户名、账号、密码、身份证号、余额，将上述信息保存到记事本文件 bank.txt。

(2) 根据用户账户号码实现用户余额的查询。

(3) 根据用户的身份证号查询用户余额。

(4) 根据用户名的相关信息(如姓氏等)，实现用户余额信息的模糊查询。

6.6 习 题

一、选择题

1. 下列有关 Swing 的叙述，错误的是(　　)。

A. Swing 是 Java 基础类(JFC)的组成部分
B. Swing 是可用来构建 GUI 的程序包
C. Swing 是 AWT 图形工具包的替代技术
D. Java 基础类(JFC)是 Swing 的组成部分

2. Swing GUI 通常由(　　)、(　　)、(　　)组成。
A. GUI 容器
B. GUI 组件
C. 布局管理器
D. GUI 事件侦听器

3. 以下关于 Swing 容器叙述,错误的是(　　)。
A. 容器是一种特殊的组件,它可用来放置其他组件
B. 容器是组成 GUI 所必需的元素
C. 容器是一种特殊的组件,它可被放置在其他容器中
D. 容器是一种特殊的组件,它可被放置在任何组件中

4. 以下关于 BorderLayout 类功能的描述,错误的是(　　)。
A. 它可以与其他布局管理器协同工作
B. 它可以对 GUI 容器中的组件完成边框式的布局
C. 它位于 java.awt 包中
D. 它是一种特殊的组件

5. JTextField 类提供的 GUI 功能是(　　)。
A. 文本区域
B. 按钮
C. 文本字段
D. 菜单

6. 将 GUI 窗口划分为东、西、南、北、中 5 个部分的布局管理器是(　　)。
A. FlowLayout
B. GridLayout
C. CardLayout
D. BorderLayout

7. 在 Swing GUI 编程中,setDefaultCloseOperation(JFrame.EXIT_ON_CLOSE)语句的作用是(　　)。
A. 当执行关闭窗口操作时,不作任何操作
B. 当执行关闭窗口操作时,调用 WindowsListener 对象并将隐藏 JFrame
C. 当执行关闭窗口操作时,退出应用程序
D. 当执行关闭窗口操作时,调用 WindowsListener 对象并隐藏和销毁 JFrame

8. (　　)、(　　)是 Swing 容器的顶层容器。
A. JPanel
B. JScrollPane
C. JWindow

D. JFrame

9. 我们要设计实现一个对话框,该对话框上有 5 个垂直方向排列的大小一致的按钮,这些按钮要充满对话框的空余空间,并且当对话框改变大小的时候按钮也要跟着改变大小,同时保证大小一致,该对话框应该选择的布局方式为(　　)。

A. FlowLayout

B. BorderLayout

C. GridLayout

D. null(no LayoutManager)

10. (　　)布局管理器是从上到下、从左到右安排组件,当移动到下一行时是居中的。

A. BorderLayout

B. FlowLayout

C. GridLayout

D. CardLayout

二、简答题

1. 创建有一个文本框和 3 个按钮的小程序,当按下每个按钮时,使不同的文字显示在文本框中。

2. 编程增加一个复选框到上题创建的程序中,使得选中或不选中复选框时插入的文字不同。

3. 编写一个程序,接受用户输入的账号和密码,给 3 次输入机会。

4. 编写成人标准身高和体重互查的程序。身高和体重在两个不同的文本框中输入,要求输入一个。输入身高,则输出体重;输入体重,则输出身高。用一个按钮启动互查,互查的公式为:

体重=身高-100

5. 阅读下面的程序,回答问题:

```
import java.awt.*;
import javax.swing.*;
public class T extends JFrame{
    public T(){
        super("GridLayout");
        Container con = this.getContentPane();
        con.setLayout(new GridLayout(2,3));
        con.add(new JButton("a"));
        con.add(new JButton("b"));
        con.add(new JButton("c"));
        con.add(new JButton("d"));
        con.add(new JButton("e"));
        con.add(new JButton("f"));
```

```
            setSize(200,80);
            setVisible(true);
        }
        public static void main(String args[]){
          new T();
        }
    }
```

画图表示程序运行后的图形界面。

学生成绩管理系统的数据库编程

项目 7 首先讲解了 MySQL 数据库及其基本 SQL 语法，并创建学生成绩数据库 S_Score；其次，介绍了 Java 数据库连接技术（JDBC），讲解了 JDBC 操作步骤、Statement 接口、ResultSet 接口、PreparedStatement 接口，并实现了学生成绩管理系统的界面功能。

工作 任务

（1）创建学生成绩数据库 S_Score。
（2）实现学生成绩管理系统界面功能。

学习 目标

（1）掌握 MySQL 的基本操作及其对应的 SQL 语法，能够独立完成 MySQL 数据库管理软件的安装，能够独立完成 MySQL 数据库的创建及管理。
（2）了解 JDBC 的概念，掌握 JDBC 访问数据库的操作步骤，掌握 Statement 接口、ResultSet 接口、PreparedStatement 接口，能够使用 DriverManager、Connection、PreparedStatement、ResultSet 对 MySQL 数据库进行增加、删除、修改、查询操作。

7.1 任务 1 创建学生成绩数据库 S_Score

小明要将学生成绩管理系统图形界面涉及的相关数据保存到数据库，必须先完成学生成绩数据库 S_Score 的创建。要完成这个工作任务，第一步，我们要完成 MySQL 数据库管理软件的安装；第二步，由于创建 MySQL 数据库要用到相关的 SQL 命令，所以在创建数据库之前我们还要复习、学习相关 SQL 基础语法，为后续的数据库操作奠定基础；第三步，利用安装的 MySQL 数据库管理软件实现学生成绩数据库 S_Score 及相关数据表格的创建。

相关知识

7.1.1 MySQL 数据库

1. MySQL 数据库简介

MySQL 是一个小型关系型数据库管理系统，开发者为瑞典 MySQL AB 公司，2008 年 1 月 MySQL 被美国的 Sun 公司收购，2009 年 4 月 Sun 公司又被美国的甲骨文(Oracle)公司收购。

MySQL 是一个单进程多线程、支持多用户、基于客户机/服务器(Client/Server，简称 C/S)的关系数据库管理系统。由于其体积小、速度快、总体拥有成本低，尤其是开放源码这一特点，许多中小型企业为了降低成本，纷纷选择了 MySQL 数据库。MySQL 的官方网站的网址是：www.mysql.com，本书使用的 MySQL 版本是 5.7.23。

2. MySQL 数据库安装

(1) 下载 MySQL 数据库。

① 打开网址 https://www.mysql.com/，进入如图 7-1 所示的 MySQL 官网主页。

图 7-1

② 点击"DOWNLOADS"，进入如图 7-2 所示的下载页面。

图 7-2

③ 在如图 7-2 所示的下载页面的下部,找到如图 7-3 所示的"Community(GPL) Downloads"超链接,点击该超链接进入如图 7-4 所示的"MySQL Community Downloads"页面。

图 7-3

图 7-4

④ 点击如图 7-4 所示"MySQL Community Downloads"页面左侧的"MySQL Community Server"选项,进入如图 7-5 所示的"Download MySQL Community Server"页面。

图 7-5

⑤ 在如图 7-5 所示的"Download MySQL Community Server"页面，点击超链接"MySQL Community Server 5.7"，进入如图 7-6 所示的"MySQL Community Server 5.7.23"页面。

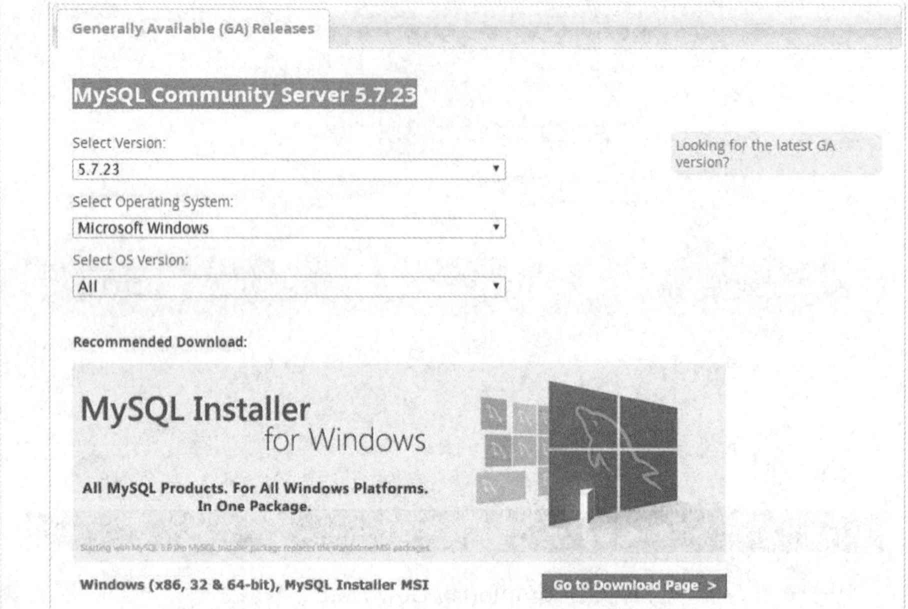

图 7-6

⑥ 在如图 7-6 所示的"MySQL Community Server 5.7.23"页面，点击"Go To Download Page"超链接，进入如图 7-7 所示的"Download MySQL Installer"页面。

图 7-7

⑦ 在如图 7-7 所示的"Download MySQL Installer"页面的下部找到如图 7-8 所示的"MySQL Installer 5.7.23"选项下载列表，点击"mysql-installer-community-5.7.23.0.msi"离线安装包对应的【Download】按钮，进入如图 7-9 所示的"Begin Your Download"页面。

图 7-8

图 7-9

⑧ 在如图 7-9 所示的"Begin Your Download"页面末尾处找到如图 7-10 所示的超链接"No thanks,just start my download.",点击该超链接后,弹出如图 7-11 所示下载页面,点击【下载】按钮,完成离线安装包的下载。

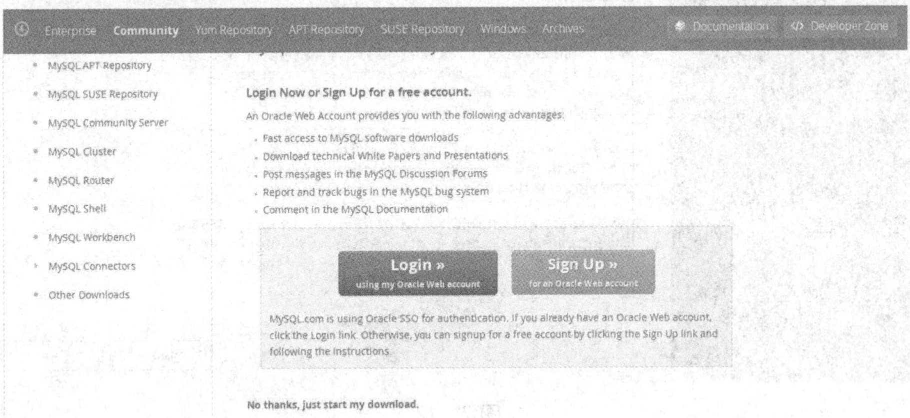

图 7-10

图 7-11

(2) 安装 MySQL 数据库。

① 双击打开下载的"mysql-installer-community-5.7.23.0.msi"离线安装包,出现如图 7-12 所示"License Agreement"(协议同意)安装界面,选中"I accept the license terms"(我同意协议)选项,点击【Next】按钮,进入如图 7-13 所示"Choosing a Setup Type"(选择安装类型)安装界面。

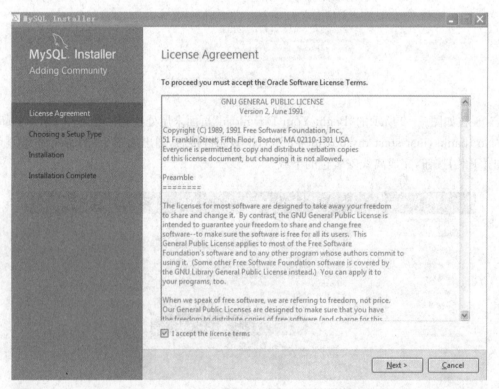

图 7-12

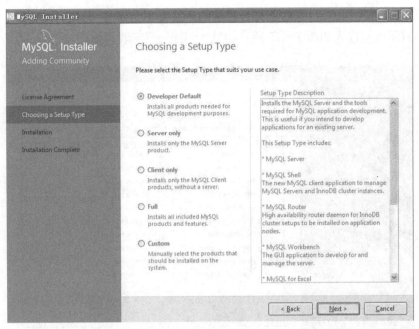

图 7-13

② 在"Choosing a Setup Type"(选择安装类型)安装界面选择"Developer Default"(开发者默认),然后点击【Next】按钮,进入如图 7-14 所示"Check Requirements"安装界面。

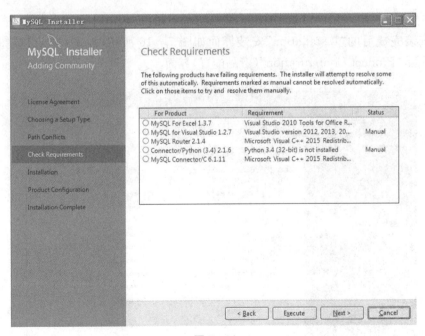

图 7-14

③ 在如图 7-14 所示的"Check Requirements"安装界面,可检查电脑当前的安装环境,并根据安装环境选择要安装的组件,然后点击【Next】按钮(点击该按钮时,会提示"一个或者

多个产品要求没有得到满足,那些不符合要求的产品将不会安装/升级,你想要继续吗?",选择【是(Yes)】即可),进入如图 7-15 所示的"Installation"安装界面。

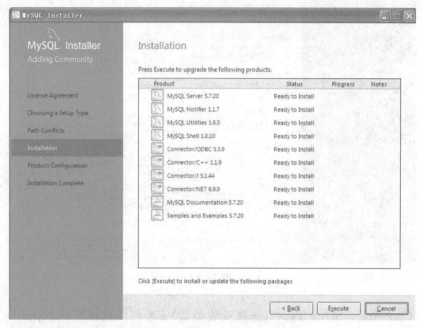

图 7-15

④ 在如图 7-15 所示的"Installation"安装界面点击【Execute】(执行)按钮,开始安装软件,全部安装完成后的"Installation"安装界面如图 7-16 所示,点击【Next】按钮,进入如图 7-17 所示"Product Configuration"(产品配置)界面。

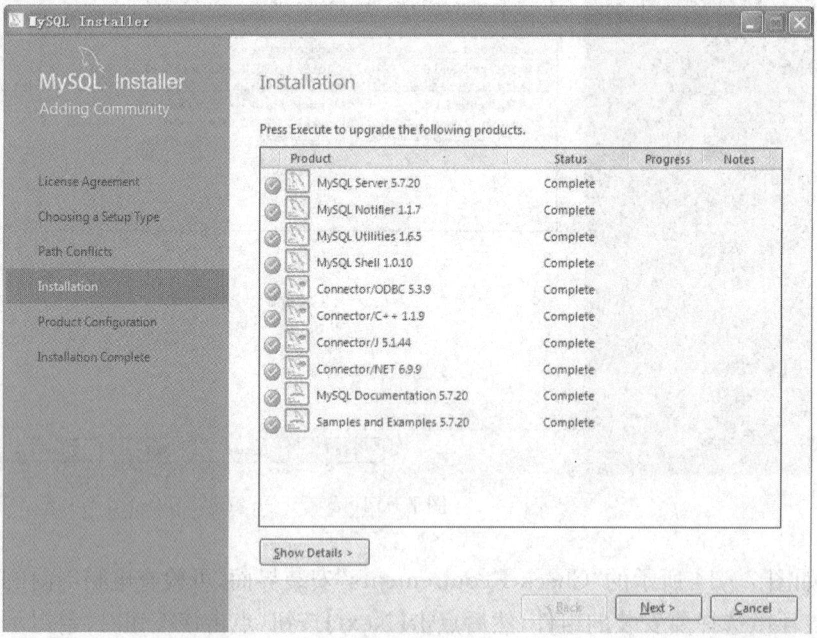

图 7-16

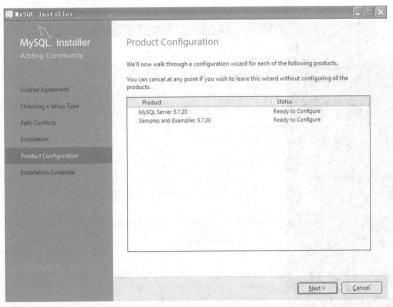

图 7-17

⑤ 在如图 7-17 所示"Product Configuration"(产品配置)界面,点击【Next】按钮,进入如图 7-18 所示"Type and Networking"(类型和网络)配置界面(可采用默认设置),选择"Standalone MySQL Server/Classic MySQL Replication"(独立的 MySQL 服务器/经典的 MySQL 复制)选项,点击【Next】按钮,在如图 7-19 所示的"Type and Networking"界面中设置"Config Type"(配置类型)项为"Development Machine",设置"Port Number"(端口号)项为"3306",然后点击【Next】按钮,进入如图 7-20 所示的"Accounts and Roles"配置界面。

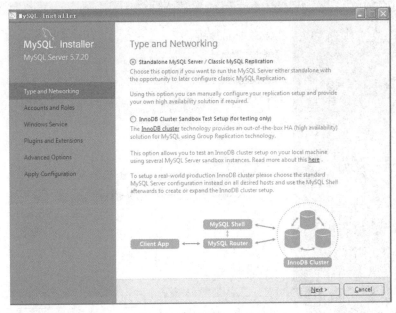

图 7-18

图 7-19

⑥ 在如图 7-20 所示的"Accounts and Roles"配置界面的"MySQL Root Password"项设置 root 密码(本书的 root 密码统一设置为 123456),在"Repeat Password"项再次输入 root 密码(用于确认),然后点击【Next】按钮,进入如图 7-21 所示"Windows Services"配置界面。

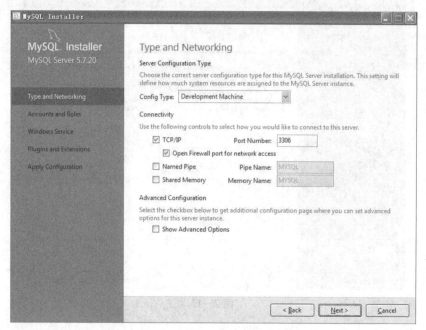

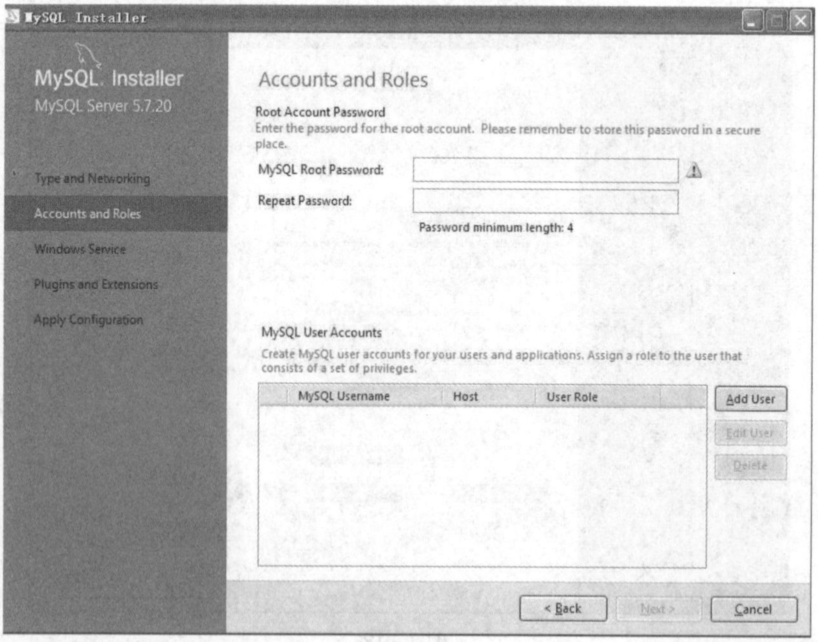

图 7-20

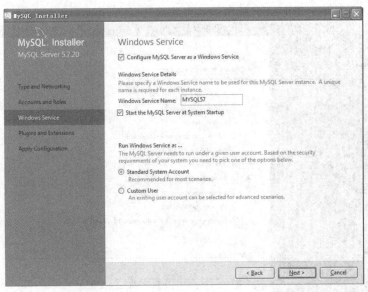

图 7-21

⑦ 在如图 7-21"Windows Services"(系统服务)配置界面中,可对"Windows Service Name"(MySQL 在 Windows 系统中的名字)、"Start the MySQL Server at System Startup"(开机启动 MySQL 服务)等项进行设置(可采用默认设置),然后点击【Next】按钮,进入如图 7-22 所示"Plugins and Extensions"(配置插件和扩展页面)配置界面。在"Plugins and Extensions"配置界面中,可采用默认配置,然后点击【Next】按钮,进入如图 7-23 所示"Apply Configuration"(应用配置页面)界面,在该界面点击【Execute】按钮,开始安装配置。配置安装完后,点击【Finish】按钮,又再次返回如图 7-17 所示的"Product Configuration"(产品配置)界面,继续点击【Next】按钮,进入如图 7-24 所示的"Connector To Server"界面。

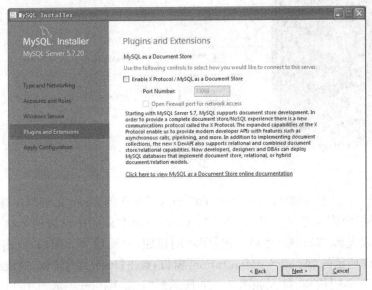

图 7-22

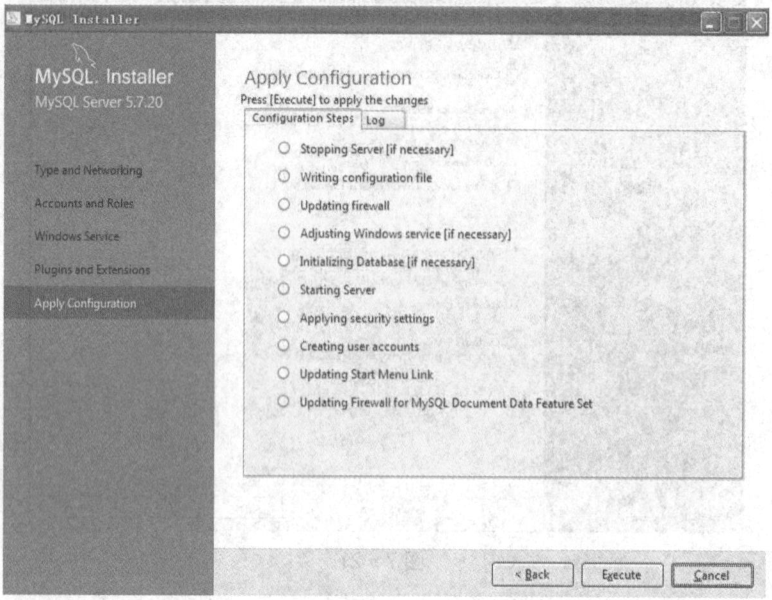

图 7-23

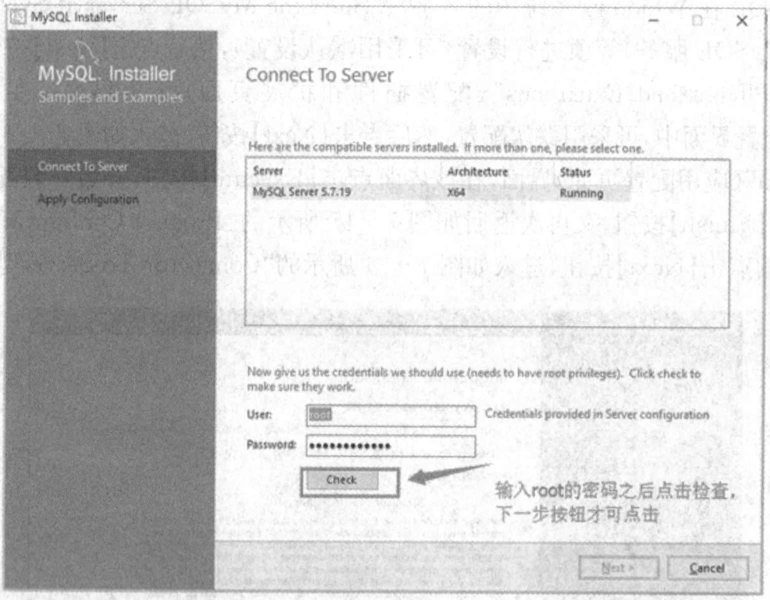

图 7-24

⑧ 在如图 7-24 所示的"Connector To Server"界面的 User 文本框中输入"root",在 Password 密码框中输入之前在图 7-20"Accounts and Roles"中设置的 root 密码(本书的 root 密码统一设置为 123456),然后点击【Check】按钮,密码检测无误后,点击【Next】按钮,进入如图 7-25 所示的"Apply Configuration"窗口,点击【Execute】按钮,完成服务器的配置应用后,点击【Finish】按钮,进入如图 7-26 所示的"Installation Complete"窗口,然后点击【Finish】按钮,完成软件安装。

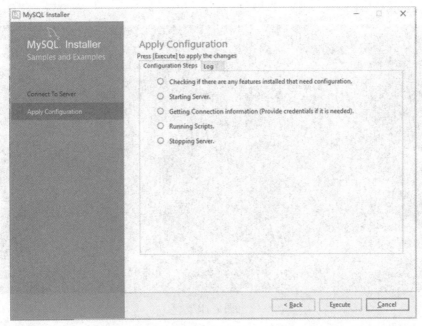

图 7-25

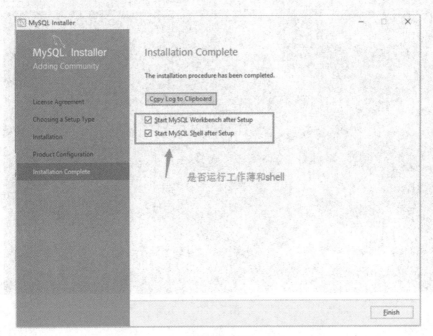

图 7-26

⑨ 安装完成后,点击"开始"—"所有程序"—"MySQL"—"MySQL Server 5.7"—"MySQL Command Line Client",打开如图 7-27 所示的 MySQL 客户端,然后输入 root 密码(123456),回车。如果出现如图 7-28 所示的界面时,表示 MySQL 软件安装成功。

图 7-27

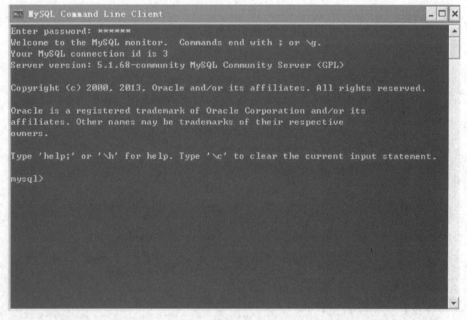

图 7-28

3. MySQL 常见命令

MySQL 数据库采用命令方式访问、操作数据库，命令行以分号作为结束符，命令不区分大小写。

(1) 连接 MySQL。

格式：mysql-h 主机地址-u 用户名-p 用户密码

① 连接到本机上的 MySQL。

首先打开 DOS 窗口，然后使用"cd"命令，进入"安装目录\bin"，如图 7-29 所示，再键入命令"mysql－u root－p123456"，回车后就连接到本机 MySQL，MySQL 的提示符是：mysql＞，如图 7-30 所示。（注：用户名前可以有空格也可以没有空格，但是密码前必须没有空格，否则需要重新输入密码。）

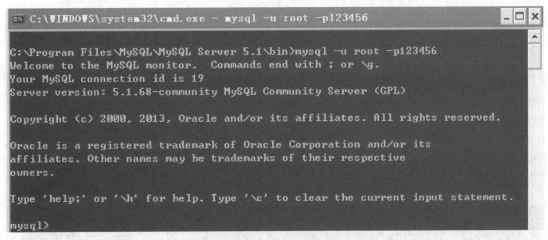

图 7-29

图 7-30

② 连接到远程主机上的 MySQL。

假设远程主机的 IP 为：120.120.120.120，用户名为 root，密码为 123456。在 DOS 下进入"安装目录\bin"后，键入以下命令：

　mysql－h 120.120.120.120－u root－p123456

（注：u 与 root 之间可以不用加空格）

③ 退出 MySQL 命令。

exit（回车），运行结果如图 7-31 所示。

(2) 修改密码。

格式：mysqladmin－u 用户名－p 旧密码 password 新密码

① 如果没有设置 root 密码，给 root 设置密码 123。

首先在 DOS 下进入"安装目录\bin"，然后键入以下命令并回车：

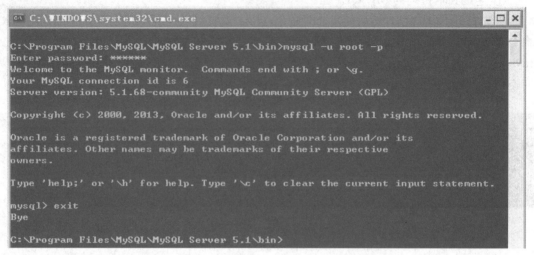

图 7-31

```
mysqladmin -u root -password 123
```

（因为开始时 root 没有密码，所以 -p 旧密码一项就可以省略了）

② 如果需要修改密码，例如将 root 的密码修改为 123，在 DOS 下进入"安装目录\bin"后，输入如下命令并回车（结果如图 7-32 所示）：

```
mysqladmin -u root -p123456 password 123
```

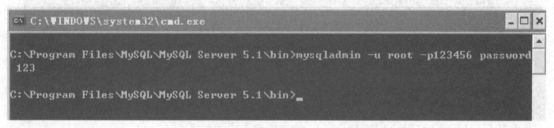

图 7-32

(3) 查看帮助。

如果想知道在 MySQL 中有哪些命令，在连接上 MySQL 后，可以输入"?"并回车，如图 7-33 所示。

(4) 创建数据库。

使用 MySQL 可以方便地创建数据库，其语法格式如下：

create database〈数据库名〉；（注：创建数据库之前要先连接 MySQL 服务器，注意以下的各种数据库相关操作都必须先连接上 MySQL 服务器）

例如，建立一个名为 S_Score 的数据库，结果如图 7-34 所示：

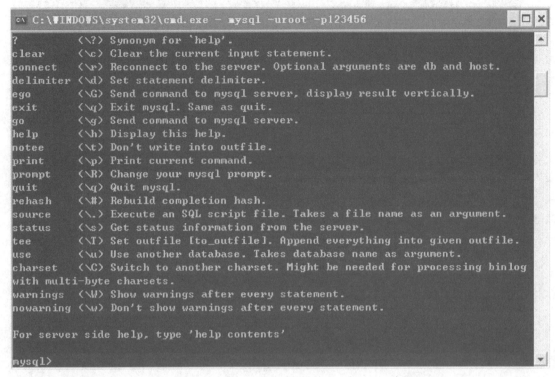

图 7-33

```
create database S_Score;
```

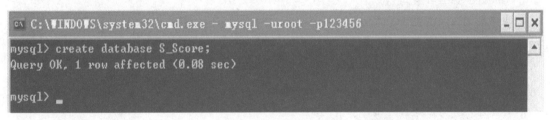

图 7-34

(5) 显示数据库。

如果想要查看 MySQL 中的数据库,其语法格式如下(结果如图 7-35 所示):

```
show databases;
```

(注:最后有个 s)

(6) 使用数据库。

命令格式:use〈数据库名〉;

例如,如果数据库 S_Score 已经存在,现在想使用它,结果如图 7-36 所示:

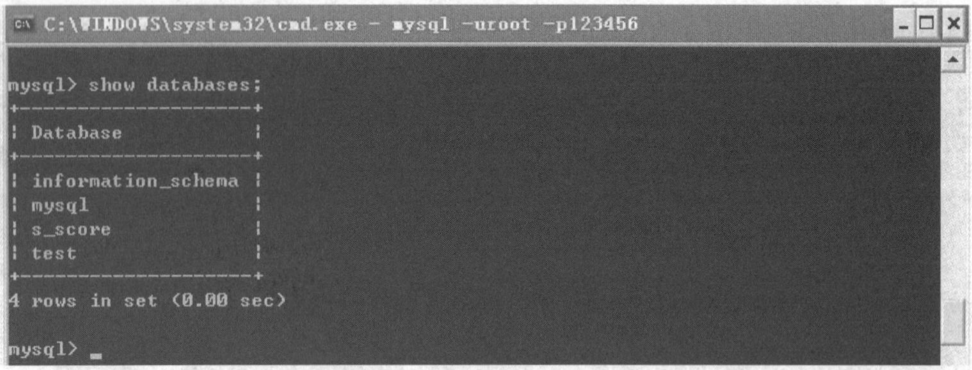

```
图 7-35
```

```
use S_Score;
```

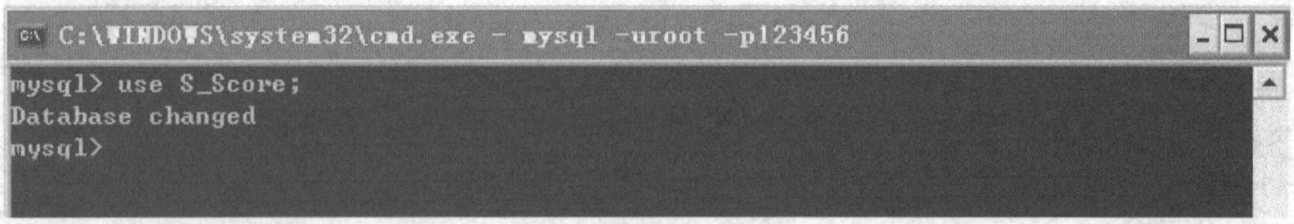

图 7-36

(7) 创建数据表。

命令格式：

```
create table<表名>(
<字段名1><类型1>[约束1],
<字段名2><类型2>[约束2],
<字段名3><类型3>[约束3],
……
<字段名n><类型n>[约束n]
);
```

注意：

创建表之前必须指定使用的数据库。

例如，在数据库 S_Score 中建立一个名为 User 的表，字段信息如表 7-1 所示。

表 7-1

字段名	数字类型	数据宽度	是否为空	是否主键	自动增加	默认值
id	int	11	否	primary key	auto_increment	
userName	varchar	50	否			
password	varchar	50	否			

```
use S_Score;
create table User(
id int(11)not null primary key auto_increment,
userName varchar(50)not null,
password varchar(50)not null
);
```

```
mysql> use  S_Score;
Database changed
mysql> create table User(
    -> id int(11) not null primary key auto_increment,
    -> userName varchar(50) not null,
    -> password varchar(50) not null
    -> );
Query OK, 0 rows affected (0.33 sec)

mysql>
```

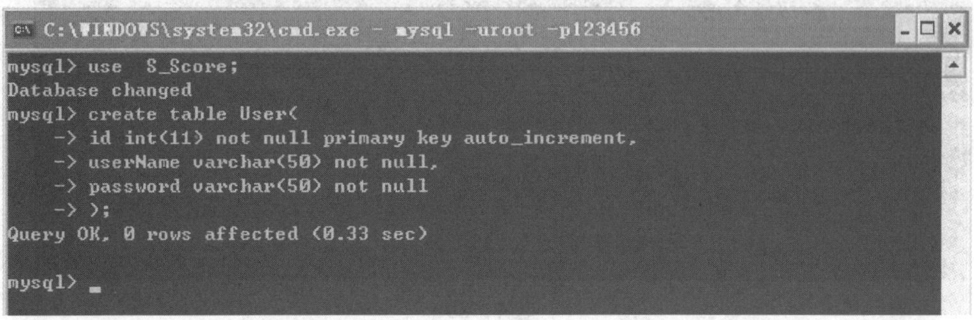

图 7-37

(8) 查看数据表和数据表的结构。

如果需要查看一个已经创建好的数据库中有哪些数据表,并且想查看指定的某一个数据表的结构,其语法格式分别如下:

① 查看数据库的全部数据表。

show tables;(注:在使用该命令前,必须已经指定使用的数据库,并且最后加 s)

例如,查看数据库 S_Score 中有哪些表,其结果如图 7-38 所示。

```
use S_Score;
show tables;
```

```
mysql> use  S_Score;
Database changed
mysql> show tables;
+------------------+
| Tables_in_s_score |
+------------------+
| user             |
+------------------+
1 row in set (0.00 sec)

mysql>
```

图 7-38

② 查看数据表结构。

desc 〈表名称〉;(注:在使用该命令前,必须已经指定使用的数据库)

例如，查看 S_Score 数据库中 user 表的结构，结果如图 7-39 所示。

```
use S_Score;
desc user;
```

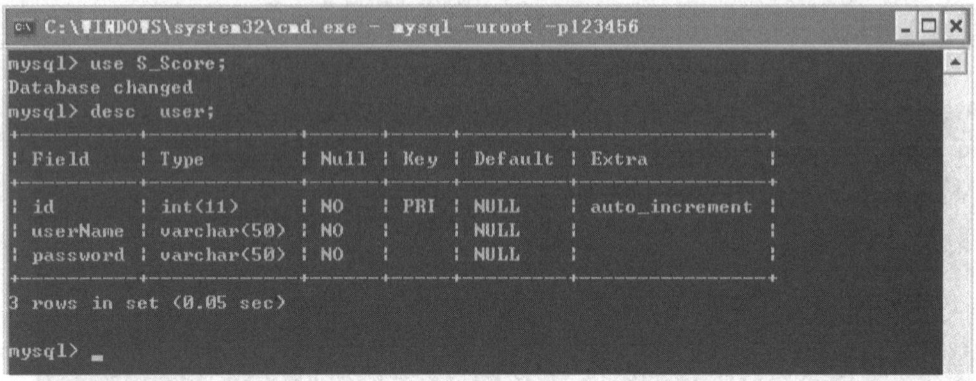

图 7-39

（9）删除表。
命令格式：drop table〈表名称〉；
例如，删除数据库 S_Score 中的 user 数据表，结果如图 7-40 所示。

```
use S_Score;
drop table user;
```

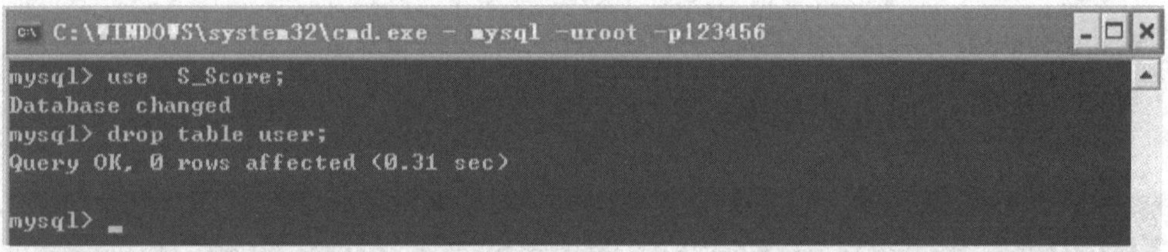

图 7-40

（10）删除数据库。
命令格式：drop database 〈数据库名称〉；
例如，删除数据 S_Score，结果如图 7-41 所示。

```
drop database S_Score;
```

7.1.2　SQL 基础语法

SQL（Structured Query Language，结构查询语言）是一种功能强大的数据库语言，通常用于与数据库的通信，是关系数据库管理系统的标准语言。

图 7-41

SQL 按功能可分为以下 3 种：
• DML(Data Manipulation Language,数据操作语言)用于检索或者修改数据。
• DDL(Data Definition Language,数据定义语言)用于定义数据的结构,如创建、修改或者删除数据库对象。
• DCL(Data Control Language,数据控制语言)用于定义数据库用户的权限。

MySQL 数据库良好地支持了标准的 SQL 语法,本模块主要涉及的是数据的增加、修改、删除、查询等相关 DDL 操作,在讲解时我们以之前创建的表 7-1 为例,完成全部相关操作讲解。

1. MySQL 的数据类型

MySQL 定义的数据字段类型,大致可以分为 3 类:数值、日期/时间和字符串(字符)类型。

(1) 数值类型。

MySQL 支持所有标准 SQL 数值数据类型,这些类型包括严格数值数据类型(INTEGER、SMALLINT、DECIMAL 和 NUMERIC),以及近似数值数据类型(FLOAT、REAL 和 DOUBLE PRECISION)。

关键字 INT 是 INTEGER 的同义词,关键字 DEC 是 DECIMAL 的同义词。

BIT 数据类型保存位字段值,并且支持 MyISAM、MEMORY、InnoDB 和 BDB 表。

作为 SQL 标准的扩展,MySQL 也支持整数类型 TINYINT、MEDIUMINT 和 BIGINT。数值类型的详细信息如表 7-2 所示。

表 7-2

类型	大小	范围(有符号)	范围(无符号)	用途
TINYINT、BIT、BOOLEAN	1 字节	(-128,127)	(0,255)	小整数值
SMALLINT	2 字节	(-32 768,32 767)	(0,65 535)	大整数值
MEDIUMINT	3 字节	(-8 388 608,8 388 607)	(0,16 777 215)	大整数值
INT、INTEGER	4 字节	(-2 147 483 648,2 147 483 647)	(0,4 294 967 295)	大整数值

续 表

类型	大小	范围(有符号)	范围(无符号)	用途
BIGINT	8字节	(−9 233 372 036 854 775 808,9 223 372 036 854 775 807)	(0,18 446 744 073 709 551 615)	极大整数值
FLOAT	4字节	(−3.402 823 466 E+38,−1.175 494 351 E−38),0,(1.175 494 351 E−38,3.402 823 466 351 E+38)	0,(1.175 494 351 E−38,3.402 823 466 E+38)	单精度浮点数值
DOUBLE	8字节	(−1.797 693 134 862 315 7 E+308,−2.225 073 858 507 201 4 E−308),0,(2.225 073 858 507 201 4 E−308,1.797 693 134 862 315 7 E+308)	0,(2.225 073 858 507 201 4 E−308,1.797 693 134 862 315 7 E+308)	双精度浮点数值
DECIMAL、DEC	对 DECIMAL(M, D),如果 M > D,为 M+2,否则为 D+2	依赖于 M 和 D 的值	依赖于 M 和 D 的值	小数值

(2) 日期和时间类型。

表示时间值的日期和时间类型为 DATETIME、DATE、TIMESTAMP、TIME 和 YEAR。每个时间类型有一个有效值范围和一个"零"值,当指定 MySQL 不能表示的不合法的值时使用"零"值。TIMESTAMP 类型有专有的自动更新特性。日期和时间类型的详细信息如表 7-3 所示。

表 7-3

类型	大小(字节)	范围	格式	用途
DATE	3	1000-01-01/9999-12-31	YYYY-MM-DD	日期值
TIME	3	'−838:59:59'/'838:59:59'	HH:MM:SS	时间值或持续时间
YEAR	1	1901/2155	YYYY	年份值
DATETIME	8	1000-01-01 00:00:00/9999-12-31 23:59:59	YYYY-MM-DD HH:MM:SS	混合日期和时间值

续 表

类型	大小(字节)	范围	格式	用途
TIMESTAMP	4	1970-01-01 00:00:00/2038 结束时间是第 2 147 483 647 秒,北京时间 2038-1-19 11:14:07,格林尼治时间 2038 年 1 月 19 日凌晨 03:14:07	YYYYMMDD HHMMSS	混合日期和时间值,时间戳

（3）字符串类型。

字符串类型指 CHAR、VARCHAR、BINARY、VARBINARY、BLOB、TEXT、ENUM 和 SET。

CHAR 和 VARCHAR 类型类似,但它们保存和检索的方式不同。它们的最大长度和是否尾部空格被保留等方面也不同。在存储或检索过程中不进行大小写转换。

BINARY 和 VARBINARY 类似于 CHAR 和 VARCHAR,不同的是它们包含二进制字符串而不要非二进制字符串。也就是说,它们包含字节字符串而不是字符字符串。这说明它们没有字符集,并且排序和比较基于列值字节的数值。

BLOB 是一个二进制大对象,可以容纳可变数量的数据。有 4 种 BLOB 类型：TINYBLOB、BLOB、MEDIUMBLOB 和 LONGBLOB。它们的区别在于可容纳存储范围不同。

有 4 种 TEXT 类型：TINYTEXT、TEXT、MEDIUMTEXT 和 LONGTEXT。对应的这 4 种 BLOB 类型,可存储的最大长度不同,可根据实际情况选择。

字符串类型的详细信息如表 7-4 所示。

表 7-4

类型	大小	用途
CHAR	0～255 字节	定长字符串
VARCHAR	0～65 535 字节	变长字符串
TINYBLOB	0～255 字节	不超过 255 个字符的二进制字符串
TINYTEXT	0～255 字节	短文本字符串
BLOB	0～65 535 字节	二进制形式的长文本数据
TEXT	0～65 535 字节	长文本数据
MEDIUMBLOB	0～16 777 215 字节	二进制形式的中等长度文本数据
MEDIUMTEXT	0～16 777 215 字节	中等长度文本数据
LONGBLOB	0～4 294 967 295 字节	二进制形式的极大文本数据
LONGTEXT	0～4 294 967 295 字节	极大文本数据

在以上数据类型中,经常使用到的是 INT、FLOAT、VARCHAR、TEXT、DATE、DATETIME,必须重点掌握。

2. 增加数据

向 MySQL 数据表中增加数据,其语法格式如下:

INSERT INTO 表名称[(字段1,字段2,…,字段n,)]VALUES(值1,值2,…,值n,);

例如,向数据库 S_Score 的 user 表添加如下信息(结果如图 7-42 所示):

userName-fanlingyun

Password-123456

SQL 命令如下:

```
use S_Score;
INSERT INTO user(userName,password)VALUES('fanlingyun','123456');
```

```
mysql> INSERT INTO user(userName,password) values('fanlingyun','123456');
Query OK, 1 row affected (0.14 sec)
```

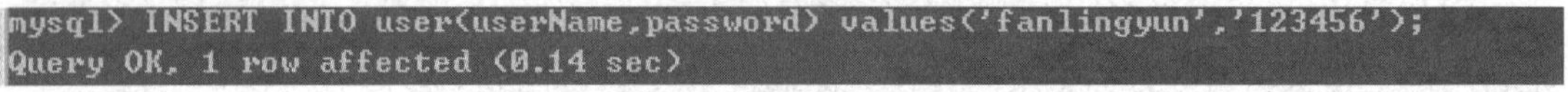

图 7-42

3. 查询数据

在 SQL 语法中,如果要查看数据表中的记录内容,可以通过查询语句完成,其语法格式如下:

```
SELECT{ * |column alias}
FROM 表名称    别名
[WHERE condition(s)];
```

(1) 查询表中全部数据。

例如,查询 user 中全部数据,SQL 命令如下(结果如图 7-43 所示):

```
select * from user;
```

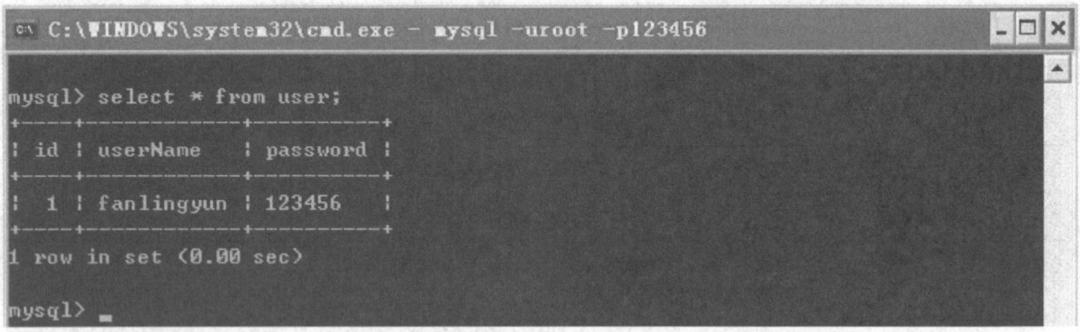

```
C:\WINDOWS\system32\cmd.exe - mysql -uroot -p123456

mysql> select * from user;
+----+-----------+----------+
| id | userName  | password |
+----+-----------+----------+
|  1 | fanlingyun| 123456   |
+----+-----------+----------+
1 row in set (0.00 sec)

mysql>
```

图 7-43

(2) 查询表格中部分列。

例如,查询 user 表中的 userName 列信息,SQL 命令如下(结果如图 7-44 所示):

```
select userName from user;
```

```
mysql> select userName from user;
+-----------+
| userName  |
+-----------+
| fanlingyun |
+-----------+
1 row in set (0.02 sec)

mysql>
```

图 7-44

(3) 查询满足条件的相关列。

例如,查询 user 表中 userName 为 fanlingyun 的用户密码,SQL 命令如下(结果如图 7-45 所示):

```
select password from user where userName = 'fanlingyun';
```

```
mysql> select password from user where userName='fanlingyun';
+----------+
| password |
+----------+
| 123456   |
+----------+
1 row in set (0.05 sec)
```

图 7-45

4. 修改数据

如果要修改数据表中的某些记录,可以使用 UPDATE 语句,语句格式如下:

```
UPDATE 表名称 SET 字段1 = 值1, …, 字段n = 值n[WHERE 更新条件];
```

例如,修改 user 表中 userName 为 fanlingyun 的用户的 password 为 123,SQL 命令如下(结果如图 7-46 所示):

```
update user set password = '123'where userName = 'fanlingyun';
```

```
mysql> update user set password='123' where userName='fanlingyun';
Query OK, 1 row affected (0.06 sec)
Rows matched: 1  Changed: 1  Warnings: 0
```

图 7-46

5. 删除数据

删除数据表中的数据，可以使用如下的语法：

```
DELETE FROM 表名称[删除条件];
```

在进行删除数据的时候，最好指定删除的条件，如果没有指定的话，则表示删除一张表中的全部数据。

例如，删除 user 表中 userName 为 fanlingyun 的记录，SQL 命令如下（结果如图 7 - 47 所示）：

```
delete from user where userName = 'fanlingyun';
```

```
mysql> delete from user where userName='fanlingyun';
Query OK, 1 row affected (0.05 sec)
```

图 7 - 47

任务 实施

7.1.3 创建学生成绩数据库 S_Score（包含 user 表、StudentScore 表）

（1）创建 S_Score 数据库代码如下：

```
create database S_Score;
use S_Score;
```

如图 7 - 48 所示。

```
mysql> create database S_Score;
Query OK, 1 row affected (0.03 sec)

mysql> use S_Score;
Database changed
mysql>
```

图 7 - 48

（2）创建 user 表代码如下：

```
create table User(
id int(11)not null primary key auto_increment,
```

```
userName varchar(50)not null,
password varchar(50)not null
);
```

(3) 向 user 表插入用户名 fan、密码 123456 的语句如下：

```
INSERT INTO user(userName,password)VALUES('fan','123456');
```

如图 7-49 所示。

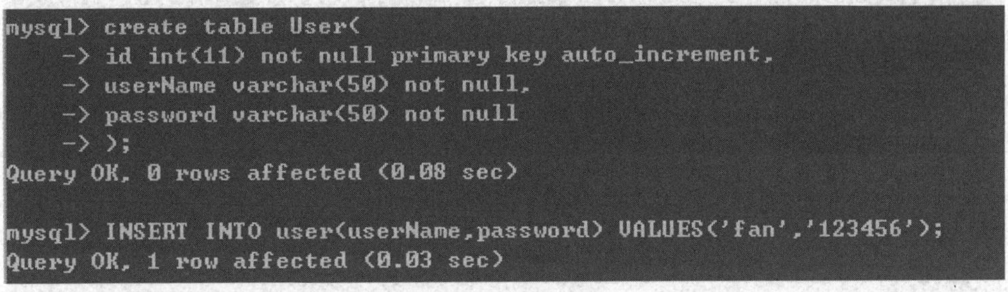

```
mysql> create table User(
    -> id int(11) not null primary key auto_increment,
    -> userName varchar(50) not null,
    -> password varchar(50) not null
    -> );
Query OK, 0 rows affected (0.08 sec)

mysql> INSERT INTO user(userName,password) VALUES('fan','123456');
Query OK, 1 row affected (0.03 sec)
```

图 7-49

(4) 创建 StudentScore 表代码如下：

```
create table StudentScore(
sno int(11)not null primary key,
sname varchar(50)not null,
sex char not null,
cname varchar(50)not null,
pingShiScore tinyint not null,
guoChengScore tinyint not null,
zhongJieScore tinyint not null,
zongHeScore double not null,
zongHeScoreJiBie varchar(20)not null,
love varchar(50)not null
);
```

如图 7-50 所示。

任务 拓展

创建自助银行服务系统的数据库 Bank，并在数据库 Bank 中创建如表 7-5 所示的用户信息表(UserInfo)：

```
mysql> create table StudentScore(
    -> sno int(11) not null primary key,
    -> sname varchar(50) not null,
    -> sex  char  not null,
    -> cname varchar(50) not null,
    -> pingShiScore tinyint not null,
    -> guoChengScore tinyint not null,
    -> zhongJieScore tinyint not null,
    -> zongHeScore double not null,
    -> zongHeScoreJiBie varchar(20) not null,
    -> love varchar(50) not null
    -> );
Query OK, 0 rows affected (0.06 sec)
```

图 7-50

表 7-5

字段名	类型	备注	属性
UserID	varchar(20)	用户卡号	primary key
Username	varchar(20)	用户名	not null
UserPW	varchar(20)	用户密码	not null
Userphone	varchar(20)	联系方式	not null
Usersex	varchar(10)	性别	not null
Useryue	int	余额	not null

任务 2　实现学生成绩管理系统界面功能

任务描述及分析

小明要通过学生成绩管理系统的图形界面实现对学生成绩数据的操作(信息添加、信息删除、信息修改、信息查询),还要通过图形界面查询数据库实现用户的安全登录。要完成这个工作任务,第一步,我们需要了解 JDBC 的概念,掌握 JDBC 访问数据库的操作步骤;第二步,我们需要掌握 JDBC 操作数据库的相关知识(DriverManager、Connection、PreparedStatement、ResultSet 等);第三步,在第一步和第二步学习的基础上,访问数据库实现用户的登录判断、信息的查询和信息的管理。

相关知识

7.2.1 JDBC 概述

JDBC(Java Database Connectivity,Java 数据库连接)提供了一种与平台无关的用于执行 SQL 语句的标准 Java API(应用程序设计接口),可以方便地实现多种关系型数据库的统一操作,它由一组用 Java 语言编写的类和接口组成。

JDBC 提供的是一套数据库操作标准,每一种数据库的生产厂商都会为其提供一个对应的 JDBC 的驱动程序,目前比较常见的 JDBC 驱动程序可以分为以下 4 类:

(1) JDBC-ODBC 桥驱动。

JDBC-ODBC 是 Sun 公司提供的标准 JDBC 操作,直接利用微软公司的 ODBC 进行数据库的连接操作。虽然 JDBC-ODBC 的一系列操作类都是最新的,但是这种操作性能较低,通常情况下不推荐使用这种方式进行操作,尤其是在实际的开发中一般也不会采用 JDBC-ODBC 桥驱动。

(2) JDBC 本地驱动。

JDBC 本地驱动是各个数据库生产商提供的 JDBC 驱动程序,JDBC 本地驱动只能应用在特定的数据库上,因此会导致程序的不可移植,但采用 JDBC 本地驱动进行数据库操作的性能较高。此外,JDBC 本地驱动程序往往是以一组 jar 包(或 Zip 包)的形式出现,需要单独配置的 classpath 环境变量。由于在实践开发中,大部分应用软件都是基于一种数据库的开发,因此使用此种模式是最多的(本书也是采用此种模式)。

(3) JDBC 网络驱动。

JDBC 网络驱动程序将 JDBC 转换为与 DBMS(数据库管理系统)无关的网络协议,之后这种协议又被某个服务器转换为一种 DBMS 协议,这种网络服务器中间件能够将它的纯 Java 客户机连接到多种不同的数据库上,因此其所用的具体协议取决于提供者。通常,这是最为灵活的 JDBC 驱动程序。

(4) 本地协议纯 JDBC 驱动。

本地协议纯 JDBC 驱动程序将 JDBC 调用直接转换为 DBMS 所使用的网络协议,这将允许从客户机上直接调用 DBMS 服务器,是互联网访问的一个很实用的解决方法。

JDBC 由 Java 语言编写的类和接口组成,通过这些类和接口可以完成数据库连接的创建、SQL 语句的执行等操作。JDBC 中所有的类和接口都保存在 java.sql 包中,在此包中规定了大量的类和接口。JDBC 的主要操作类及接口如表 7-6 所示。

表 7-6

序号	类及接口	描述
1	java.sql.DriverManager	用于管理 JDBC 驱动程序
2	java.sql.Connection	用于建立与特定数据库的连接,一个连接就是一个会话,建立连接后便可以执行 SQL 语句和获得检索结果

续表

序号	类及接口	描述
3	java.sql.Statement	一个 Statement 对象用于执行静态 SQL 语句,并获得语句执行后产生的结果
4	java.sql.PreparedStatement	创建一个可以编译的 SQL 语句对象,该对象可以被多次运行,以提高执行的效率,接口是 Statement 的子接口
5	java.sql.ResultSet	用于创建表示 SQL 语句检索结果的结果集,用户通过结果集完成对数据库的访问
6	java.sql.Date	该类是标准 java.util.Date 的一个子集,用于表示与 SQL DATE 相同的日期类型,该日期不包括时间
7	java.sql.Timestamp	标准 java.util.Date 类的扩展,用于表示 SQL 时间戳,并增加了一个能表示纳秒的时间域
8	java.sql.CallableStatement	用于执行 SQL 存储过程
9	java.sql.DatabaseMetaData	与 java.sql.ResultSetMetaData 一同用于访问数据库的元信息
10	java.sql.Driver	定义一个数据库驱动程序的接口
11	java.sql.DataTruncation	在 JDBC 遇到数据截断的异常时,报告一个警告(读数据时)或产生一个异常(写数据时)
12	java.sql.DriverPropertyInfo	高级程序设计人员通过 DriverPropertyInfo 与 Driver 进行交流,可使用 getDriverPropertyInfo 获取或提供驱动程序的信息
13	java.sql.Time	该类是标准 java.util.Date 的一个子集,用于表示时、分、秒
14	java.sql.SQLException	对数据库访问时产生的错误的描述信息
15	java.sql.SQLWarning	对数据库访问时产生的警告的描述信息
16	java.sql.Types	定义了表示 SQL 类型的常量

JDBC 访问数据库的操作中最常用的类和接口如下:
- DriverManager:是一个最常用的类,使用此类可以取得一个数据库的连接。
- Connection:每一个 Connection 的实例化对象都表示一个数据库连接。
- 数据库的更改:Statement、PreparedStatement。
- 数据库的查询:ResultSet。

7.2.2 JDBC 访问数据库的操作步骤

MySQL 数据库安装并配置完成后,就可以按照如下步骤完成数据库的操作了:

(1) 加载数据库驱动程序:各种数据库都会提供 JDBC 的驱动程序开发包(一般为 *.jar 或 *.Zip),只须将 JDBC 操作所需要的开发包配置到 classpath 路径即可。

(2) 连接数据库:根据各个数据库的不同,连接的地址也不同,此连接地址将由数据库厂商提供,一般在使用 JDBC 连接数据库的时候都要求用户输入数据库连接的用户名和密码。本模块使用的是 MySQL 数据库。用户在取得连接之后才可以对数据库进行查询或更改的操作。

(3) 使用语句进行数据库操作：数据库操作分为更改和查询两种操作，除了可以使用标准的 SQL 语句之外，对于各种数据库也可以使用其自己提供的各种命令。

(4) 关闭数据库连接：数据库操作完成之后需要关闭连接以释放资源。

注意：

要想连接数据库，就要使用连接地址，但数据库的连接是非常有限的，所以打开之后一定要关闭。

1. 配置 MySQL 数据库驱动程序

如果要采用 JDBC 本地驱动模式完成 MySQL 数据库操作，必须先将 MySQL 数据库的驱动程序配置到本机的环境变量 classpath 中。例如，假设 MySQL 数据库驱动程序在本机位置的路径是"E:\mysql-connector-java-5.1.21-bin.jar"，配置环境变量 classpath 的步骤如下：

(1) 在计算机桌面上右击"我的电脑"图标，在弹出的快捷菜单中选择"属性"，打开系统属性窗口，在系统属性窗口中点击"高级"选项，再点击【环境变量】按钮，如图 7-51 所示，此时打开"环境变量"对话框，如图 7-52 所示。

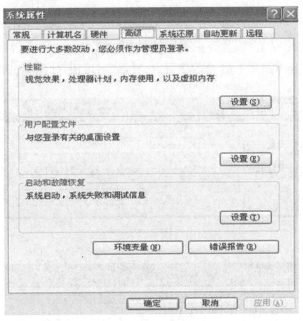

图 7-51

(2) 在如图 7-52 所示的"环境变量"对话框中的"Administrator 的用户变量"里寻找"classpath"这一项。如果没有这一项，就单击【新建】按钮，在弹出的"新建用户变量"窗口中选中"变量名"项并输入"classpath"，然后选中"变量值"项并输入"E:\mysql-connector-java-5.1.21-bin.jar;"（请注意别忘记输入最后的分号），然后点击【确定】按钮，如图 7-53 所示。

反之，若找到了，就选中它，然后单击【编辑】按钮，在弹出的"编辑系统变量"对话框中，

图 7-52

选中"变量值"项,在"变量值"项的开头添加如下语句"E:\mysql-connector-java-5.1.21-bin.jar;"(请注意别忘记输入最后的分号),然后点击【确定】按钮,如图 7-54 所示。

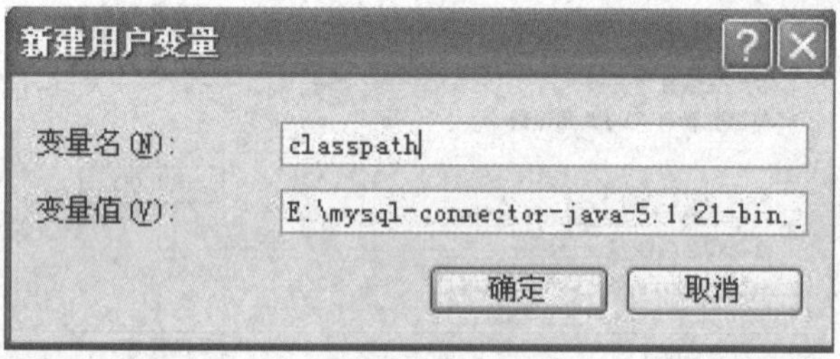

图 7-53

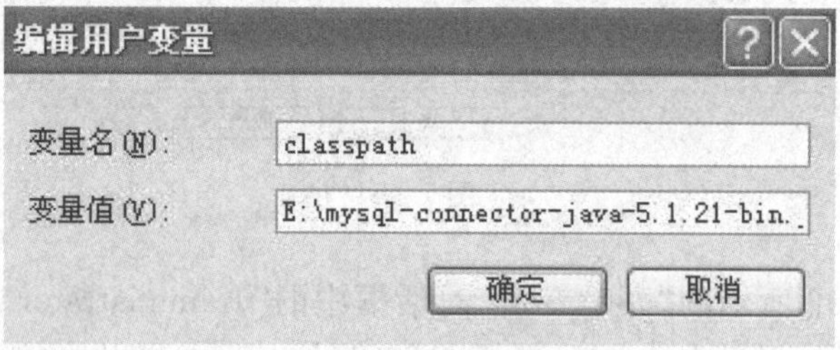

图 7-54

注:MySQL 数据库的驱动程序下载地址为"https://dev.mysql.com/downloads/connector/"。

2. 加载 MySQL 驱动程序

MySQL 驱动程序配置完成后,可通过 Class.forName()语句加载驱动程序,由于不同数据库的驱动程序不一样,MySQL 驱动程序提供的类名是"com.mysql.jdbc.Driver"(或 org.gjt.mm.mysql.Driver),因此其对应的加载驱动程序语句为:

```
Class.forName("com.mysql.jdbc.Driver");
```

或

```
Class.forName("org.gjt.mm.mysql.Driver");
```

注意:

"org.gjt.mm.mysql.Driver"是早期的驱动名称,后来改名为"com.mysql.jdbc.Driver",在最新版本的 MySQL JDBC 驱动中,为了保持对老版本的兼容,仍然保留了"org.gjt.mm.mysql.Driver",但实际上"org.gjt.mm.mysql.Driver"调用了"com.mysql.jdbc.Driver",二者没有本质区别,现在一般都推荐使用"com.mysql.jdbc.Driver"。

【例 7.1】 加载 MySQL 驱动程序,如果加载成功则输出"MySQL 驱动程序加载成功!",否则输出错误信息和"MySQL 驱动程序加载失败!"。

```java
public class LoadMysql{
    public static void main(String args[]){
        try{
            Class.forName("com.mysql.jdbc.Driver");//加载 MySQL 驱动程序
            System.out.println("MySQL 驱动程序加载成功!");//成功提示信息
        }
        catch(ClassNotFoundException e){
            e.printStackTrace();//输出错误信息
            System.out.println("MySQL 驱动程序加载失败!");//失败提示信息
        }
    }
};
```

如图 7-55 所示。

注意:

加载驱动程序,应使用异常处理结构。

图 7-55

3. 连接、关闭数据库

MySQL 数据库驱动程序实现正常加载后,就可以使用 DriverManager 类连接 MySQL 数据库了。DriverManager 类中的常用方法如表 7-7 所示:

表 7-7

序号	方法	描述
1	public static Connection getConnection(String url) throws SQLException	通过连接地址连接数据库
2	public static Connection getConnection(String url, String user, String password) throws SQLException	通过连接地址连接数据库,同时输入用户名和密码

由表 7-7 可知,DriverManager 的主要作用是得到一个数据库的连接,DriverManager 类通过 getConnection() 方法取得连接对象,此方法返回的类型是 Connection 对象。虽然有两种 getConnection() 方法,但不管使用哪种方式连接,都需要提供一个数据库的连接地址 url。此外,如果在连接数据库的时候需要用户名 user 和密码 password,则还需要将用户名和密码设置上。

MySQL 数据库的连接地址 url 格式如下:

```
jdbc:mysql://IP 地址:端口号/数据库名称
```

- JDBC 协议:JDBC URL 中的协议总是 JDBC;
- 子协议:驱动程序名或数据库连接机制(这种机制可由一个或多个驱动程序支持)的名称,如 mysql;
- 子名称:一种标识数据库的方法,必须遵循"//主机名:端口/子协议"的标准 URL 命名约定,如//localhost:3306/S_Score。

通过 DriverManager 成功取得 Connection 接口对象,实际上就表示已连接上 MySQL 数据库,此后就可以通过该接口实现 MySQL 数据库的更新及查询等所有操作。Connection 接口的常用方法如表 7-8 所示。

表 7-8

序号	方法	描述
1	Statement createStatement() throws SQLException	创建一个 Statement 对象
2	Statement createStatement(int resultSetType, int resultSetConcurrency) throws SQLException	创建一个 statement 对象,该对象将生成具有给定类型和并发性的 ResultSet 对象
3	PreparedStatement prepareStatement(String sql) throws SQLException	创建一个 PreparedStatement 类型的对象

续表

序号	方　　法	描　　述
4	PreparedStatement prepareStatement（String sql, int resultSetType, int resultSetConcurrency）throws SQLException	创建一个 PreparedStatement 对象，该对象将生成具有给定类型和并发性的 ResultSet 对象
5	boolean isClosed()throws SQLException	判断连接是否已关闭
6	void close()throws SQLException	关闭数据库

上述方法在 JDBC 的操作会经常用到，我们在后续的课程内容中会陆续讲到，下面我们先举例说明如何连接数据库。

【例 7.2】实现 MySQL 数据库 S_Score 的连接。

```java
import java.sql.Connection;
import java.sql.DriverManager;
import java.sql.SQLException;
public class Conn_S_Score{
    public static void main(String args[]){
    Connection conn = null;//数据库连接
    try{
    Class.forName("com.mysql.jdbc.Driver");//加载驱动程序
    }
    catch(ClassNotFoundException e){
    e.printStackTrace();//输出加载驱动错误信息
    }
    try{
    conn = DriverManager.getConnection("jdbc:mysql://localhost:3306/S_Score","root","123456");//通过连接地址，用户名，密码使用 getConnection()方法创建数据库连接 conn
    System.out.println("数据库连接 con 创建成功!");
    }
    catch(SQLException e){
    e.printStackTrace();//输出创建连接错误信息
    }
    }
};
```

如图 7-56 所示。

上例只实现了数据库的连接，但是没有关闭数据库的连接。在实际的应用开发中，由于数据库资源有限，要求数据库在操作完毕后，必须关闭连接，其语法格式如下：

图 7-56

```
数据库连接名.close();
```

【例7.3】实现上例中数据库连接 conn 的关闭。

```
import java.sql.Connection;
import java.sql.DriverManager;
import java.sql.SQLException;
public class Conn_S_Score{
    public static void main(String args[]){
    Connection conn = null;//数据库连接
    try{
    Class.forName("com.mysql.jdbc.Driver");//加载驱动程序
    }
    catch(ClassNotFoundException e){
    e.printStackTrace();//输出加载驱动错误信息
    }
    try{
    conn = DriverManager.getConnection("jdbc:mysql://localhost:3306/S_Score","root","123456");//通过连接地址,用户名,密码使用 getConnection()方法创建数据库连接 conn
        System.out.println("数据库连接 conn 创建成功!");
    }
    catch(SQLException e){
    e.printStackTrace();//输出创建连接错误信息
    }
    try{
    conn.close();//数据库关闭
    System.out.println("数据库连接 conn 关闭成功!");
    }catch(SQLException e){
    e.printStackTrace();
    }
```

```
            }
        };
```

如图 7-57 所示。

```
E:\>javac Conn_S_Score.java

E:\>java Conn_S_Score
数据库连接conn创建成功!
数据库连接conn关闭成功!

E:\>
```

图 7-57

上例中,我们成功创建并关闭了数据库的连接,但在实际的软件开发中,需要对数据库进行多种操作。如果每次创建连接都要重复输入连接地址、用户名、密码的详细信息,这样会非常烦琐,我们可以通过创建常量解决这个问题。

【例 7.4】通过创建常量创建并关闭数据库的连接。

```java
import java.sql.Connection;
import java.sql.DriverManager;
import java.sql.SQLException;
public class Conn_S_Score{
//定义 MySQL 的数据库驱动程序
public static final String DBDRIVER = "com.mysql.jdbc.Driver";
//定义 MySQL 数据库的连接地址
public static final String DBURL = "jdbc:mysql://localhost:3306/S_Score";
//MySQL 数据库的连接用户名
public static final String DBUSER = "root";
//MySQL 数据库的连接密码
public static final String DBPASS = "123456";
    public static void main(String args[]){
    Connection conn = null;//数据库连接
    try{
    Class.forName(DBDRIVER);//加载驱动程序
    }
    catch(ClassNotFoundException e){
    e.printStackTrace();//输出加载驱动错误信息
    }
    try{
```

```
            conn = DriverManager.getConnection(DBURL,DBUSER,DBPASS);//通过连接地
址,用户名,密码使用 getConnection()方法创建数据库连接 conn
        System.out.println("数据库连接conn创建成功!");
        }
        catch(SQLException e){
        e.printStackTrace();//输出创建连接错误信息
        }
        try{
        conn.close();//数据库关闭
        System.out.println("数据库连接conn关闭成功!");
        }catch(SQLException e){
        e.printStackTrace();
        }
        }
    };
```

如图 7-58 所示。

图 7-58

3. 执行数据库更改操作

MySQL 数据库连接之后,就可以进行数据库的具体操作了。而要实现操作数据库,就要使用 Statement 接口或 PreparedStatement 接口完成。

(1) Statement 接口。

Statement 接口可以使用 Connection 接口中提供的 createStatement()方法实例化。createStatement()方法可以创建一个 Statement 对象,用于发送 SQL 语句到数据库,其语法格式如下:

```
Statement  对象名=连接名.createStatement();
```

例如:

```
Statement stmt = con.createStatement();
```

Statement 对象用于执行一个静态的 SQL 语句并返回它产生的结果,没有参数的 SQL

语句通常使用 Statement 对象执行。在缺省情况下,任一时刻每个 Statement 对象只产生一个 ResultSet 集,因此对数据库希望有不同操作得到结果集时,需要创建不同的 Statement 对象。Statement 的常用方法如表 7-9 所示:

表 7-9

序号	方法	描述
1	int executeUpdate(String sql) throws SQLException	执行数据库更改的 SQL 语句,如 INSERT、UPDATE、DELETE 等,返回更新的记录数
2	ResultSet executeQuery(String sql) throws SQLException	执行数据库查询操作,返回一个结果集对象
3	void addBatch(String sql) throws SQLException	增加一个待执行的 SQL 语句
4	int[] executeBatch() throws SQLException	批量执行 SQL 语句
5	void close() throws SQLException	关闭 Statement 操作
6	boolean execute(String sql) throws SQLException	执行 SQL 语句

注意:

executeUpdate()方法和 executeQuery()方法是用于执行数据库操作的方法,其中 executeUpdate()方法用于执行数据库更改的 SQL 语句,如 INSERT、UPDATE、DELETE 等,并返回更新的记录数;而 executeQuery()方法只用于执行数据库查询操作并返回一个结果集对象。此外,由于数据库资源有限,数据库操作完成后,Statement 接口对象必须使用 close()方法关闭。

接下来,我们首先举例讲解使用 executeUpdate()方法,实现数据库的插入、更新、删除操作。

① 实现数据库插入操作。

【例 7.5】 要求在之前创建的 S_Score 数据库的 user 表中插入如下信息(结果如图 7-59 所示):

userName-admin

Password-123456

程序代码如下:

```
import java.sql.Connection;
import java.sql.DriverManager;
import java.sql.Statement;
public class State_Insert1{
//定义 MySQL 的数据库驱动程序
public static final String DBDRIVER = "com.mysql.jdbc.Driver";
//定义 MySQL 数据库的连接地址
```

```java
public static final String DBURL = "jdbc:mysql://localhost:3306/S_Score";
//MySQL 数据库的连接用户名
public static final String DBUSER = "root";
//MySQL 数据库的连接密码
public static final String DBPASS = "123456";
public static void main(String args[])    {
Connection conn = null;//数据库连接
Statement stmt = null;//数据库操作
try{
Class.forName(DBDRIVER);//加载驱动程序
String sql = " INSERT INTO user ( userName, password)" + " VALUES ('admin', '123456')";
conn = DriverManager.getConnection(DBURL,DBUSER,DBPASS);
stmt = conn.createStatement();//实例化 Statement 对象
stmt.executeUpdate(sql);//执行数据库更新操作
stmt.close();//关闭操作
conn.close();//数据库关闭
}
catch(Exception e){
e.printStackTrace();//输出错误信息
}
}
}
```

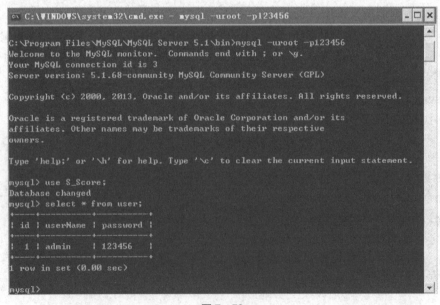

图 7-59

在以上的程序中可以发现，执行的 SQL 语句是一条标准的 SQL 语句，但是程序中的 SQL 语句中的数据是固定的（admin、123456），实际开发中执行的 SQL 语句会根据需求发生不同变化。如果要将具体的内容换成一个变量，则可以将代码修改如下：

```java
import java.sql.Connection;
import java.sql.DriverManager;
import java.sql.Statement;
public class State_Insert2{
//定义MySQL的数据库驱动程序
public static final String DBDRIVER = "com.mysql.jdbc.Driver";
//定义MySQL数据库的连接地址
public static final String DBURL = "jdbc:mysql://localhost:3306/S_Score";
//MySQL数据库的连接用户名
public static final String DBUSER = "root";
//MySQL数据库的连接密码
public static final String DBPASS = "123456";
public static void main(String args[])  {
Connection conn = null;//数据库连接
Statement stmt = null;//数据库操作
try{
Class.forName(DBDRIVER);//加载驱动程序
String name = "admin1";//姓名
String password = "123456";//密码
String sql = "INSERT INTO user(userName,password)" + "VALUES('" + name + "','" + password + "')";
conn = DriverManager.getConnection(DBURL,DBUSER,DBPASS);
stmt = conn.createStatement();//实例化Statement对象
stmt.executeUpdate(sql);//执行数据库更新操作
stmt.close();//关闭操作
conn.close();//数据库关闭
}
catch(Exception e){
e.printStackTrace();
}
}
}
```

② 实现数据库更新操作。

【例 7.6】要求将 S_Score 数据中的 userName 为 admin 的用户的密码改为"123123",结果如图 7-60 所示。

```java
import java.sql.Connection;
import java.sql.DriverManager;
import java.sql.Statement;
public class State_Update{
//定义 MySQL 的数据库驱动程序
public static final String DBDRIVER = "com.mysql.jdbc.Driver";
//定义 MySQL 数据库的连接地址
public static final String DBURL = "jdbc:mysql://localhost:3306/S_Score";
//MySQL 数据库的连接用户名
public static final String DBUSER = "root";
//MySQL 数据库的连接密码
public static final String DBPASS = "123456";
public static void main(String args[])   {
Connection conn = null;//数据库连接
Statement stmt = null;//数据库操作
try{
Class.forName(DBDRIVER);//加载驱动程序
String name = "admin";//姓名
String password = "123123";//密码
String sql = "update user set password = '" + password + "'where userName = '" + name + "'";
conn = DriverManager.getConnection(DBURL,DBUSER,DBPASS);
stmt = conn.createStatement();//实例化 Statement 对象
stmt.executeUpdate(sql);//执行数据库更新操作
stmt.close();//关闭操作
conn.close();//数据库关闭
}
catch(Exception e){
e.printStackTrace();
}
}
}
```

图 7-60

③ 实现数据库删除操作。

【例 7.7】删除 S_Score 数据库中 userName 为 admin 的记录,结果如图 7-61 所示。

```
import java.sql.Connection;
import java.sql.DriverManager;
import java.sql.Statement;
public class State_Delete{
//定义 MySQL 的数据库驱动程序
public static final String DBDRIVER = "com.mysql.jdbc.Driver";
//定义 MySQL 数据库的连接地址
public static final String DBURL = "jdbc:mysql://localhost:3306/S_Score";
//MySQL 数据库的连接用户名
public static final String DBUSER = "root";
//MySQL 数据库的连接密码
public static final String DBPASS = "123456";
public static void main(String args[])    {
Connection conn = null;//数据库连接
Statement stmt = null;//数据库操作
try{
Class.forName(DBDRIVER);//加载驱动程序
String name = "admin";//姓名
String sql = "delete from user where userName = '" + name + "'";
conn = DriverManager.getConnection(DBURL,DBUSER,DBPASS);
stmt = conn.createStatement();//实例化 Statement 对象
stmt.executeUpdate(sql);//执行数据库更新操作
stmt.close();//关闭操作
conn.close();//数据库关闭
}
catch(Exception e){
e.printStackTrace();
```

```
        }
    }
}
```

```
mysql> select * from user;
+----+----------+----------+
| id | userName | password |
+----+----------+----------+
|  2 | admin1   | 123456   |
+----+----------+----------+
1 row in set (0.00 sec)
```

图 7-61

(2) PreparedStatement 接口。

PreparedStatement 属于预处理操作，与直接使用 Statement 不同的是，PreparedStatement 在操作时，将一个 SQL 语句预编译后存储到 PreparedStatement 对象中，但是此 SQL 语句的具体内容暂时不设置，而是之后再进行设置，PreparedStatement 对象可以多次执行该 SQL 语句。PreparedStatement 继承于 Statement，扩展了 Statement 的用途，提高了 Statement 的执行效率。它与 Statement 对象有两点不同：

- 同一个对象可以多次使用。
- 它的 SQL 语句可以带输入(IN)参数。

PreparedStatement 在程序中执行的 SQL 语句中的输入参数使用占位符"?"来实现，且必须使用 PreparedStatement 提供的相应设置方法按照"?"出现的顺序来设置占位符的具体值，才能执行语句。PreparedStatement 除了继承了 Statement 的所有操作方法外，还增加了很多新操作方法，其基本操作方法如表 7-10 所示。

表 7-10

序号	方　　法	描　　述
1	int executeUpdate() throws SQLException	执行设置的预处理 SQL 语句
2	ResultSet executeQuery (String sql) throws SQLException	执行数据库查询操作，返回一个结果集对象
3	void setInt (int parameterIndex, int x) throws SQLException	指定要设置的索引编号，并设置整数内容
4	void setFloat (int parameterIndex, float x) throws SQLException	指定要设置的索引编号，并设置浮点数内容
5	void setString(int parameterIndex, String x) throws SQLException	指定要设置的索引编号，并设置字符串内容
6	void setDate (int parameterIndex, Date x) throws SQLException	指定要设置的索引编号，并设置 java.sql.Date 类型的日期内容

注意:

- PreparedStatement 接口的实例对象可以通过 Connection 接口的 prepareStatement() 方法得到,其格式如下:

```
public PreparedStatement prepareStatement(String sql)throws SQLException
```

- 创建 PreparedStatement 对象语法格式如下:

```
PreparedStatement 对象名 = 连接名.prepareStatement(SQL 语句);
```

例如:

```
PreparedStatement pstmt = con.prepareStatement("UPDATE user SET password = ? WHERE userName = ?");
```

- PreparedStatement 对象创建好后,就可以用相应的 setXXX 方法,按照"?"的顺序设置具体的参数值。例如上例中有两个"?",而且其数据类型都为 varchar,所以应该使用 setString() 方法设置输入参数值为:

```
pstmt.setString(1,"123456");//第一个"?"对应的值
pstmt.setString(2,"admin");//第二个"?"对应的值
```

也可以使用变量表示参数值,例如:

```
String pwd = "123456";
String pwd = "admin";
pstmt.setString(1,pwd);//第一个"?"对应的值
pstmt.setString(2,name);//第二个"?"对应的值
```

- 虽然 Statement 和 PreparedStatement 都可以实现数据库操作,但在实际的开发应用中很少使用 Statement 接口,因为其采用的是拼凑的 SQL 语句形式,这样很有可能造成 SQL 的注入漏洞,所以强烈建议在开发中采用 PreparedStatement。

① 实现数据库插入操作。

【例 7.8】在 S_Score 数据库的 user 表中插入如下数据:

userName	password
admin2	123456

```java
import java.sql.Connection;
import java.sql.DriverManager;
import java.sql.PreparedStatement;
public class Pre_Insert{
```

```java
//定义MySQL的数据库驱动程序
public static final String DBDRIVER = "com.mysql.jdbc.Driver";
//定义MySQL数据库的连接地址
public static final String DBURL = "jdbc:mysql://localhost:3306/S_Score";
//MySQL数据库的连接用户名
public static final String DBUSER = "root";
//MySQL数据库的连接密码
public static final String DBPASS = "123456";
public static void main(String args[])   {
Connection conn = null;//数据库连接
PreparedStatement pstmt = null;//数据库操作
try{
Class.forName(DBDRIVER);//加载驱动程序
String name = "admin2";//姓名
String pwd = "123456";//密码
String sql = "INSERT INTO user(userName,password)VALUES(?,?)";
conn = DriverManager.getConnection(DBURL,DBUSER,DBPASS);
pstmt = conn.prepareStatement(sql);//实例化PreparedStatement对象
pstmt.setString(1,name);
pstmt.setString(2,pwd);
pstmt.executeUpdate();//执行数据库更新操作
pstmt.close();//关闭操作
conn.close();//数据库关闭
}
catch(Exception e){
e.printStackTrace();
}
}
}
```

如图 7-62 所示。

```
mysql> select * from user;
+----+----------+----------+
| id | userName | password |
+----+----------+----------+
|  2 | admin1   | 123456   |
|  3 | admin2   | 123456   |
+----+----------+----------+
2 rows in set (0.00 sec)
```

图 7-62

② 实现数据库更新操作。

【例 7.9】将数据库 S_Score 的 user 表中用户名为 admin2 的密码改为"123123",结果如图 7-63 所示。

```java
import java.sql.Connection;
import java.sql.DriverManager;
import java.sql.PreparedStatement;
public class Pre_Update{
//定义 MySQL 的数据库驱动程序
public static final String DBDRIVER = "com.mysql.jdbc.Driver";
//定义 MySQL 数据库的连接地址
public static final String DBURL = "jdbc:mysql://localhost:3306/S_Score";
//MySQL 数据库的连接用户名
public static final String DBUSER = "root";
//MySQL 数据库的连接密码
public static final String DBPASS = "123456";
public static void main(String args[])  {
Connection conn = null;//数据库连接
PreparedStatement pstmt = null;//数据库操作
try{
Class.forName(DBDRIVER);//加载驱动程序
String name = "admin2";//姓名
String pwd = "123123";//密码
String sql = "update user set password = ?where userName = ?";
conn = DriverManager.getConnection(DBURL,DBUSER,DBPASS);
pstmt = conn.prepareStatement(sql);//实例化 PreparedStatement 对象
pstmt.setString(1,pwd);
pstmt.setString(2,name);
pstmt.executeUpdate();//执行数据库更新操作
pstmt.close();//关闭操作
conn.close();//数据库关闭
}
catch(Exception e){
e.printStackTrace();
}
}
}
```

```
mysql> select * from user;
+----+----------+----------+
| id | userName | password |
+----+----------+----------+
|  2 | admin1   | 123456   |
|  3 | admin2   | 123123   |
+----+----------+----------+
2 rows in set (0.00 sec)
```

图 7-63

③ 实现数据库删除操作。

【例 7.10】删除 S_Score 数据库的 user 表中 userName 为"admin2"的记录。

```java
import java.sql.Connection;
import java.sql.DriverManager;
import java.sql.PreparedStatement;
public class Pre_Delete{
//定义 MySQL 的数据库驱动程序
public static final String DBDRIVER = "com.mysql.jdbc.Driver";
//定义 MySQL 数据库的连接地址
public static final String DBURL = "jdbc:mysql://localhost:3306/S_Score";
//MySQL 数据库的连接用户名
public static final String DBUSER = "root";
//MySQL 数据库的连接密码
public static final String DBPASS = "123456";
public static void main(String args[])   {
Connection conn = null;//数据库连接
PreparedStatement pstmt = null;//数据库操作
try{
Class.forName(DBDRIVER);//加载驱动程序
String name = "admin2";//姓名
String sql = "delete from user where userName = ?";
conn = DriverManager.getConnection(DBURL,DBUSER,DBPASS);
pstmt = conn.prepareStatement(sql);//实例化 PreparedStatement 对象
pstmt.setString(1,name);
pstmt.executeUpdate();//执行数据库更新操作
pstmt.close();//关闭操作
conn.close();//数据库关闭
}
catch(Exception e){
e.printStackTrace();
```

```
        }
    }
}
```

如图 7-64 所示。

```
mysql> select * from user;
+----+----------+----------+
| id | userName | password |
+----+----------+----------+
|  2 | admin1   | 123456   |
+----+----------+----------+
1 row in set (0.00 sec)
```

图 7-64

4. ResultSet 接口

数据库的操作主要分为更改和查询操作,查询操作会将满足条件的全部查询结果返回给用户,在 JDBC 的操作中数据库的所有查询记录将使用 ResultSet 进行接收,并使用 ResultSet 显示内容,其常用操作方法如表 7-11 所示。

表 7-11

序号	方　　法	描述
1	boolean next() throws SQLException	将指针移到下一行
2	int getInt(int columnIndex) throws SQLException	以整数形式按列编号取得指定列的内容
3	int getInt(String columnLabel) throws SQLException	以整数形式取得指定列的内容
4	float getFloat(int columnIndex) throws SQLException	以浮点数形式按列编号取得指定列的内容
5	float getFloat(String columnLabel) throws SQLException	以浮点数形式取得指定列的内容
6	String getString(int columnIndex) throws SQLException	以字符串形式按列编号取得指定列的内容
7	String getString(String columnLabel) throws SQLException	以字符串形式取得指定列的内容
8	Date getDate(int columnIndex) throws SQLException	以 Date 形式按列编号取得指定列的内容
9	Date getDate(String columnLabel) throws SQLException	以 Date 形式取得指定列的内容

在进行查询操作时,一般使用 PreparedStatement 接口的 executeQuery()方法完成数据库的查询,此方法的返回值就是 ResultSet 对象,其语法格式如下:

ResultSet 对象名 = PreparedStatement 对象名.executeQuery(sql 查询语句);

例如:

```
ResultSet rs = pstmt.executeQuery(sql);
```

ResultSet 对象代表数据库查询结果集,因此也可将其看成一张数据表。ResultSet 对象持有一个游标,该游标指向当前数据行,初始化时游标定位到第一行之前,next()方法将游标移动到下一行,当对象行完时,返回错误。通常使用循环来完成每行的遍历。例如:

```
while(rs.next())
{ … …
}
```

注意:

● ResultSet 接收到的查询结果存放在内存中,可以使用 next()方法依次取出,然后进行判断。如果有结果,则根据要取出的数据类型使用相应的 getXxx()语句的形式将内容取出。例如:

```
String name = rs.getString("name");//取出 name 列的内容
```

● getString()方法可以取出任何类型的数据。

● 在使用 getXxx()方法取出数据时,可以使用列的名称指定要取出的内容,也可以使用查询时取值顺序编号指定要取出的内容。例如,

```
String name = rs.getString(2);//取出 name 列的内容
```

(1) 使用列名取出数据。

【例 7.11】S_Score 数据库中 user 表的数据记录如图 7-65 所示,要求查询表中密码为"123456"的记录对应的 **id** 和 **userName**。

图 7-65

```
import java.sql.Connection;
import java.sql.DriverManager;
import java.sql.ResultSet;
import java.sql.PreparedStatement;

public class Rs_Select1{
```

```java
//定义 MySQL 的数据库驱动程序
public static final String DBDRIVER = "com.mysql.jdbc.Driver";
//定义 MySQL 数据库的连接地址
public static final String DBURL = "jdbc:mysql://localhost:3306/S_Score";
//MySQL 数据库的连接用户名
public static final String DBUSER = "root";
//MySQL 数据库的连接密码
public static final String DBPASS = "123456";
public static void main(String args[]){
Connection conn = null;//数据库连接
PreparedStatement pstmt = null;//数据库操作
String pwd = "123456";//设置查询关键字
ResultSet rs = null;//接收查询结果
String sql = "SELECT id,userName FROM user where password = ?";
try{
Class.forName(DBDRIVER);//加载驱动程序
conn = DriverManager.getConnection(DBURL,DBUSER,DBPASS);
pstmt = conn.prepareStatement(sql);//实例化 PreparedStatement 对象
pstmt.setString(1,pwd);
rs = pstmt.executeQuery();//执行查询
while(rs.next()){
int id = rs.getInt("id");
String name = rs.getString("userName");
System.out.print("编号:" + id);
System.out.println("   姓名:" + name);
}
rs.close();
pstmt.close();
conn.close();//数据库关闭
}
catch(Exception e){
e.printStackTrace();
}
}
}
```

如图 7 - 66 所示。

```
E:\>javac Rs_Select1.java

E:\>java Rs_Select1
编号: 1   姓名: admin2
编号: 2   姓名: admin
编号: 3   姓名: admin3
编号: 4   姓名: admin1
```

图 7 - 66

(2) 使用编号取出数据。

【例 7.12】S_Score 数据库中 user 表的数据记录如图 7 - 65 所示,要求查询表中密码为 "123456" 的记录对应的 id 和 userName。

```java
import java.sql.Connection;
import java.sql.DriverManager;
import java.sql.ResultSet;
import java.sql.PreparedStatement;

public class Rs_Select2{
//定义 MySQL 的数据库驱动程序
public static final String DBDRIVER = "com.mysql.jdbc.Driver";
//定义 MySQL 数据库的连接地址
public static final String DBURL = "jdbc:mysql://localhost:3306/S_Score";
//MySQL 数据库的连接用户名
public static final String DBUSER = "root";
//MySQL 数据库的连接密码
public static final String DBPASS = "123456";
public static void main(String args[]){
Connection conn = null;//数据库连接
PreparedStatement pstmt = null;//数据库操作
String pwd = "123456";//设置查询关键字
ResultSet rs = null;//接收查询结果
String sql = "SELECT id,userName FROM user where password = ?";
try{
Class.forName(DBDRIVER);//加载驱动程序
conn = DriverManager.getConnection(DBURL,DBUSER,DBPASS);
pstmt = conn.prepareStatement(sql);//实例化 PreparedStatement 对象
pstmt.setString(1,pwd);
rs = pstmt.executeQuery();//执行查询
while(rs.next()){
```

```
        int id = rs.getInt(1);
        String name = rs.getString(2);
        System.out.print("编号:" + id);
        System.out.println("   姓名:" + name);
    }
    rs.close();
    pstmt.close();
    conn.close();//数据库关闭
    }
    catch(Exception e){
        e.printStackTrace();
    }
    }
}
```

如图 7-67 所示。

```
E:\>javac Rs_Select2.java

E:\>java Rs_Select2
编号: 1  姓名: admin2
编号: 2  姓名: admin
编号: 3  姓名: admin3
编号: 4  姓名: admin1
```

图 7-67

任务 实施

7.2.3 学生成绩管理系统功能实现

1. 实现登录功能

(1) 数据库连接模块 DBUtil 代码如下:

```
import java.sql.Connection;
import java.sql.DriverManager;
import java.sql.SQLException;
import javax.swing.JOptionPane;

public class DBUtil{
    private static String DriverName = "com.mysql.jdbc.Driver";
    private static String url = "jdbc:mysql://127.0.0.1:3306/S_Score";
```

```java
            private static String user = "root";
            private static String password = "123456";
            private static Connection conn = null;
        //创建连接
        public static Connection getConnection(){
            try{
                Class.forName(DriverName);
                conn = DriverManager.getConnection(url,user,password);
            }catch(ClassNotFoundException e){
                JOptionPane.showMessageDialog(null,e.toString(),"加载驱动错误",JOptionPane.ERROR_MESSAGE);
            }catch(SQLException e){
                JOptionPane.showMessageDialog(null,e.toString(),"数据库连接错误!",JOptionPane.ERROR_MESSAGE);
            }
            return conn;
        }
        //关闭连接
        public static void colseConnction(){
            if(conn! = null){
              try{
                    conn.close();
              }catch(SQLException e){
                    JOptionPane.showMessageDialog(null,e.toString(),"关闭数据库错误!",JOptionPane.ERROR_MESSAGE);
                }
            }
        }
    }
```

(2) 封装 User 表操作的 DAO 程序 UserDAO 代码如下：

```java
import java.sql.Connection;
import java.sql.PreparedStatement;
import java.sql.ResultSet;
import java.sql.SQLException;
import javax.swing.JDialog;
```

```java
import javax.swing.JOptionPane;

public class UserDAO{
    private Connection conn = null;
    private String sql = null;
    private PreparedStatement ps = null;
    private ResultSet rs = null;

    public UserDAO(){
        conn = DBUtil.getConnection();
    }

    //判断是否是合法的用户
    public boolean login(String userName,String password){
        boolean find = false;
        sql = "select * from User where userName = ?and password = ?";
        try{
            ps = conn.prepareStatement(sql);
            ps.setString(1,userName);
            ps.setString(2,password);
            rs = ps.executeQuery();
            if(rs.next()){
                find = true;
            }
        }
        catch(SQLException e){
            JOptionPane.showMessageDialog(null,e.toString());
        }
        return find;
    }
}
```

（3）登录窗口 LoginJFrame 代码如下：

```java
import java.awt.event.ActionEvent;
import java.awt.event.ActionListener;
import javax.swing.JButton;
import javax.swing.JFrame;
```

```java
import javax.swing.JLabel;
import javax.swing.JPasswordField;
import javax.swing.JTextField;
import javax.swing.JOptionPane;

/**
 *用户登录界面
 *
 */
public class LoginJFrame extends JFrame implements ActionListener{
    private JTextField uNameText = new JTextField();//用户名文本组件
    private JPasswordField uPswPwd = new JPasswordField();//密码文本组件
    private JLabel uNameLabel = new JLabel();//用户名标签组件
    private JLabel uPswLabel = new JLabel();//密码标签组件
    private JButton uLoginButton = new JButton("登录");//定义按钮对象
    private JButton uResetButton = new JButton("重置");//定义按钮对象
    private UserDAO userdao;
    public static MainJFrame mainFrame;//定义主窗体对象

    /**
     *构造方法
     *
     */
    public LoginJFrame(){
        uNameLabel.setText("用户名:");
        uNameLabel.setBounds(16,14,54,24);
        uNameText.setBounds(72,14,150,24);
        uPswLabel.setText("密  码:");
        uPswLabel.setBounds(16,50,54,24);
        uPswPwd.setEchoChar('*');
        uPswPwd.setBounds(72,50,150,24);
        uLoginButton.setBounds(65,90,60,24);
        uResetButton.setBounds(140,90,60,24);
        uLoginButton.addActionListener(this);
        uResetButton.addActionListener(this);
        this.add(uNameLabel);
        this.add(uNameText);
        this.add(uPswLabel);
```

```java
        this.add(uPswPwd);
        this.add(uLoginButton);
        this.add(uResetButton);
        this.setTitle("登录窗口");
        this.setLayout(null);
        this.setDefaultCloseOperation(JFrame.EXIT_ON_CLOSE);
        this.setSize(250,170);
        this.setLocationRelativeTo(null);
        this.setResizable(false);
        this.setVisible(true);
    }

    /**
     * 在文本框上回车、单击按钮激发的事件
     *
     * @param curr
     */
    public void actionPerformed(ActionEvent e){
        if(e.getSource() == uLoginButton){//点击登录按钮
            //(1)判断登录名和密码不能为空
            String name = uNameText.getText().trim();//获取用户名
            String password = new String(uPswPwd.getPassword());//获取密码
            if(name.length() == 0){
                JOptionPane.showMessageDialog(this,"请输入用户名!");
                return;
            }
            if(password.length() == 0){
                JOptionPane.showMessageDialog(this,"请输入密码!");
                return;
            }
            userdao = new UserDAO();
            //(2)判断登录名和密码是否正确
            if(userdao.login(name,password)){
                mainFrame = new MainJFrame();
                this.dispose();
            }else{
                JOptionPane.showMessageDialog(this,"用户名或密码不存在,请重新输入!");
```

```java
            uNameText.setText("");
            uPswPwd.setText("");
            return;
        }
    }
            if(e.getSource() == uResetButton){///重置按钮
        uNameText.setText("");
        uPswPwd.setText("");
        }
    }

    /**
     * 主函数
     *
     * @param args
     */
    public static void main(String[] args){
        new LoginJFrame();
    }
}
```

(4) 主窗口 MainJFrame 代码如下：

```java
import java.awt.event.ActionEvent;
import java.awt.event.ActionListener;
import javax.swing.JFrame;
import javax.swing.JMenu;
import javax.swing.JMenuBar;
import javax.swing.JMenuItem;

public class MainJFrame extends JFrame implements ActionListener{
    //创建菜单栏
    JMenuBar jMenuBar = new JMenuBar();
    //创建菜单
    JMenu jMenu01 = new JMenu("信息管理");
    //创建菜单项
    JMenuItem jMenuItem01_00 = new JMenuItem("信息添加");
    JMenuItem jMenuItem01_01 = new JMenuItem("信息修改");
```

```java
        JMenuItem jMenuItem01_02 = new JMenuItem("信息删除");
    //创建菜单
    JMenu jMenu02 = new JMenu("信息查询");
    //创建菜单项
    JMenuItem jMenuItem02_00 = new JMenuItem("成绩查询");
    //创建菜单
    JMenu jMenu03 = new JMenu("系统退出");
    //创建菜单项
    JMenuItem jMenuItem03_00 = new JMenuItem("退出");

    public MainJFrame(){
        super("学生成绩管理系统");
        //菜单栏
        jMenu01.add(jMenuItem01_00);
        jMenu01.addSeparator();
        jMenu01.add(jMenuItem01_01);
        jMenu01.addSeparator();
        jMenu01.add(jMenuItem01_02);
        jMenu02.add(jMenuItem02_00);
        jMenu03.add(jMenuItem03_00);
        jMenuBar.add(jMenu01);
        jMenuBar.add(jMenu02);
        jMenuBar.add(jMenu03);

        //在窗口上设置菜单栏
        this.setJMenuBar(jMenuBar);
        jMenuItem01_00.addActionListener(this);
        jMenuItem01_01.addActionListener(this);
        jMenuItem01_02.addActionListener(this);
        jMenuItem02_00.addActionListener(this);
        jMenuItem03_00.addActionListener(this);
        //设置窗口
        this.setDefaultCloseOperation(JFrame.EXIT_ON_CLOSE);
        this.setSize(700,580);
        this.setLocationRelativeTo(null);
        this.setVisible(true);
    }
```

```java
public void actionPerformed(ActionEvent e){
    if(e.getSource() = = jMenuItem01_00){
    new AddJFrame();
    }
    if(e.getSource() = = jMenuItem01_01){
    new SnoInputJFrame();
    }
    if(e.getSource() = = jMenuItem01_02){
    new DeleteJFrame();
    }
    if(e.getSource() = = jMenuItem02_00){
    new SelectJFrame();
    }
    if(e.getSource() = = jMenuItem03_00){
    System.exit(0);
    }
}

public static void main(String args[]){
    new MainJFrame();
}
}
```

如图 7-68、图 7-69 所示。

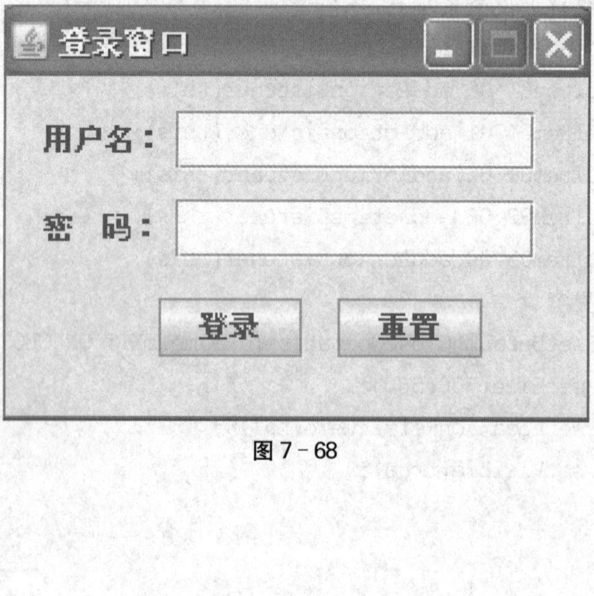

图 7-68

图 7-69

2. 实现学生成绩信息管理(添加、修改、删除)、查询功能
(1) 学生成绩信息的封装类 StudentScoreBean 代码如下：

```java
public class StudentScoreBean{
    private int sno;
    private String sname;
    private char sex;
    private String cname;
    private byte pingShiScore;
    private byte guoChengScore;
    private byte zhongJieScore;
    private double zongHeScore;
    private String zongHeScoreJiBie;
    private String love;
    public int getSno(){
        return sno;
    }
    public void setSno(int sno){
        this.sno = sno;
    }
    public String getSname(){
        return sname;
    }
    public void setSname(String sname){
```

```java
        this.sname = sname;
    }
    public char getSex(){
        return sex;
    }
    public void setSex(char sex){
        this.sex = sex;
    }
    public String getCname(){
        return cname;
    }
    public void setCname(String cname){
        this.cname = cname;
    }
    public byte getPingShiScore(){
        return pingShiScore;
    }
    public void setPingShiScore(byte pingShiScore){
        this.pingShiScore = pingShiScore;
    }
    public byte getGuoChengScore(){
        return guoChengScore;
    }
    public void setGuoChengScore(byte guoChengScore){
        this.guoChengScore = guoChengScore;
    }
    public byte getZhongJieScore(){
        return zhongJieScore;
    }
    public void setZhongJieScore(byte zhongJieScore){
        this.zhongJieScore = zhongJieScore;
    }
    public double getZongHeScore(){
        return zongHeScore;
    }
    public void setZongHeScore(double zongHeScore){
        this.zongHeScore = zongHeScore;
    }
```

```java
    public String getZongHeScoreJiBie(){
        return zongHeScoreJiBie;
    }
    public void setZongHeScoreJiBie(String zongHeScoreJiBie){
        this.zongHeScoreJiBie = zongHeScoreJiBie;
    }
    public String getLove(){
        return love;
    }
    public void setLove(String love){
        this.love = love;
    }
}
```

（2）学生成绩表 StudentScore 的操作封装程序 StudentScoreDAO 代码如下：

```java
import java.sql.Connection;
import java.sql.PreparedStatement;
import java.sql.ResultSet;
import java.sql.SQLException;
import javax.swing.JDialog;
import javax.swing.JOptionPane;

public class StudentScoreDAO{
    private Connection conn = null;
    private String sql = null;
    private PreparedStatement ps = null;
    private ResultSet rs = null;
    
    public StudentScoreDAO(){
        conn = DBUtil.getConnection();
    }
    
    //按学号查找学生成绩信息
    public StudentScoreBean searchBySno(int sno){
        StudentScoreBean S_S_Bean = null;
        sql = "select * from StudentScore where sno = ?";
        try{
```

```java
            ps = conn.prepareStatement(sql);
            ps.setInt(1,sno);
            rs = ps.executeQuery();
            if(rs.next()){
                S_S_Bean = new StudentScoreBean();
                S_S_Bean.setSno(rs.getInt("sno"));
                S_S_Bean.setSname(rs.getString("sname"));
                S_S_Bean.setSex(rs.getString("sex").charAt(0));
                S_S_Bean.setCname(rs.getString("cname"));
                S_S_Bean.setPingShiScore(rs.getByte("pingShiScore"));
                S_S_Bean.setGuoChengScore(rs.getByte("guoChengScore"));
                S_S_Bean.setZhongJieScore(rs.getByte("zhongJieScore"));
                S_S_Bean.setZongHeScore(rs.getDouble("zongHeScore"));
        S_S_Bean.setZongHeScoreJiBie(rs.getString("zongHeScoreJiBie"));
                S_S_Bean.setLove(rs.getString("love"));
            }
        }
        catch(SQLException e){
        JOptionPane.showMessageDialog(null,e.toString(),"查找学生成绩出错!",JOptionPane.ERROR_MESSAGE);
        }
        return S_S_Bean;
    }

    //按姓名查找学生成绩信息
    public StudentScoreBean searchBySname(String sname){
        StudentScoreBean S_S_Bean = null;
        sql = "select * from StudentScore where sname = ?";
        try{
            ps = conn.prepareStatement(sql);
            ps.setString(1,sname);
            rs = ps.executeQuery();
            if(rs.next()){
                S_S_Bean = new StudentScoreBean();
                S_S_Bean.setSno(rs.getInt("sno"));
                S_S_Bean.setSname(rs.getString("sname"));
                S_S_Bean.setSex(rs.getString("sex").charAt(0));
                S_S_Bean.setCname(rs.getString("cname"));
```

```java
                    S_S_Bean.setPingShiScore(rs.getByte("pingShiScore"));
                    S_S_Bean.setGuoChengScore(rs.getByte("guoChengScore"));
                    S_S_Bean.setZhongJieScore(rs.getByte("zhongJieScore"));
                    S_S_Bean.setZongHeScore(rs.getDouble("zongHeScore"));
                S_S_Bean.setZongHeScoreJiBie(rs.getString("zongHeScoreJiBie"));
                    S_S_Bean.setLove(rs.getString("love"));
                }
            }
            catch(SQLException e){
                JOptionPane.showMessageDialog(null,e.toString(),"查找学生成绩出错!",JOptionPane.ERROR_MESSAGE);
            }
            return S_S_Bean;
    }

    //查找所有学生成绩信息
        public ResultSet searchAll(){
            sql = "select * from StudentScore";
            try{
                ps = conn.prepareStatement(sql);
                rs = ps.executeQuery();
            }
            catch(SQLException e){
                JOptionPane.showMessageDialog(null,e.toString(),"查找学生成绩出错!",JOptionPane.ERROR_MESSAGE);
            }
            return rs;
        }

    //添加学生成绩信息
        public void addScore(StudentScoreBean S_S_Bean){
        sql = "insert into StudentScore(sno,sname,sex,cname,pingShiScore,guoChengScore,zhongJieScore,zongHeScore,zongHeScoreJiBie,love)values(?,?,?,?,?,?,?,?,?,?)";
            try{
                ps = conn.prepareStatement(sql);
                ps.setInt(1,S_S_Bean.getSno());
                ps.setString(2,S_S_Bean.getSname());
```

```java
            ps.setString(3,String.valueOf(S_S_Bean.getSex()));
            ps.setString(4,S_S_Bean.getCname());
            ps.setByte(5,S_S_Bean.getPingShiScore());
            ps.setByte(6,S_S_Bean.getGuoChengScore());
            ps.setByte(7,S_S_Bean.getZhongJieScore());
            ps.setDouble(8,S_S_Bean.getZongHeScore());
            ps.setString(9,S_S_Bean.getZongHeScoreJiBie());
            ps.setString(10,S_S_Bean.getLove());
            ps.executeUpdate();
            JOptionPane.showMessageDialog(null,"学生成绩信息已添加到数据库!");
        }
        catch(SQLException e){
            JOptionPane.showMessageDialog(null,e.toString(),"添加学生成绩出错!",JOptionPane.ERROR_MESSAGE);
        }
    }

    //修改学生成绩信息
    public void updateScore(StudentScoreBean S_S_Bean,int number){
        sql = "UPDATE StudentScore SET sno = ?,sname = ?,sex = ?,cname = ?,pingShiScore = ?,guoChengScore = ?,zhongJieScore = ?,zongHeScore = ?,zongHeScoreJiBie = ?,love = ? WHERE sno = ?";
        try{
            ps = conn.prepareStatement(sql);
            ps.setInt(1,S_S_Bean.getSno());
            ps.setString(2,S_S_Bean.getSname());
            ps.setString(3,String.valueOf(S_S_Bean.getSex()));
            ps.setString(4,S_S_Bean.getCname());
            ps.setByte(5,S_S_Bean.getPingShiScore());
            ps.setByte(6,S_S_Bean.getGuoChengScore());
            ps.setByte(7,S_S_Bean.getZhongJieScore());
            ps.setDouble(8,S_S_Bean.getZongHeScore());
            ps.setString(9,S_S_Bean.getZongHeScoreJiBie());
            ps.setString(10,S_S_Bean.getLove());
            ps.setInt(11,number);
            ps.executeUpdate();
            JOptionPane.showMessageDialog(null,"学生成绩信息修改成功!");
```

```
            }
            catch(SQLException e){
                JOptionPane.showMessageDialog(null,e.toString(),"修改学生成绩出错!",JOptionPane.ERROR_MESSAGE);
            }
        }

        //删除学生成绩信息
        public void deleteScore(int number){
            sql = "delete from StudentScore where sno = ?";
            try{
                ps = conn.prepareStatement(sql);
                ps.setInt(1,number);
                ps.executeUpdate();
                JOptionPane.showMessageDialog(null,"学生成绩信息删除成功!");
            }catch(SQLException e){
                JOptionPane.showMessageDialog(null,e.toString(),"删除学生成绩出错!",JOptionPane.ERROR_MESSAGE);
            }
        }
    }
```

(3) 学生成绩添加窗口 AddJFrame 代码如下:

```
import javax.swing.JButton;
import javax.swing.JCheckBox;
import javax.swing.JComboBox;
import javax.swing.JFrame;
import javax.swing.JLabel;
import javax.swing.JOptionPane;
import javax.swing.JPanel;
import javax.swing.JRadioButton;
import javax.swing.ButtonGroup;
import javax.swing.JTextField;
import javax.swing.BorderFactory;
import java.awt.event.ActionEvent;
import java.awt.event.ActionListener;
```

```java
public class AddJFrame extends JFrame implements ActionListener{
    private JPanel jp1;
    private JPanel jp2;
    private JLabel snoLabel;
    private JTextField snoText;
    private JLabel snameLabel;
    private JTextField snameText;
    private JLabel sexLabel;
    private JRadioButton sexJRB[];
    private JLabel cnameLabel;
    private JComboBox cnameJCB;
    String strCname = "==请选择==,Java程序设计,网络数据库技术与应用,JSP动态网页设计,网页设计与制作,图形图像处理";
    private JLabel pingShiLabel;
    private JTextField pingShiText;
    private JLabel guoChengLabel;
    private JTextField guoChengText;
    private JLabel zhongJieLabel;
    private JTextField zhongJieText;
    private JLabel zongHeLabel;
    private JTextField zongHeText;
    private JLabel zongHeJiBieLabel;
    private JTextField zongHeJiBieText;
    private JLabel loveLabel;
    private JCheckBox loveJCK[];
    private JButton addButton;
    private JButton resetButton;
    private JButton returnButton;

    public AddJFrame(){
        this.setTitle("添加学生成绩信息");
        this.setLayout(null);
        this.setSize(520,450);
        jp1 = new JPanel(null);

        //第一行
        snoLabel = new JLabel("学号:");
        jp1.add(snoLabel);
```

```java
snoLabel.setBounds(10,20,40,30);

snoText = new JTextField();
jp1.add(snoText);
snoText.setBounds(60,20,100,30);

snameLabel = new JLabel("姓名:");
jp1.add(snameLabel);
snameLabel.setBounds(170,20,40,30);

snameText = new JTextField();
snameText.setBounds(220,20,100,30);
jp1.add(snameText);

sexLabel = new JLabel("性别:");
sexLabel.setBounds(330,20,40,30);
jp1.add(sexLabel);

sexJRB = new JRadioButton[2];
ButtonGroup bg = new ButtonGroup();
sexJRB[0] = new JRadioButton("男");
sexJRB[0].setSelected(true);
jp1.add(sexJRB[0]);
sexJRB[0].setBounds(380,20,40,30);
sexJRB[1] = new JRadioButton("女");
jp1.add(sexJRB[1]);
sexJRB[1].setBounds(440,20,40,30);
bg.add(sexJRB[0]);
bg.add(sexJRB[1]);

//第二行
cnameLabel = new JLabel("课程名称:");
cnameLabel.setBounds(10,80,80,30);
jp1.add(cnameLabel);

cnameJCB = new JComboBox();
//将字符串分割成字符串数组 split
String[]cnameArray = strCname.split(",");
```

```java
for(int i = 0; i<cnameArray.length; i++){
    cnameJCB.addItem(cnameArray[i]);
}
cnameJCB.setBounds(80,80,85,30);
jp1.add(cnameJCB);

pingShiLabel = new JLabel("平时成绩:");
pingShiLabel.setBounds(165,80,70,30);
jp1.add(pingShiLabel);

pingShiText = new JTextField();
pingShiText.setBounds(235,80,80,30);
jp1.add(pingShiText);

guoChengLabel = new JLabel("过程考核成绩:");
guoChengLabel.setBounds(315,80,100,30);
jp1.add(guoChengLabel);

guoChengText = new JTextField();
guoChengText.setBounds(410,80,80,30);
jp1.add(guoChengText);

//第三行
zhongJieLabel = new JLabel("终结考核成绩:");
zhongJieLabel.setBounds(10,140,100,30);
jp1.add(zhongJieLabel);

zhongJieText = new JTextField();
zhongJieText.setBounds(105,140,80,30);
jp1.add(zhongJieText);

zongHeLabel = new JLabel("综合成绩:");
zongHeLabel.setBounds(185,140,80,30);
jp1.add(zongHeLabel);

zongHeText = new JTextField();
zongHeText.setBounds(255,140,80,30);
zongHeText.setEnabled(false);
```

```
jp1.add(zongHeText);

zongHeJiBieLabel = new JLabel("综合成绩级别:");
zongHeJiBieLabel.setBounds(335,140,95,30);
jp1.add(zongHeJiBieLabel);

zongHeJiBieText = new JTextField();
zongHeJiBieText.setBounds(430,140,60,30);
zongHeJiBieText.setEnabled(false);
jp1.add(zongHeJiBieText);

//第四行
loveLabel = new JLabel("爱好:");
loveLabel.setBounds(10,200,40,30);
jp1.add(loveLabel);

loveJCK = new JCheckBox[5];
loveJCK[0] = new JCheckBox("篮球");
loveJCK[0].setBounds(60,200,80,30);
jp1.add(loveJCK[0]);

loveJCK[1] = new JCheckBox("足球");
loveJCK[1].setBounds(150,200,80,30);
jp1.add(loveJCK[1]);

loveJCK[2] = new JCheckBox("乒乓球");
loveJCK[2].setBounds(240,200,80,30);
jp1.add(loveJCK[2]);

loveJCK[3] = new JCheckBox("羽毛球");
loveJCK[3].setBounds(330,200,80,30);
jp1.add(loveJCK[3]);

loveJCK[4] = new JCheckBox("排球");
loveJCK[4].setBounds(420,200,60,30);
jp1.add(loveJCK[4]);

this.add(jp1);
```

```java
        jp1.setBounds(5,5,500,260);
        jp1.setBorder(BorderFactory.createTitledBorder("学生成绩信息"));//设
置标题

        jp2 = new JPanel(null);
        //将相关操作的按钮组件添加到 jp2 上
        //设置窗口的大小
        addButton = new JButton("添   加");
        addButton.setBounds(40,40,100,40);
        jp2.add(addButton);
        addButton.addActionListener(this);

        resetButton = new JButton("重   置");
        resetButton.setBounds(200,40,100,40);
        resetButton.addActionListener(this);
        jp2.add(resetButton);

        returnButton = new JButton("返   回");
        returnButton.setBounds(360,40,100,40);
        returnButton.addActionListener(this);
        jp2.add(returnButton);

        this.add(jp2);
        jp2.setBounds(5,280,500,120);
        jp2.setBorder(BorderFactory.createTitledBorder("添加界面相关操作"));
//设置标题
        this.setLocationRelativeTo(null);
        this.setResizable(false);
        this.setDefaultCloseOperation(JFrame.DO_NOTHING_ON_CLOSE);
        this.setVisible(true);
    }

    public void actionPerformed(ActionEvent e){
        if(e.getSource() == addButton){
            //判断组件内容不允许为空
            if(!isNullForm()){
                return;
            }
            //判断学号是否已存在,避免学生成绩信息重复
```

```java
            StudentScoreDAO ssd = new StudentScoreDAO();
            String sno = snoText.getText().trim();
            try{
                StudentScoreBean ssb = ssd.searchBySno(Integer.parseInt(sno));
                if(ssb! = null){
                    JOptionPane.showMessageDialog(this,"对不起,【学号】已经存在,请重新输入学号!");
                    return;
                }else{
                    ssb = getScore();
                    ssd.addScore(ssb);
                }
            }
            catch(Exception e1){
                JOptionPane.showMessageDialog(this,e1.toString());
            }
        }
        if(e.getSource() = = resetButton){
            snoText.setText("");
            snameText.setText("");
            pingShiText.setText("");
            guoChengText.setText("");
            zhongJieText.setText("");
            zongHeText.setText("");
            zongHeJiBieText.setText("");
            cnameJCB.setSelectedIndex(0);
            for(int i = 0;i<5;i + +){
                loveJCK[i].setSelected(false);
            }
        }
        if(e.getSource() = = returnButton){
            this.dispose();
        }
    }
    /**
     *判断组件的内容是否为空
     *
```

```java
 * @return
 */
private boolean isNullForm(){
if(snoText.getText().trim().length() == 0){
    JOptionPane.showMessageDialog(this,"【学号】不能为空!");
    snoText.requestFocus(true);
    return false;
}
if(snameText.getText().trim().length() == 0){
    JOptionPane.showMessageDialog(this,"【姓名】不能为空!");
    snameText.requestFocus(true);
    return false;
}
if(pingShiText.getText().trim().length() == 0){
    JOptionPane.showMessageDialog(this,"【平时成绩】不能为空!");
    pingShiText.requestFocus(true);
    return false;
}
if(guoChengText.getText().trim().length() == 0){
    JOptionPane.showMessageDialog(this,"【过程考核成绩】不能为空!");
    guoChengText.requestFocus(true);
    return false;
}
if(zhongJieText.getText().trim().length() == 0){
    JOptionPane.showMessageDialog(this,"【终结考核成绩】不能为空!");
    zhongJieText.requestFocus(true);
    return false;
}
    return true;
}

/**
 * 获取组件上的信息
 *
 * @return
 */
public StudentScoreBean getScore(){
    StudentScoreBean ssb = new StudentScoreBean();
```

```java
try{
    int sno = Integer.parseInt(snoText.getText());
    ssb.setSno(sno);
    String sname = snameText.getText();
    ssb.setSname(sname);
    char sex;
    if(sexJRB[0].isSelected()){
    sex = '男';
    }else{
    sex = '女';}
    ssb.setSex(sex);
    String cname = cnameJCB.getSelectedItem().toString();
    ssb.setCname(cname);
    byte pingShiScore = Byte.parseByte(pingShiText.getText());
    ssb.setPingShiScore(pingShiScore);
    byte guoChengScore = Byte.parseByte(guoChengText.getText());
    ssb.setGuoChengScore(guoChengScore);
    byte zhongJieScore = Byte.parseByte(zhongJieText.getText());
    ssb.setZhongJieScore(zhongJieScore);
    double
zongHeScore = pingShiScore * 0.1 + guoChengScore * 0.4 + zhongJieScore * 0.5;
    zongHeText.setText(String.valueOf(zongHeScore));
    ssb.setZongHeScore(zongHeScore);
    String zongHeScoreJiBie = null;
    if(zongHeScore<60){
    zongHeScoreJiBie = "不及格";
    }else if(zongHeScore<70){
    zongHeScoreJiBie = "及格";
    }else if(zongHeScore<80){
    zongHeScoreJiBie = "中";
    }
    else if(zongHeScore<90){
    zongHeScoreJiBie = "良";
    }
    else if(zongHeScore< = 100){
    zongHeScoreJiBie = "优";
    }else{
    zongHeScoreJiBie = "初始成绩有误!";
```

```
            }
            zongHeJiBieText.setText(zongHeScoreJiBie);
            ssb.setZongHeScoreJiBie(zongHeScoreJiBie);
            String love = "";
            if(loveJCK[0].isSelected())
            love = love + "篮球" + "   ";
            if(loveJCK[1].isSelected())
            love = love + "足球" + "   ";
            if(loveJCK[2].isSelected())
            love = love + "乒乓球" + "   ";
            if(loveJCK[3].isSelected())
            love = love + "羽毛球" + "   ";
            if(loveJCK[4].isSelected())
            love = love + "排球" + "   ";
            if(love.trim().length() = = 0)
            love = "未选爱好";
            ssb.setLove(love);
            }
            catch(Exception e){
            JOptionPane.showMessageDialog(this,e.toString());
            }
            return ssb;
        }

    public static void main(String[]args){
        new AddJFrame();
    }
}
```

如图 7-70 至图 7-72 所示。

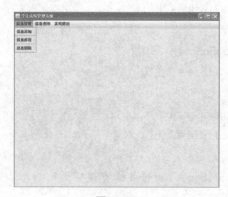

图 7-70

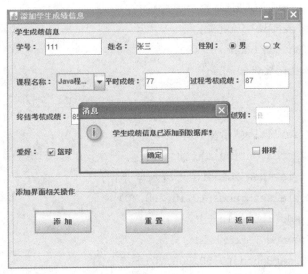

图 7 – 71

图 7 – 72

（4）实现学生成绩信息修改功能。

① 学号输入窗口 SnoInputJFrame 代码如下：

```java
import java.awt.event.ActionEvent;
import java.awt.event.ActionListener;
import java.awt.event.WindowAdapter;
import java.awt.event.WindowEvent;
import javax.swing.JButton;
import javax.swing.JDialog;
import javax.swing.JFrame;
import javax.swing.JLabel;
import javax.swing.JOptionPane;
import javax.swing.JTextField;

/**
 *学号输入界面
 */
public class SnoInputJFrame extends JDialog implements ActionListener{
```

```java
        private JLabel lbLTitle;
        private JTextField txtNo;
        private JButton btnNext;
        public static int xuehao;

        public SnoInputJFrame(){
            this.setTitle("请输入学生学号!");
            this.setLayout(null);
            lbLTitle = new JLabel("请输入学号:");
            lbLTitle.setBounds(10,10,200,30);
            this.add(lbLTitle);
            txtNo = new JTextField();
            txtNo.setBounds(5,50,200,20);
            this.add(txtNo);
            btnNext = new JButton("下一步");
            btnNext.setBounds(60,90,80,20);
            btnNext.addActionListener(this);
            this.add(btnNext);
            //匿名内部类
            this.addWindowListener(new WindowAdapter(){
                public void windowClosing(WindowEvent arg0){
                    dispose();
                }
            });
            this.setSize(220,150);
            this.setLocationRelativeTo(null);
            this.setVisible(true);
        }

        public void actionPerformed(ActionEvent e){
            String input = txtNo.getText().trim();
            if(input.length() == 0){
            JOptionPane.showMessageDialog(this,"请输入学号!");
            return;
            }
            try{
            StudentScoreDAO ssd = new StudentScoreDAO();
            StudentScoreBean ssb = ssd.searchBySno(Integer.parseInt(input));
```

```java
        if(ssb! = null){
            xuehao = ssb.getSno();
            new UpdateJFrame();
            this.dispose();
        }else{
            txtNo.setText("");
            JOptionPane.showMessageDialog(this,"对不起,【学号】不存在,请重新输入学号!");
            return;
        }
    }
    catch(Exception e1){
        JOptionPane.showMessageDialog(this,e1.toString());
    }
    }
}
```

② 学生成绩修改窗口 UpdateJFrame 代码如下:

```java
import java.awt.event.ActionEvent;
import java.awt.event.ActionListener;
import javax.swing.JButton;
import javax.swing.JCheckBox;
import javax.swing.JComboBox;
import javax.swing.JFrame;
import javax.swing.JLabel;
import javax.swing.JOptionPane;
import javax.swing.JPanel;
import javax.swing.JRadioButton;
import javax.swing.ButtonGroup;
import javax.swing.JTextField;
import javax.swing.BorderFactory;

public class UpdateJFrame extends JFrame implements ActionListener{
    private JPanel jp1;
    private JPanel jp2;
    private JLabel snoLabel;
    private JTextField snoText;
```

```java
        private JLabel snameLabel;
        private JTextField snameText;
        private JLabel sexLabel;
        private JRadioButton sexJRB[];
        private JLabel cnameLabel;
        private JComboBox cnameJCB;
        String strCname = "= =请选择= =,Java程序设计,网络数据库技术与应用,JSP动态网页设计,网页设计与制作,图形图像处理";
        private JLabel pingShiLabel;
        private JTextField pingShiText;
        private JLabel guoChengLabel;
        private JTextField guoChengText;
        private JLabel zhongJieLabel;
        private JTextField zhongJieText;
        private JLabel zongHeLabel;
        private JTextField zongHeText;
        private JLabel zongHeJiBieLabel;
        private JTextField zongHeJiBieText;
        private JLabel loveLabel;
        private JCheckBox loveJCK[];
        private JButton updateButton;
        private JButton resetButton;
        private JButton returnButton;
        int xuehao1;
        SnoInputJFrame s;

        public UpdateJFrame(){
            this.setTitle("修改学生成绩信息");
            this.setLayout(null);
            this.setSize(520,450);
            jp1 = new JPanel(null);

            s = new SnoInputJFrame();
            xuehao1 = s.xuehao;
            s.dispose();
            StudentScoreDAO ssd1 = new StudentScoreDAO();
            StudentScoreBean ssb1 = ssd1.searchBySno(xuehao1);
```

```java
snoLabel = new JLabel("学号:");
jp1.add(snoLabel);
snoLabel.setBounds(10,20,40,30);

snoText = new JTextField();
jp1.add(snoText);
snoText.setBounds(60,20,100,30);
snoText.setText(String.valueOf(ssb1.getSno()));

snameLabel = new JLabel("姓名:");
jp1.add(snameLabel);
snameLabel.setBounds(170,20,40,30);

snameText = new JTextField();
snameText.setBounds(220,20,100,30);
snameText.setText(ssb1.getSname());
jp1.add(snameText);

sexLabel = new JLabel("性别:");
sexLabel.setBounds(330,20,40,30);
jp1.add(sexLabel);

sexJRB = new JRadioButton[2];
ButtonGroup bg = new ButtonGroup();
sexJRB[0] = new JRadioButton("男");
jp1.add(sexJRB[0]);
sexJRB[0].setBounds(380,20,40,30);
sexJRB[1] = new JRadioButton("女");
jp1.add(sexJRB[1]);
sexJRB[1].setBounds(440,20,40,30);
bg.add(sexJRB[0]);
bg.add(sexJRB[1]);
if(String.valueOf(ssb1.getSex()).equals("男")){
sexJRB[0].setSelected(true);
}else{
sexJRB[1].setSelected(true);
}
```

```java
//第二行
cnameLabel = new JLabel("课程名称:");
cnameLabel.setBounds(10,80,80,30);
jp1.add(cnameLabel);

cnameJCB = new JComboBox();
//将字符串分割成字符串数组split
String[]cnameArray = strCname.split(",");
for(int i = 0;i<cnameArray.length;i++){
    cnameJCB.addItem(cnameArray[i]);
}
cnameJCB.setBounds(80,80,85,30);
jp1.add(cnameJCB);

pingShiLabel = new JLabel("平时成绩:");
pingShiLabel.setBounds(165,80,70,30);
jp1.add(pingShiLabel);

pingShiText = new JTextField();
pingShiText.setBounds(235,80,80,30);
pingShiText.setText(String.valueOf(ssb1.getPingShiScore()));
jp1.add(pingShiText);

guoChengLabel = new JLabel("过程考核成绩:");
guoChengLabel.setBounds(315,80,100,30);
jp1.add(guoChengLabel);

guoChengText = new JTextField();
guoChengText.setBounds(410,80,80,30);
guoChengText.setText(String.valueOf(ssb1.getGuoChengScore()));
jp1.add(guoChengText);

//第三行
zhongJieLabel = new JLabel("终结考核成绩:");
zhongJieLabel.setBounds(10,140,100,30);
jp1.add(zhongJieLabel);

zhongJieText = new JTextField();
```

```
zhongJieText.setBounds(105,140,80,30);
zhongJieText.setText(String.valueOf(ssb1.getZhongJieScore()));
jp1.add(zhongJieText);

zongHeLabel = new JLabel("综合成绩:");
zongHeLabel.setBounds(185,140,80,30);
jp1.add(zongHeLabel);

zongHeText = new JTextField();
zongHeText.setBounds(255,140,80,30);
zongHeText.setText(String.valueOf(ssb1.getZongHeScore()));
zongHeText.setEnabled(false);
jp1.add(zongHeText);

zongHeJiBieLabel = new JLabel("综合成绩级别:");
zongHeJiBieLabel.setBounds(335,140,95,30);
jp1.add(zongHeJiBieLabel);

zongHeJiBieText = new JTextField();
zongHeJiBieText.setBounds(430,140,60,30);
zongHeJiBieText.setText(ssb1.getZongHeScoreJiBie());
zongHeJiBieText.setEnabled(false);
jp1.add(zongHeJiBieText);

//第四行
loveLabel = new JLabel("爱好:");
loveLabel.setBounds(10,200,40,30);
jp1.add(loveLabel);

loveJCK = new JCheckBox[5];
loveJCK[0] = new JCheckBox("篮球");
loveJCK[0].setBounds(60,200,80,30);
jp1.add(loveJCK[0]);

loveJCK[1] = new JCheckBox("足球");
loveJCK[1].setBounds(150,200,80,30);
jp1.add(loveJCK[1]);
```

```java
        loveJCK[2] = new JCheckBox("乒乓球");
        loveJCK[2].setBounds(240,200,80,30);
        jp1.add(loveJCK[2]);

        loveJCK[3] = new JCheckBox("羽毛球");
        loveJCK[3].setBounds(330,200,80,30);
        jp1.add(loveJCK[3]);

        loveJCK[4] = new JCheckBox("排球");
        loveJCK[4].setBounds(420,200,60,30);
        jp1.add(loveJCK[4]);

        this.add(jp1);
        jp1.setBounds(5,5,500,260);
        jp1.setBorder(BorderFactory.createTitledBorder("学生成绩信息"));//设置标题

        jp2 = new JPanel(null);
        //将相关操作的按钮组件添加到jp2上
        //设置窗口的大小
        updateButton = new JButton("修  改");
        updateButton.setBounds(40,40,100,40);
        jp2.add(updateButton);
        updateButton.addActionListener(this);

        resetButton = new JButton("重  置");
        resetButton.setBounds(200,40,100,40);
        resetButton.addActionListener(this);
        jp2.add(resetButton);

        returnButton = new JButton("返  回");
        returnButton.setBounds(360,40,100,40);
        returnButton.addActionListener(this);
        jp2.add(returnButton);

        this.add(jp2);
        jp2.setBounds(5,280,500,120);
```

```java
        jp2.setBorder(BorderFactory.createTitledBorder("添加界面相关操
作"));//设置标题

        this.setLocationRelativeTo(null);
        this.setResizable(false);
        this.setDefaultCloseOperation(JFrame.DO_NOTHING_ON_CLOSE);
        this.setVisible(true);
    }

    public void actionPerformed(ActionEvent e){
        if(e.getSource() == updateButton){
            //判断组件内容不允许为空
            if(!isNullForm()){
                return;
            }
            StudentScoreDAO ssd = new StudentScoreDAO();
            StudentScoreBean ssb = getScore();
            try{
                ssd.updateScore(ssb,xuehao1);
            }
            catch(Exception e1){
                JOptionPane.showMessageDialog(this,e1.toString());
            }
        }
        if(e.getSource() == resetButton){
            snoText.setText("");
            snameText.setText("");
            pingShiText.setText("");
            guoChengText.setText("");
            zhongJieText.setText("");
            zongHeText.setText("");
            zongHeJiBieText.setText("");
            cnameJCB.setSelectedIndex(0);
            for(int i = 0;i<5;i++){
                loveJCK[i].setSelected(false);
            }
        }
        if(e.getSource() == returnButton){
```

```java
            this.dispose();
        }
    }


    /**
     * 获取组件上的信息
     *
     * @return
     */
    public StudentScoreBean getScore(){
        StudentScoreBean ssb = new StudentScoreBean();
        try{
            int sno = Integer.parseInt(snoText.getText());
            ssb.setSno(sno);
            String sname = snameText.getText();
            ssb.setSname(sname);
            char sex;
            if(sexJRB[0].isSelected()){
            sex = '男';
            }else{
            sex = '女';}
            ssb.setSex(sex);
            String cname = cnameJCB.getSelectedItem().toString();
            ssb.setCname(cname);
            byte pingShiScore = Byte.parseByte(pingShiText.getText());
            ssb.setPingShiScore(pingShiScore);
            byte guoChengScore = Byte.parseByte(guoChengText.getText());
            ssb.setGuoChengScore(guoChengScore);
            byte zhongJieScore = Byte.parseByte(zhongJieText.getText());
            ssb.setZhongJieScore(zhongJieScore);
            double
zongHeScore = pingShiScore * 0.1 + guoChengScore * 0.4 + zhongJieScore * 0.5;
            zongHeText.setText(String.valueOf(zongHeScore));
            ssb.setZongHeScore(zongHeScore);
            String zongHeScoreJiBie = null;
            if(zongHeScore<60){
            zongHeScoreJiBie = "不及格";
```

```java
        }else if(zongHeScore<70){
        zongHeScoreJiBie = "及格";
        }else if(zongHeScore<80){
        zongHeScoreJiBie = "中";
        }
        else if(zongHeScore<90){
        zongHeScoreJiBie = "良";
        }
        else if(zongHeScore<=100){
        zongHeScoreJiBie = "优";
        }else{
        zongHeScoreJiBie = "初始成绩有误!";
        }
        zongHeJiBieText.setText(zongHeScoreJiBie);
        ssb.setZongHeScoreJiBie(zongHeScoreJiBie);
        String love = "";
        if(loveJCK[0].isSelected())
        love = love + "篮球" + "   ";
        if(loveJCK[1].isSelected())
        love = love + "足球" + "   ";
        if(loveJCK[2].isSelected())
        love = love + "乒乓球" + "   ";
        if(loveJCK[3].isSelected())
        love = love + "羽毛球" + "   ";
        if(loveJCK[4].isSelected())
        love = love + "排球" + "   ";
        if(love.trim().length() == 0)
        love = "未选爱好";
        ssb.setLove(love);
        }
        catch(Exception e){
        JOptionPane.showMessageDialog(this,e.toString());
        }
        return ssb;
    }

    /**
    * 判断组件的内容是否为空
```

```java
     *
     * @return
     */
    private boolean isNullForm(){
        if(snoText.getText().trim().length()==0){
            JOptionPane.showMessageDialog(this,"【学号】不能为空!");
            snoText.requestFocus(true);
            return false;
        }
        if(snameText.getText().trim().length()==0){
            JOptionPane.showMessageDialog(this,"【姓名】不能为空!");
            snameText.requestFocus(true);
            return false;
        }
        if(pingShiText.getText().trim().length()==0){
            JOptionPane.showMessageDialog(this,"【平时成绩】不能为空!");
            pingShiText.requestFocus(true);
            return false;
        }
        if(guoChengText.getText().trim().length()==0){
            JOptionPane.showMessageDialog(this,"【过程考核成绩】不能为空!");
            guoChengText.requestFocus(true);
            return false;
        }
        if(zhongJieText.getText().trim().length()==0){
            JOptionPane.showMessageDialog(this,"【终结考核成绩】不能为空!");
            zhongJieText.requestFocus(true);
            return false;
        }
        return true;
    }

    public static void main(String[]args){
        new UpdateJFrame();
    }
}
```

如图 7-73 至图 7-76 所示。

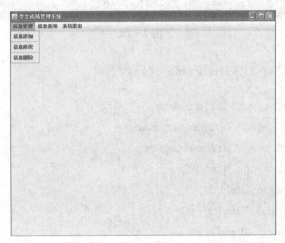

图 7-73

图 7-74

图 7-75

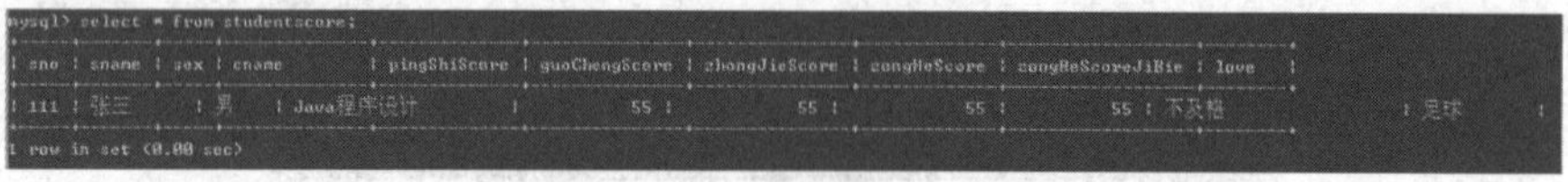

图 7-76

（5）学生成绩删除窗口 DeleteJFrame 代码如下：

```java
import java.awt.event.ActionEvent;
import java.awt.event.ActionListener;
import java.awt.event.WindowAdapter;
import java.awt.event.WindowEvent;
import javax.swing.JButton;
import javax.swing.JDialog;
import javax.swing.JFrame;
import javax.swing.JLabel;
import javax.swing.JOptionPane;
import javax.swing.JTextField;

public class DeleteJFrame extends JDialog implements ActionListener{
    private JLabel lbLTitle;
    private JTextField txtNo;
    private JButton btnNext;

    public DeleteJFrame(){
        this.setTitle("请输入学生学号！");
        this.setLayout(null);
        lbLTitle = new JLabel("请输入学号：");
        lbLTitle.setBounds(10,10,200,30);
        this.add(lbLTitle);
        txtNo = new JTextField();
        txtNo.setBounds(5,50,200,20);
        this.add(txtNo);
        btnNext = new JButton("下一步");
        btnNext.setBounds(60,90,80,20);
        btnNext.addActionListener(this);
        this.add(btnNext);
        //匿名内部类
        this.addWindowListener(new WindowAdapter(){
            public void windowClosing(WindowEvent arg0){
```

```java
                    dispose();
                }
            });
            this.setSize(220,150);
            this.setLocationRelativeTo(null);
            this.setVisible(true);
        }

        public void actionPerformed(ActionEvent e){
            String input = txtNo.getText().trim();
            if(input.length() = = 0){
                JOptionPane.showMessageDialog(this,"请输入学号!");
                return;
            }
            try{
                int number = Integer.parseInt(input);
                StudentScoreDAO ssd = new StudentScoreDAO();
                StudentScoreBean ssb = ssd.searchBySno(number);
                if(ssb! = null){
                    int n = JOptionPane.showConfirmDialog(this,"您确认删除该学生的成绩信息吗?","确认对话框",JOptionPane.YES_NO_OPTION,JOptionPane.QUESTION_MESSAGE);
                    if(n = = JOptionPane.YES_OPTION){
                        ssd.deleteScore(number);
                        this.dispose();
                    }
                    if(n = = JOptionPane.NO_OPTION){
                        this.dispose();
                    }
                }else{  txtNo.setText("");
                    JOptionPane.showMessageDialog(this,"对不起,【学号】不存在,请重新输入学号!");
                    return;
                }
            }
            catch(Exception e1){
                JOptionPane.showMessageDialog(this,e1.toString());
            }
```

```
        }
    }
```

如图 7-77 至图 7-81 所示。

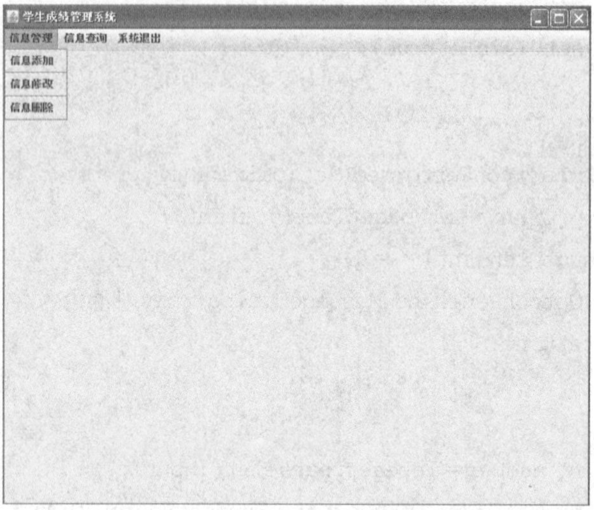

图 7-77

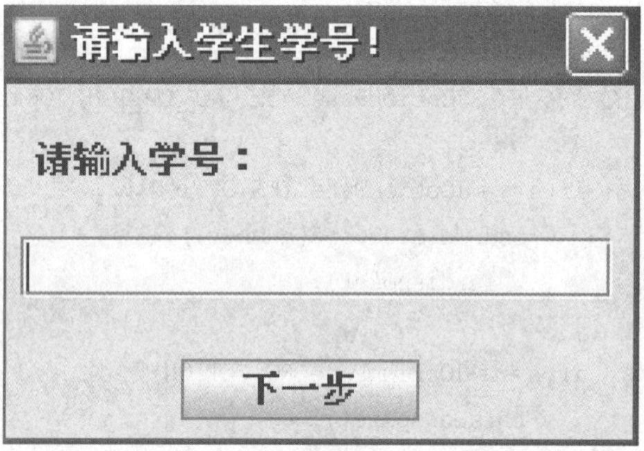

图 7-78

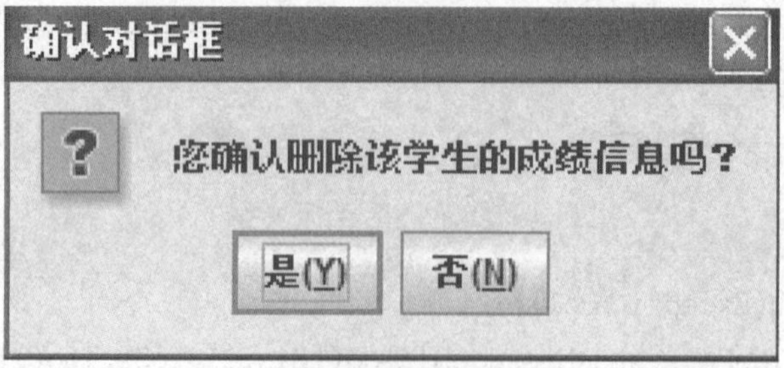

图 7-79

图 7-80

图 7-81

(6) 学生成绩查询窗口 SelectJFrame 代码如下：

```
import java.awt.event.ActionEvent;
import java.awt.event.ActionListener;
import javax.swing.JButton;
import javax.swing.JComboBox;
import javax.swing.JFrame;
import javax.swing.JLabel;
import javax.swing.JOptionPane;
import javax.swing.JPanel;
import javax.swing.JTextField;
import javax.swing.BorderFactory;
import javax.swing.JScrollPane;
import javax.swing.JScrollPane;
import javax.swing.JTable;
import javax.swing.table.DefaultTableModel;
import javax.swing.table.TableColumnModel;
import javax.swing.table.TableColumn;
import java.sql.ResultSet;

public class SelectJFrame extends JFrame implements ActionListener{
    private JPanel jp1;
    private JPanel jp2;
    private JLabel selectLabel;
    private JComboBox selectJCB;
    String strSelect = "==请选择==,学  号,姓  名,所  有";
```

```java
            private JTextField selectText;
            private JButton selectButton;
            private JButton returnButton;
            //表格列标题
            String[]titles = {"学号","姓名","性别","课程名称","平时成绩","过程考核成绩","终结考核成绩","综合成绩","综合成绩级别","爱好"};
            //表格显示内容
            Object[][]cj;
            DefaultTableModel DTM;
            JTable JT;
            JScrollPane JSP1;

            public SelectJFrame(){
            this.setTitle("查询学生成绩信息");
            this.setLayout(null);
            this.setSize(580,450);
            jp1 = new JPanel(null);
            selectLabel = new JLabel("请选择关键字并输入查询信息：");
            selectLabel.setBounds(10,20,190,30);
            jp1.add(selectLabel);
            selectJCB = new JComboBox();
            //将字符串分割成字符串数组 split
            String[]selectArray = strSelect.split(",");
            for(int i = 0;i<selectArray.length;i++){
                selectJCB.addItem(selectArray[i]);
            }
            selectJCB.setBounds(210,20,90,30);
            jp1.add(selectJCB);
            selectJCB.addActionListener(this);
            selectText = new JTextField();
            selectText.setBounds(310,20,90,30);
            jp1.add(selectText);
            selectButton = new JButton("查询");
            selectButton.setBounds(410,20,70,30);
            jp1.add(selectButton);
            selectButton.addActionListener(this);
            returnButton = new JButton("返回");
            returnButton.setBounds(490,20,70,30);
```

```java
        jp1.add(returnButton);
        returnButton.addActionListener(this);
        this.add(jp1);
        jp1.setBounds(5,5,565,60);
        jp1.setBorder(BorderFactory.createTitledBorder("查询方式选择"));//设置标题

        jp2 = new JPanel(null);
        //将相关操作的按钮组件添加到jp2上
        //设置窗口的大小
        DTM = new DefaultTableModel(cj,titles);
        JT = new JTable(DTM);
        JSP1 = new JScrollPane(JT);
        jp2.add(JSP1);
        JSP1.setBounds(10,30,550,300);
        this.add(jp2);
        jp2.setBounds(5,70,565,340);
        jp2.setBorder(BorderFactory.createTitledBorder("学生成绩信息"));//设置标题
        this.setLocationRelativeTo(null);
        this.setResizable(false);
        this.setDefaultCloseOperation(JFrame.DO_NOTHING_ON_CLOSE);
        this.setVisible(true);
    }

    public void actionPerformed(ActionEvent e){
        if(e.getSource() == selectButton){
            try{
                DTM.setRowCount(0);
                String selected = selectJCB.getSelectedItem().toString();
                if(selected.equals("学   号")){
                    String input = selectText.getText().trim();
                    if(input.length() == 0){
                        JOptionPane.showMessageDialog(this,"请输入查询信息!");
                        return;
                    }
                    int number = Integer.parseInt(selectText.getText());
                    StudentScoreDAO ssd = new StudentScoreDAO();
```

```java
            StudentScoreBean ssb = ssd.searchBySno(number);
            if(ssb!=null){
                DTM.addRow(new Object[]{ssb.getSno(),ssb.getSname(),ssb.getSex(),ssb.getCname(),ssb.getPingShiScore(),ssb.getGuoChengScore(),ssb.getZhongJieScore(),ssb.getZongHeScore(),ssb.getZongHeScoreJiBie(),ssb.getLove()});

                JOptionPane.showMessageDialog(this,"查询完毕!");
                selectJCB.setSelectedIndex(0);
                selectText.setText("");
            }else{
                selectJCB.setSelectedIndex(0);
                selectText.setText("");
                JOptionPane.showMessageDialog(this,"未找到该学生信息!");
            }
        }
        if(selected.equals("姓   名")){
            String input = selectText.getText().trim();
            if(input.length()==0){
                JOptionPane.showMessageDialog(this,"请输入查询信息!");
                return;
            }
            String name = selectText.getText();
            StudentScoreDAO ssd = new StudentScoreDAO();
            StudentScoreBean ssb = ssd.searchBySname(name);
            if(ssb!=null){
                DTM.addRow(new Object[]{ssb.getSno(),ssb.getSname(),ssb.getSex(),ssb.getCname(),ssb.getPingShiScore(),ssb.getGuoChengScore(),ssb.getZhongJieScore(),ssb.getZongHeScore(),ssb.getZongHeScoreJiBie(),ssb.getLove()});

                JOptionPane.showMessageDialog(this,"查询完毕!");
                selectJCB.setSelectedIndex(0);
                selectText.setText("");
            }else{
                selectJCB.setSelectedIndex(0);
                selectText.setText("");
```

```java
                JOptionPane.showMessageDialog(this,"未找到该学生信息!");
            }
        }
        if(selected.equals("所  有")){
            StudentScoreDAO ssd = new StudentScoreDAO();
            ResultSet rs = ssd.searchAll();
            while(rs.next()){
                DTM.addRow(new Object[]{rs.getInt("sno"),
rs.getString("sname"),rs.getString("sex").charAt(0),rs.getString("cname"),rs.
getByte("pingShiScore"), rs.getByte("guoChengScore"), rs.getByte("
zhongJieScore"),rs.getDouble("zongHeScore"),rs.getString("zongHeScoreJiBie"),
rs.getString("love")});
            }
            selectJCB.setSelectedIndex(0);
            selectText.setText("");
            JOptionPane.showMessageDialog(this,"查询完毕!");
        }
        if(selected.equals("==请选择==")){
            JOptionPane.showMessageDialog(this,"请选择查询方式!");
        }
    }
    catch(Exception e1){
        JOptionPane.showMessageDialog(this,e1.toString());
    }
}

if(e.getSource() == selectJCB){
    String selected = selectJCB.getSelectedItem().toString();
    if(selected.equals("所  有")){
        selectText.setText("");
        selectText.setEditable(false);
    }else{
        selectText.setText("");
        selectText.setEditable(true);
    }
}

if(e.getSource() == returnButton){
```

```
            this.dispose();
        }
    }
    public static void main(String[ ]args){
        new SelectJFrame();
    }
}
```

如图 7-82 至图 7-86 所示。

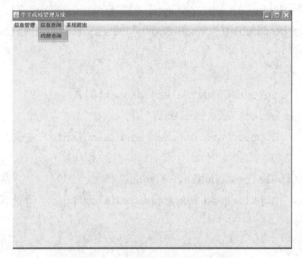

图 7-82

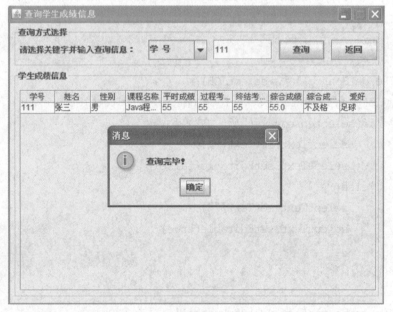

图 7-83

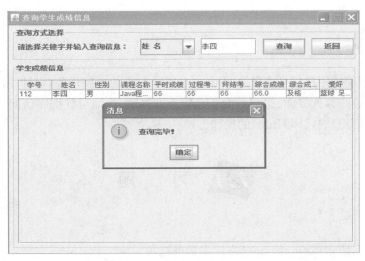

图 7-84

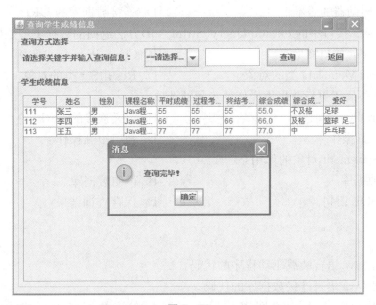

图 7-85

图 7-86

任务拓展

根据 7.1 创建的用户信息表(UserInfo)，参考项目 6 的拓展任务，设计用户登录界面、主界面、开户界面、存/取款界面、销户界面和余额查询界面，并编写数据库访问程序，实现相应功能模块对用户信息表(UserInfo)的操作。

7.3 习题

一、单选题

1. 请选出不是 getConnection()方法参数的选项。（   ）
   A. 数据库用户名　　　　　　　　　B. 数据库的访问密码
   C. JDBC 驱动器的版本　　　　　　D. 连接数据库的 URL
2. Statement 接口中的 executeQuery(String sql)方法返回的数据类型是（   ）。
   A. Statement 接口实例　　　　　　B. Connection 接口实例
   C. DatabaseMetaData 类的对象　　D. ResultSet 类的对象
3. 请选出不属于更新数据库操作的步骤的选项。（   ）
   A. 加载 JDBC 驱动程序　　　　　　B. 定义连接的 URL
   C. 执行查询操作　　　　　　　　　D. 执行更新操作
4. 建立 Statement 对象的作用是（   ）。
   A. 连接数据库　　　　　　　　　　B. 声明数据库
   C. 执行 SQL 语句　　　　　　　　D. 保存查询结果

二、简答题

1. 请写出 Java 语言装载驱动程序的代码。
2. 请写出 Java 语言连接数据库的代码。
3. 请使用 SQL 命令创建图书数据库 Books，在该数据库中创建表格 BookInfo，BookInfo 表包含书号、书名、作者、出版社、出版日期、数量字段。然后使用 Java 语言实现如下操作：
   （1）对图书信息进行插入操作。
   （2）对图书信息进行修改操作。
   （3）对图书信息进行删除操作。
   （4）对图书信息进行查询操作。

项目 8

简单网络聊天软件

项目 8 首先介绍了 IP 地址、InetAddress 类、URL、URLConnection、TCP 及 UDP 程序设计等 Java 网络程序设计相关知识;其次,讲解了 Java 多线程的实现方式、状态变化、主要操作方法、同步、生命周期等多线程程序设计相关知识,并设计实现了具有聊天功能的简单网络聊天软件。

工作任务

(1) 一对一网络聊天软件的单机模拟实现。
(2) 一对一网络聊天软件的多线程实现。

学习目标

(1) 理解 IP 地址与 InetAddress 的关系,了解 URL、URLConnection 定位网络资源的方法,掌握 TCP、UDP 程序设计知识,能够根据需要独立完成简单的 Java 网络程序设计。

(2) 熟悉线程的基本概念,了解线程的生命周期,掌握线程的创建方法,学会线程同步的处理方法,并能通过线程处理聊天信息的发送和接收。

任务 1 一对一网络聊天软件的单机模拟实现

任务描述及分析

小明接到一个开发任务:模拟实现网络聊天软件(一对一)。要完成这个工作任务,第一步,我们需要系统学习 Java 网络程序设计相关知识;第二步,利用项目 6 所学的 Java 图形界面知识创建简单网络聊天软件的服务器和客户端界面,并结合 Java 网络程序设计相关知识,单机模拟实现服务器端和客户端的一对一通信。

相关知识

8.1.1　IP 地址和 InetAddress 类

网络使不同物理位置上的计算机实现了资源共享和相互通信,而网络上不同位置的两台计算机如何在数量众多的计算机中找到对方,从而实现相互之间的通信呢？答案是通过计算机的 IP 地址。

1. IP 地址

互联网中的每一台计算机都有一个唯一表示自己的 IP 地址,IP 地址是一个 32 位长度的二进制数据,实际上我们通常把它表示成由 4 个点号"."隔开的十进制整数(每个整数的取值范围是 0 到 255)。例如在 Windows 系统中,右键单击"网络连接"中的"本地连接",在弹出的"本地连接属性"窗口中,选择"Internet 协议(TCP/IP)"并点击【属性】(如图 8-1 所示),在弹出的如图 8-2 所示的"Internet 协议(TCP/IP)"中可以看到 IP 地址的十进制表示形式。

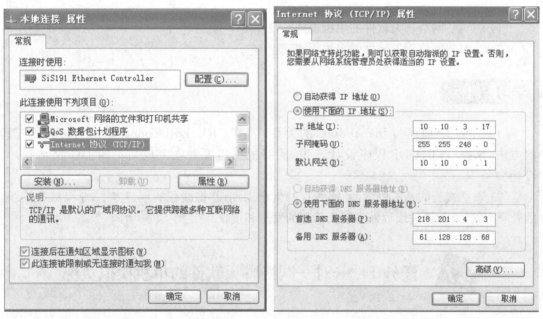

图 8-1　　　　　　　　　　　图 8-2

图 8-2 所示的 IP 地址是 IPv4 版的,也是使用最广泛的版本。IPv4 版的 IP 地址分为 5 类,如表 8-1 所示。

表 8-1

IPv4 地址分类		
地址分类	地址范围	用途
A 类地址	1.0.0.1～126.255.255.254	保留给政府机构
B 类地址	128.0.0.1～191.255.255.254	分配给中等规模的公司
C 类地址	192.0.0.1～223.255.255.254	分配给任何需要的人
D 类地址	224.0.0.1～239.255.255.254	用于组播
E 类地址	240.0.0.1～255.255.255.254	用于实验

注意：

IP 地址 127.0.0.1 表示本机地址，还可使用 localhost 代表本机。

由于 IP 地址使用数字作为计算机的标识，人们不容易记忆。由此，我们为互联网上的每一台计算机分配了一串用点"."分隔的字符组成的计算机名称（即域名，如 www.baidu.com），用于标识计算机。在实际使用中，我们可以用简便、容易记忆的域名代替 IP 地址，计算机会通过域名服务器将域名自动转换为对应的 IP 地址。例如，在 Windows 系统的 DOS 窗口输入命令：ping www.baidu.com，可以查到域名为 www.baidu.com 的计算机对应的 IP 地址，如图 8-3 所示。

```
C:\Documents and Settings\Administrator>ping www.baidu.com

Pinging www.a.shifen.com [183.232.231.172] with 32 bytes of data:

Reply from 183.232.231.172: bytes=32 time=178ms TTL=52
Reply from 183.232.231.172: bytes=32 time=137ms TTL=52
Reply from 183.232.231.172: bytes=32 time=96ms TTL=52
Reply from 183.232.231.172: bytes=32 time=135ms TTL=52

Ping statistics for 183.232.231.172:
    Packets: Sent = 4, Received = 4, Lost = 0 (0% loss),
Approximate round trip times in milli-seconds:
    Minimum = 96ms, Maximum = 178ms, Average = 136ms
```

图 8-3

注意：

互联网上的计算机使用 IP 地址标识自己，域名只是方便人的使用，人们输入的域名可以通过域名服务器自动转换为 IP 地址，计算机之间通过转换生成的 IP 地址相互确认位置。

2. InetAddress 类

现在，我们已经知道 Internet 上的主机有两种方式表示地址：

(1) 域名，如 www.baidu.com。

(2) IP 地址，如 183.232.231.172。

Java 语言又如何描述计算机的域名和 IP 地址呢？Java 语言用 InetAddress 类来封装

(表示)主机的 IP 地址及其域名,即 InetAddress 类对象含有一个互联网主机的域名和 IP 地址信息,如 www.baidu.com/183.232.231.172。

InetAddress 类位于 java.net 包下(即使用时需要导入头文件 java.net.InetAddress),且 InetAddress 类的构造方法被私有化了,不能通过构造方法对其产生实例对象。因此,InetAddress 类的实例化,我们一般使用静态方法 getLocalHost()、getByName()等。使用上述静态方法创建 InetAddress 类对象时,如果找不到本地机器的地址,这些方法通常会抛出 UnknownHostException 异常,所以应该在程序中进行异常处理。此外,InetAddress 类的对象还有一些常用的非静态方法用于获取主机名、IP 地址等,详见表 8-2。

表 8-2

方 法	描 述
public static InetAddress getLocalHost() throws UnknownHostException	通过本机得到 InetAddress 对象
public static InetAddress getByName(String host) throws UnknownHostException	通过主机名称得到 InetAddress 对象
public String getHostName()	返回代表与 InetAddress 对象相关的主机名的字符串
public String getHostAddress()	返回代表与 InetAddress 对象相关的主机地址的字符串
public boolean isReachable(int timeout) throws IOException	判断地址是否可达,同时指定超时时间
boolean equals(Object other)	如果对象具有和 other 相同的 Internet 地址则返回 true

【例 8.1】利用 InetAddress 类(注意测试本例时,请确保互联网已连接):
(1) 获取本机的域名及 IP 地址信息;
(2) 获取域名为 www.baidu.com 的主机信息(域名、IP 地址)。

```
import java.net.InetAddress;
public class InetAddressTest
{
public static void main(String args[])throws Exception{//抛出异常
InetAddress loc_Address = InetAddress.getLocalHost();//得到本机 InetAddress 对象
    System.out.println("本机的 InetAddress 对象信息:" + loc_Address);//输出本机 InetAddress 对象信息
    System.out.println("本机的域名:" + loc_Address.getHostName());//输出本机的域名
```

```
        System.out.println("本机的 IP 地址:" + loc_Address.getHostAddress());//输出
本机的 IP 地址
        InetAddress baidu_Address = InetAddress.getByName("www.baidu.com");//得到
域名为 www.baidu.com 的 InetAddress 对象
        System.out.println("百度的 InetAddress 对象信息:" + baidu_Address);//输出域
名为 www.baidu.com 的主机 InetAddress 对象信息
        System.out.println("百度的域名:" + baidu_Address.getHostName());//输出百度
主机名
        System.out.println("百度的 IP 地址:" + baidu_Address.getHostAddress());//输
出百度 IP 地址
    }
}
```

如图 8-4 所示。

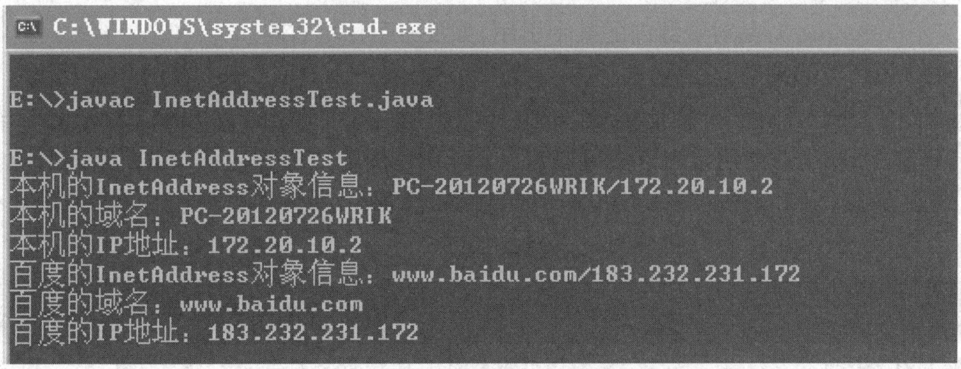

```
C:\WINDOWS\system32\cmd.exe

E:\>javac InetAddressTest.java

E:\>java InetAddressTest
本机的InetAddress对象信息：PC-20120726WRIK/172.20.10.2
本机的域名：PC-20120726WRIK
本机的IP地址：172.20.10.2
百度的InetAddress对象信息：www.baidu.com/183.232.231.172
百度的域名：www.baidu.com
百度的IP地址：183.232.231.172
```

图 8-4

8.1.2 URL 类和 URLConnection 类

当前,我们所用到的网络程序总体可分为两大类:

(1) B/S 架构(Browser/Server,浏览器/服务器)的网络程序,如淘宝、天猫等商务网站。对于万维网(WWW)的网站等基于 B/S 架构的网络程序,我们可以使用 URL(Uniform Resource Locator,即统一资源定位符)表示(定位)网络上资源的位置。在 Java 的网络类库中,URL 类、URLConnection 类为使用 URL 在互联网上获取信息提供了一个简单的、简洁的用户编程接口(API),接下来我们将详细讲解。

(2) C/S 架构(Client/Server,客户端/服务器)的网络程序,如 QQ 聊天软件、下载软件等。对于 C/S 架构的网络程序,Java 语言利用 Socket 实现通信。但是,Java 语言的 Socket 通信机制提供了两种通信方式:面向连接方式(TCP)、无连接方式(UDP)。其中,面向连接方式由 java.net 包中的 ServerSocket、Socket 类实现,而无连接方式由 java.net 包中的 DatagramSocket 和 DatagramPacket 类实现,我们将在 8.1.3 中详细讲解。

1. URL 类

(1) URL 的基本结构。

我们通过 URL 可以访问互联网上的各种网络资源，URL 的基本结构由如下 5 部分组成：

〈传输协议〉://〈主机名〉:〈端口号〉/〈文件名〉#〈引用〉

例如：http://www.baidu.com:80/index.html#chapter1.ppt

传输协议(protocol)：有 HTTP、FTP、File 等，默认为 HTTP 协议。

主机名(hostname)：指定资源所在的主机名，主机名可以是 IP 地址，也可以是主机的名字或者域名。

端口号(port)：端口号用来区分一个计算机中提供的不同服务，如 Web 服务、FTP 服务等，每一种服务都用一个端口号，范围是 0～65535。在 URL 中，hostname 后面的冒号及端口号是可以省略的，HTTP 的默认端口号是 80。

文件名(filename)：文件名包括该文件的完整路径。在 HTTP 协议中，有一个默认的文件名是 index.html，因此，http://www.baidu.com 与 http://www.baidu.com/index.html 两者等价。

引用(reference)：是对资源内的某个引用，如没有引用可省略。

(2) URL 类的常用方法。

Java 的 URL 类有多个构造函数，每个都能引发一个 MalformedURLException 异常。此外，通过 URL 类对象的常用方法可以打开该 URL 的输入流和由该 URL 标识的位置的连接，还可获得 URL 的协议名、主机名、端口号、文件名等，构造方法及常用方法见表 8-3。

表 8-3

方 法	功 能	类型
public URL(String spec)	使用 URL 字符串 spec 构造一个 URL 对象	构造
public URL(String protocol, String host, String file)	指定的协议 protocol、主机名 host、文件路径及文件名 file，创建一个 URL 对象	构造
public URL(String protocol, String host, int port, String file)	用指定的协议 protocol、主机名 host、端口号 port、文件路径及文件名 file，创建一个 URL 对象	构造
InputStream openStream()	打开该 URL 的输入流	普通
URLConnection openConnection()	打开由该 URL 标识的位置的连接	普通
public String getProtocol()	获取该 URL 的协议名	普通
public String getHost()	获取该 URL 的主机名	普通
public int getPort()	获取该 URL 的端口号。若无端口，返回 -1	
public String getFile()	获取该 URL 中的文件名	
public String getContent()	获取传输协议	
public String getPath()	获取该 URL 的路径	

续　表

方　　法	功　　能	类型
void set (string protocol, string host, int prot, string file, string ref)	设置该 URL 的各域的值	
public String toString()	将 URL 转化为字符串	

(3) 利用 URL 对象获取网络资源。

利用 URL 对象获取网络资源的步骤如下：

① 创建 URL 对象；

② 使用 URL 对象的 openStream()方法，返回一个 InputStream；

③ 从 InputStream 读入即可。

【例 8.2】使用 URL"www. baidu. com"找到对应主机上的页面资源，并下载显示(注意调试本程序时，请保持互联网的连接)。

```
import java.net.URL;
import java.io.InputStream;
public class URLReader{
public static void main(String[ ]args)throws Exception{
    URL u = new URL("http://www.baidu.com/");
    InputStream in = u.openStream();
    byte[ ]b = new byte[in.available()];
    in.read(b);
    System.out.println(new String(b));
    in.close();
    }
}
```

如图 8－5 所示。

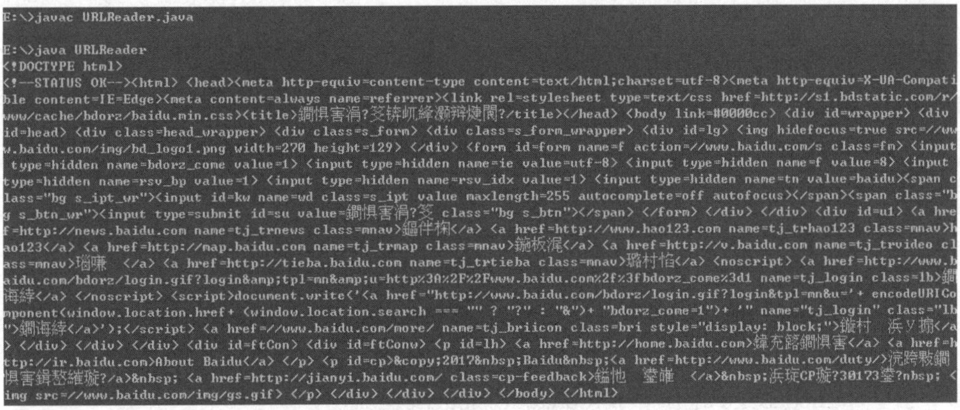

```
E:\>javac URLReader.java

E:\>java URLReader
<!DOCTYPE html>
<!--STATUS OK--><html> <head><meta http-equiv=content-type content=text/html;charset=utf-8><meta http-equiv=X-UA-Compati
ble content=IE=Edge><meta content=always name=referrer><link rel=stylesheet type=text/css href=http://s1.bdstatic.com/r/
www/cache/bdorz/baidu.min.css><title>鐧惧害涓?笅锛屼綘灏辩煡閬?/title></head> <body link=#0000cc> <div id=wrapper> <div
id=head> <div class=head_wrapper> <div class=s_form> <div class=s_form_wrapper> <div id=lg> <img hidefocus=true src=//ww
w.baidu.com/img/bd_logo1.png width=270 height=129> </div> <form id=form name=f action=//www.baidu.com/s class=fm> <input
 type=hidden name=bdorz_come value=1> <input type=hidden name=ie value=utf-8> <input type=hidden name=f value=8> <input
type=hidden name=rsv_bp value=1> <input type=hidden name=rsv_idx value=1> <input type=hidden name=tn value=baidu><span c
lass="bg s_ipt_wr"><input id=kw name=wd class=s_ipt value maxlength=255 autocomplete=off autofocus></span><span class="b
g s_btn_wr"><input type=submit id=su value=鐧惧害涓?笅 class="bg s_btn"></span> </form> </div> </div> <div id=u1> <a hre
f=http://news.baidu.com name=tj_trnews class=mnav>鏂伴椈</a> <a href=http://www.hao123.com name=tj_trhao123 class=mnav>h
ao123</a> <a href=http://map.baidu.com name=tj_trmap class=mnav>鍦板浘</a> <a href=http://v.baidu.com name=tj_trvideo cl
ass=mnav>瑙嗛</a> <a href=http://tieba.baidu.com name=tj_trtieba class=mnav>璐村惂</a> <noscript> <a href=http://www.b
aidu.com/bdorz/login.gif?login&amp;tpl=mn&amp;u=http%3A%2F%2Fwww.baidu.com%2f%3fbdorz_come%3d1 name=tj_login class=lb>鐧
诲綍</a> </noscript> <script>document.write('<a href="http://www.baidu.com/bdorz/login.gif?login&tpl=mn&u='+ encodeURICo
mponent(window.location.href+ (window.location.search === "" ? "?" : "&")+ "bdorz_come=1")+ '" name="tj_login" class="lb
">鐧诲綍</a>');</script> <a href=//www.baidu.com/more/ name=tj_briicon class=bri style="display: block;">鏇村 浜у搧</a
> </div> </div> </div> <div id=ftCon> <div id=ftConw> <p id=lh> <a href=http://home.baidu.com>鍏充簬鐧惧害</a> <a href=h
ttp://ir.baidu.com>About Baidu</a> </p> <p id=cp>&copy;2017 Baidu <a href=http://www.baidu.com/duty/>浣跨敤鐧
惧害鍓嶅繀璇?/a>  <a href=http://jianyi.baidu.com/ class=cp-feedback>鎰忚 鍙嶉</a> 浜琁CP璇?30173鍙?nbsp; <
img src=//www.baidu.com/img/gs.gif> </p> </div> </div> </div> </body> </html>
```

图 8－5

2. URLConnection 类

虽然通过 URL 类的 openStream()方法能够读取网络上资源中的数据,但是 Java 提供的 URLConnection 类中包含了更加丰富的方法,可以对网络上的资源进行更多的处理。例如,通过 URLConnection 类,既可以从 URL 中读取数据,也可以向 URL 中的资源发送数据。URLConnection 类表示在应用程序和 URL 所标识的资源之间的一个通信连接,它是一个抽象类。

注意:

创建 URLConnection 对象之前必须先创建一个 URL 对象,然后通过调用 URL 类提供的 openConnection()方法,就可以获得一个 URLConnection 类的对象。URLConnection 类对象的常用方法如表 8-4 所示。

表 8-4

方法名	功能说明	方法名	功能说明
void connect()	建立 URL 连接	long getExpiration()	获取响应数据的终止时间
Object getContent()	获取该 URL 的内容	InputStream getInputStream()	获取该连接的输入流
int getContentLength()	获取响应数据的内容长度	long getLastModified()	获取响应数据的最后修改时间
String getContentType()	获取响应数据的内容类型	OutputStream getOutputStream()	获取该连接的输出流
long getDate()	获取响应数据的创建时间		

【例 8.3】 使用 URLConnection 获取 URL"www.baidu.com"的响应数据的内容类型及对应的页面资源(注意调试本程序,请保持互联网的连接)。

```java
import java.io.InputStream;
import java.net.URL;
import java.net.URLConnection;
public class URLC_Test{
    public static void main(String args[])throws Exception{
        int c;
        URL url = new URL("http://www.baidu.com");
        URLConnection urlcon = url.openConnection();
        System.out.println("响应数据的内容类型:" + urlcon.getContentType());
        System.out.println("页面资源的代码为:");
        InputStream in = urlcon.getInputStream();
```

```
        while((c = in.read())! = -1){
            System.out.print((char)c);
        }
        in.close();
    }
}
```

如图 8-6 所示。

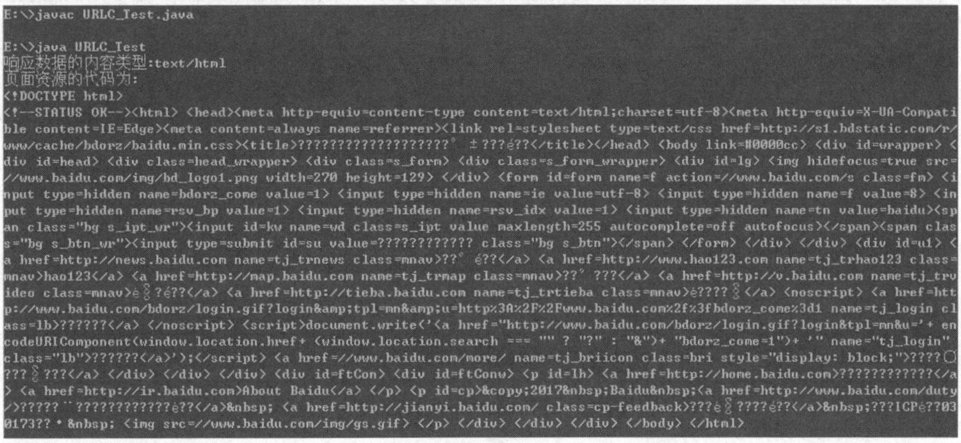

图 8-6

8.1.3　TCP 和 UDP 程序设计

互联网上的两台计算机，一台计算机发送服务请求，并接收对方的回应数据；另一台计算机等待来自对方的要求，并处理要求、传回结果。发送请求的计算机我们称之为客户端，等待请求并提供处理服务的计算机我们称之为服务器。Java 语言利用 Socket 类的相关方法，实现服务器和客户端之间的请求和处理结果等相关数据的传输。

Socket 有两种主要的操作方式：面向连接的和无连接的。面向连接的操作使用 TCP 协议，发送数据前必须先建立连接。面向连接的操作比无连接的操作效率更低，但是数据的安全性更高。无连接的操作使用 UDP 协议，发送数据之前不需要建立连接，其特点是快速和高效，但是数据安全性不佳。

1. TCP 程序设计

Java 语言使用 ServerSocket 类和 Socket 类设计实现面向连接的 TCP 程序。其中，ServerSocket 类主要用于服务器端程序设计，用于监听指定端口上的客户端连接请求，并为每个新的连接创建一个 Socket 对象，之后客户端便可以与服务器端开始通信了；而客户端程序主要使用 Socket 类设计实现。

TCP 程序设计的一般步骤如下：

① 服务器程序创建一个 ServerSocket 对象，然后调用 accept()方法等待客户端建立连接；

② 客户端程序创建一个 Socket 对象，并请求与服务器建立连接；

③ 建立连接后，可以用 Socket 类的 getInputStream() 和 getOutputStream() 方法获得读写数据的输入/输出流；

④ 通信结束后，双方调用 Socket 类的 close() 方法断开连接。

(1) ServerSocket 类。

ServerSocket 类主要用于服务器端，其常用操作方法如表 8-5 所示。

表 8-5

方法	功　　能	类型
ServerSocket(int port)	在指定端口 port 上创建一个 ServerSocket 类对象	构造
ServerSocket(int port, int backlog)	在指定端口 port 上创建一个 ServerSocket 类对象，并进入监听状态，第二个参数 backlog 是服务器忙时保持连接请求的等待客户数量	构造
ServerSocket(int port, int backlog, InetAddress bindAddr)	使用指定的端口 port 和要绑定到的服务器 IP 地址 bindAddr 创建一个 ServerSocket 类对象，并进入监听状态，参数 backlog 同上	构造
Socket accept()	接收该连接并返回该连接的 Socket 对象	普通
void close()	关闭此服务器的 ServerSocket	普通
InetAddress getInetAddress()	获取该服务器 Socket 所绑定的地址	普通
boolean isClosed()	返回 ServerSocket 的关闭状态	普通
int getLocalPort()	获取该服务器 Socket 所监听的端口号	普通

注意：

① 服务器端程序每次运行时，都要使用 aceept() 方法等待客户端连接，此方法执行之后服务器端将进入阻塞状态，直到客户端连接之后，服务器端程序才可以向下继续执行。aceept() 方法的返回值类型是 Socket，每一个 Socket 都表示一个客户端对象。

② aceept() 方法返回 Socket 对象后，就可以使用该 Socket 对象的 getInputStream() 方法获取客户端发送过来的信息，还可以使用 getOutputStream() 将信息发送给客户端。

③ Java 网络程序需要使用输入、输出流的形式完成数据传输，因此需要导入 java.io 包。

ServerSocket 类设计实现的 TCP 服务器端程序应依次执行如下操作：

① 创建 ServerSocket 对象，绑定并监听端口；

② 调用 ServerSocket 的 accept() 方法监听客户端的请求，如果接收到客户端请求，得到一个 Socket；

③ 调用 Socket 的 getInputStream() 和 getOutputStream() 方法获取和客户端相连的 IO 流（输入流可以读取客户端输出流写出的数据，输出流可以写出数据到客户端的输入流）；

④ 关闭相关资源。

【例 8.4】使用 ServerSocket 类及 Socket 类完成一个服务器的程序开发,此服务器端程序向客户端程序输出"你好,客户端!我是服务器"的字符串信息。

```java
import java.net.ServerSocket;
import java.net.Socket;
import java.io.PrintStream;
public class TcpServer{
public static void main(String args[])throws Exception{//所有异常抛出
//定义 ServerSocket 类对象 server,服务器在 8888 端口上监听
ServerSocket server = new ServerSocket(8888);
Socket client = null;//表示客户端
PrintStream out = null;//打印流输出最方便,用于向客户端输出信息
System.out.println("服务器运行,等待客户端连接。");//输出提示信息
client = server.accept();//程序进入到阻塞状态,等待连接
String str = "你好,客户端!我是服务器。";//表示要输出的信息
out = new PrintStream(client.getOutputStream());
out.println(str);//向客户端输出信息
client.close();
server.close();
}
}
```

如图 8-7 所示。

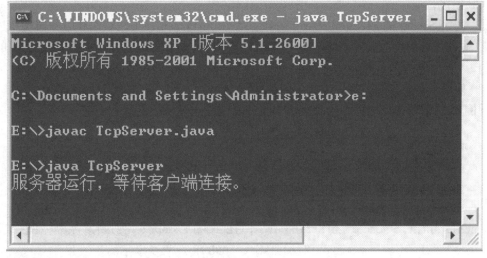

图 8-7

从上图的运行结果可以知道,服务器端程序执行到 accept()方法时进入阻塞状态,阻塞状态会在客户端连接之后改变。

(2) Socket 类。

客户端程序可以通过构造一个 Socket 类的对象来建立与服务器的连接，Socket 类的常用操作方法如表 8-6 所示。

表 8-6

方　　法	功　　能	类型
Socket(String host, int port)	构造 Socket 对象，同时指定要连接的服务器的主机名称 host 及连接端口 port	构造
InputStream getInputStream()	获取 Socket 对应的输入流	普通
OutputStream getOutputStream()	获取该 Socket 的输出流	普通
void close()	关闭 Socket 连接	普通
InetAddress getInetAddress()	获取当前连接的远程主机的互联网地址	普通
InetAddress getLocalAddress()	获取本地主机的互联网地址	普通
int getLocalPort()	获取本地连接的端口号	普通
int getPort()	获取远程主机端口号	普通

注意：

客户端程序创建 Socket 类对象后，就可以使用该 Socket 对象的 getInputStream()方法获取服务器端程序发送过来的信息，还可以使用 getOutputStream()将信息发送给服务器端程序。

Socket 类设计实现的 TCP 客户端程序应依次执行如下操作：

① 创建一个 Socket 实例，连接到服务器端：构造函数向指定的远程主机和端口建立一个 TCP 连接；

② 调用 Socket 的 getInputStream()和 getOutputStream()方法获取和服务器端相连的 IO 流（输入流可以读取服务器端输出流写出的数据，输出流可以写出数据到服务器端的输入流）；

③ 使用 Socket 类的 close 方法关闭连接。

【例 8.5】 编写客户端程序接收上例中服务器端发送过来的信息。

```
import java.net.Socket;
import java.io.BufferedReader;
import java.io.InputStreamReader;
public class TcpClient{
public static void main(String args[])throws Exception{//所有异常抛出
Socket client = new Socket("localhost",8888);//表示客户端
BufferedReader buf = null;//一次性接收完成
buf = new BufferedReader(new InputStreamReader(client.getInputStream()));
String str = buf.readLine();
```

```
        System.out.println("客户端接收到的内容:" + str);
        buf.close();
        client.close();
    }
}
```

如图 8-8 所示。

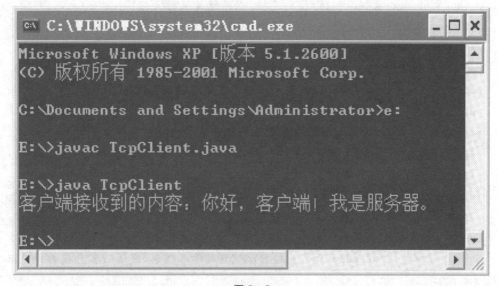

图 8-8

通过图 8-8 会发现客户端和服务器端只能连接一次,程序就退出。这是由于服务器端只能执行一次 accept()方法。我们可以通过循环使用 accept()方法,这样就可以实现服务器端、客户端的一对一连续通信。

【例 8.6】服务器端、客户端一对一连续通信的实现。

① 服务器端程序设计。

```
import java.net.ServerSocket;
import java.net.Socket;
import java.io.BufferedReader;
import java.io.PrintStream;
import java.io.InputStreamReader;
public class TcpServer2{
public static void main(String args[])throws Exception{//所有异常抛出
ServerSocket server = new ServerSocket(8888);//定义 ServerSocket 类,服务器在 8888 端口上监听
Socket client = null;//表示客户端
```

```java
BufferedReader buf = null;//接收输入流
PrintStream out = null;//打印流输出最方便
BufferedReader input = new BufferedReader(new InputStreamReader(System.in));//接收键盘数据
boolean f = true;//定义标志位
while(f){
System.out.println("服务器运行,等待客户端连接。");
client = server.accept();//得到连接,程序进入阻塞状态
out = new PrintStream(client.getOutputStream());
//准备接收客户端的输入信息
buf = new BufferedReader(new InputStreamReader(client.getInputStream()));
boolean flag = true;//标志位,表示可以一直接收并回应信息
while(flag){
String str = buf.readLine();//接收客户端发送的内容
if(str.equals("exit")){//如果输入的内容为 exit,表示结束
flag = false;
}else{
System.out.println("客户端传来的内容为:" + str);
System.out.print("请输入服务器端信息:");
String echo = input.readLine();//接收键盘的输入信息,发送给客户端的回应信息
out.println(echo);
}
}
client.close();
}
server.close();
}
}
```

② 客户端程序设计。

```java
import java.net.Socket;
import java.io.BufferedReader;
import java.io.InputStreamReader;
import java.io.PrintStream;
public class TcpClient2{
public static void main(String args[])throws Exception{//所有异常抛出
Socket client = new Socket("localhost",8888);//表示客户端
```

```
        BufferedReader buf = new BufferedReader(new InputStreamReader(client.
getInputStream()));//一次性接收完成
    PrintStream out = new PrintStream(client.getOutputStream());//发送数据
    BufferedReader input = new BufferedReader(new InputStreamReader(System.
in));//接收键盘数据
    boolean flag = true;//定义标志位
    while(flag){
    System.out.print("请输入客户端信息:");
    String str = input.readLine();//接收键盘的输入信息
    out.println(str);
    if(str.equals("exit")){
    flag = false;
    }else{
    String echo = buf.readLine();//接收返回结果
    System.out.println("服务器端传来的内容为:" + echo);//输出回应信息
    }
    }
    buf.close();
    client.close();
    }
    }
```

如图 8-9、图 8-10 所示。

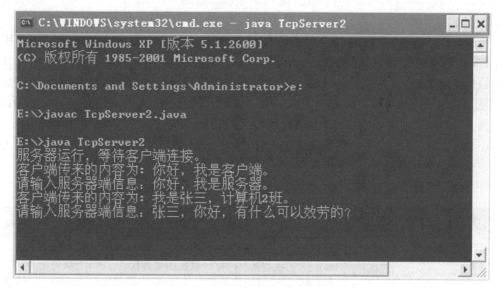

图 8-9

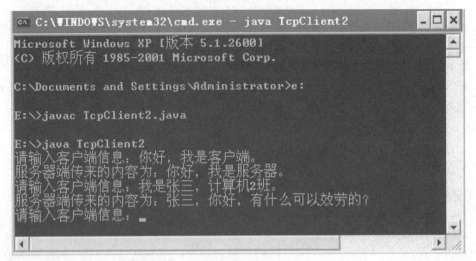

图 8-10

通过程序运行结果可知,服务器端和客户端程序已经实现了一对一的连续通信,而且当一个客户端程序结束后,服务器端不结束,而是等待下一个客户端的到来,以便再次开始一对一的连续通信。

注意:

在该案例程序的运行过程中,服务器端每次只能连接一个客户端。如果要同时实现多个客户端的连接,可以使用多线程机制,详见工作任务 8.2 多人网络聊天实现。

2. UDP 程序设计

在 Java 网络程序设计过程中,无连接的 UDP 程序可以用 DatagramSocket 类和 DatagramPacket 类实现。在 UDP 程序开发中使用 DatagramPacket 类包装一条要发送的信息,之后使用 DatagramSocket 类用于完成信息的发送操作。

(1) DatagramPacket 类。

使用 UDP 协议传输数据,首先应将数据打包,java.net 包中的 DatagramPacket 类用来创建数据包对象。但是,创建的数据包有两种,一种数据包用于发送数据,该数据包有要传递到的目的地址;另一种数据包用来接收传递过来的数据包中的数据。要创建接收的数据包,常用 DatagramPacket 类的构造方法 public DatagramPacket(byte ibuft[],int ilength),而创建发送数据包则常用构造方法 public DatagramPacket(byte ibuf[], int ilength, InetAddrss iaddr, int port)。此外,DatagramPacket 类创建的数据包对象还有一些常用方法用来获取数据包的信息等,详见表 8-7。

表 8-7

方 法	功 能	类型
public DatagramPacket (byte ibuft[],int ilength)	实例化 DatagramPacket 对象,并指定接收数据长度。参数 ibuf[]为接收数据包的存储数据的缓冲区,ilength 为从传递过来的数据包中读取的字节数,接收到的数据从 ibuft[0]开始存放,直到整个数据包接收完毕或者将 ilength 的字节写入 ibuft 为止	构造

续 表

方 法	功 能	类型
public DatagramPacket(byte ibuf[], int ilength, InetAddrss iaddr, int port)	实例化 DatagramPacket 对象时指定发送的数据、数据的长度、目标地址及端口。参数 iaddr 为数据包要传递到的目标地址,port 为目标地址的程序接收数据包的端口号(即目标地址的计算机上运行的客户程序是在哪一个端口接收服务器发送过来的数据包),ibuf[]为要发送数据的存储区,从 ibuf 数组的 0 位置开始填充数据包 ilength 字节	构造
public byte[]getData()	返回一个字节数组,其中是数据包的数据(即接收的数据)	普通
public int getLength()	返回要发送或接收数据的长度	普通
public InetAddress getAddress()	如果是发送数据包,则获得数据包要发送的目标地址;但是如果是接收数据包,则返回发送此数据包的源地址	普通
pubic int getPort()	返回数据包中的目标地址的主机端口号	普通

(2) DatagramSocket 类。

DatagramSocket 类用于发送和接收数据包,DatagramSocket 类的常用方法如表 8-8 所示。

表 8-8

方 法	功 能	类型
public DatagramSocket(int port) throws SocketException	创建 DatagramSocket 对象,并指定监听的端口	构造
public void receive (DatagramPacket p)	从网络上接收数据包并将其存储在 DatagramPacket 对象 p 中。p 中的数据缓冲区必须足够大,receive()把尽可能多的数据存放在 p 中。如果装不下,就把其余的部分丢弃	普通
public Void Send(DatagramPacket p)	发送数据包	普通
public int getLocalPort()	返回本地 DatagramSocket 对象正在监听的端口号	普通
public void close()	关闭套接字	普通

(3) UDP 程序实现。

在 UDP 程序设计过程中,一般先编写客户端程序,再编写服务器端程序。Java 语言编写的 UDP 客户端程序应依次执行如下操作:

① 使用构造方法 DatagramSocket(int port)建立 socket(套间字)服务,参数 port 指定监听端口。

② 使用构造方法 public DatagramPacket(byte ibuft[], int ilength)定义一个数据包(DatagramPacket),用于存储接收到的数据。

③ 通过 DatagramSocket 类的 receive()方法将接收到的数据存入上面定义的数据包中。

④ 使用 DatagramPacket 类的 getData()等方法从数据包中提取传送的内容等数据。

⑤ 关闭资源。

典型的 UDP 服务器端程序应依次执行如下操作：

① 使用构造方法 DatagramSocket(int port)建立 socket(套间字)服务。

② 使用构造方法 public DatagramPacket(byt ibuf[], int ilength, InetAddrss iaddr, int port)定义一个数据包(DatagramPacket)，将数据打包到 DatagramPacket 中去，用于发送。

③ 通过 DatagramSocket 类的 send()方法发送数据。

④ 关闭资源。

【例 8.7】UDP 客户端程序设计。

```
import java.net.DatagramPacket;
import java.net.DatagramSocket;
public class UDPClient{
public static void main(String args[])throws Exception{//所有异常抛出
DatagramSocket ds = null;//定义接收数据包的对象
byte[]buf = new byte[1024];//开辟空间,以接收数据
DatagramPacket dp = null;//声明 DatagramPacket 对象
ds = new DatagramSocket(8000);//客户端在 8000 端口上等待服务器发送信息
dp = new DatagramPacket(buf,1024);//所有的信息使用 buf 保存
ds.receive(dp);//接收数据
String str = new String(dp.getData(), 0, dp.getLength()) + " from" + dp.getAddress().getHostAddress() + ":" + dp.getPort();
System.out.println(str);//输出内容
}
}
```

【例 8.8】UDP 服务器端程序设计。

```
import java.net.DatagramPacket;
import java.net.DatagramSocket;
import java.net.InetAddress;
public class UDPServer{
public static void main(String args[])throws Exception{//所有异常抛出
DatagramSocket ds = null;//定义发送数据包的对象
DatagramPacket dp = null;//声明 DatagramPacket 对象
ds = new DatagramSocket(3000);//服务器端在 3000 端口上等待服务器发送信息
String str = "hello World!!!";
dp = new
```

```
DatagramPacket ( str. getBytes ( ), str. length ( ), InetAddress. getByName ( "
localhost"),8000);//所有的信息使用 buf 保存
    System.out.println("发送信息.");
    ds.send(dp);//发送信息出去
    ds.close();
    }
    }
```

如图 8-11、图 8-12 所示。

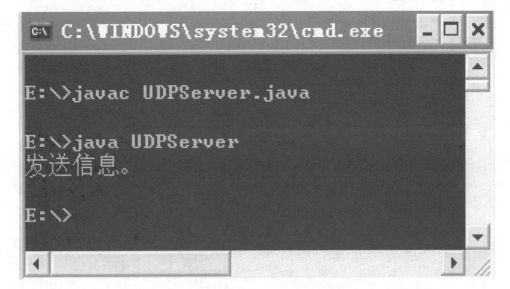

图 8-11

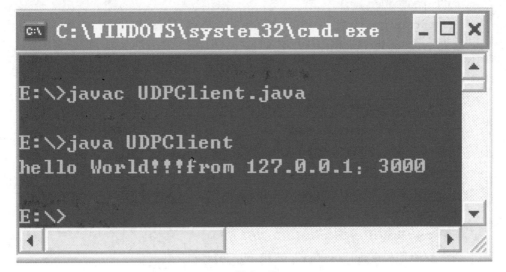

图 8-12

> **任务 实施**

1. 服务器端程序设计实现

```java
import javax.swing.*;
import java.awt.*;
import java.awt.event.*;
import java.io.*;
import java.net.*;
import java.text.SimpleDateFormat;
import java.util.Date;
public class ChatServer extends JFrame{
    JLabel ipJLabel = new JLabel("IP:");
    JLabel portJLabel = new JLabel("端口:");
    JTextField ipField    = new JTextField();
    JTextField portField = new JTextField();
    JTextField msgField   = new JTextField();
    JButton listenBtn    = new JButton("侦听");
    JButton disConBtn    = new JButton("断开");
    JButton sendBtn      = new JButton("发送");
    JTextArea msgRecArea = new JTextArea();
    JScrollPane scrollPanel = new JScrollPane(msgRecArea);
    ServerSocket server;
    Socket client;
    BufferedReader br;
    PrintWriter pw;
    BufferedReader in;

    public ChatServer(){
        super("服务器");
        this.setSize(500,500);
        this.setLayout(null);
        this.setVisible(true);
        ipJLabel.setBounds(20,30,20,30);
        ipField.setBounds(50,30,150,30);
        portJLabel.setBounds(210,30,50,30);
        portField.setBounds(260,30,60,30);
```

```java
    listenBtn.setBounds(330,30,60,30);
    disConBtn.setBounds(400,30,60,30);
    this.add(ipJLabel);
    this.add(ipField);
    this.add(portJLabel);
    this.add(portField);
    this.add(listenBtn);
    this.add(disConBtn);
    msgField.setBounds(50,70,300,30);
    sendBtn.setBounds(400,70,60,30);
    this.add(msgField);
    this.add(sendBtn);
    scrollPanel.setBounds(50,120,360,300);
    this.add(scrollPanel);
    msgRecArea.setEditable(false);
    listenBtn.setEnabled(false);
disConBtn.setEnabled(false);
    ipField.setEditable(false);
    
    InetAddress inetAddr;
try{
    inetAddr = InetAddress.getLocalHost();
    ipField.setText(inetAddr.getHostAddress());
    portField.setText("8888");
    
    }catch(UnknownHostException e){
        e.printStackTrace();
    }
    ButtonListener listener = new ButtonListener();
    sendBtn.addActionListener(listener);
    this.addWindowListener(new WindowAdapter(){
        public void windowClosing(WindowEvent e){
            System.exit(0);
        }
    });
    
    //等待客户连接+接受消息
    try{
```

```java
            server = new ServerSocket(8888);
            msgRecArea.append("服务器已启动,等待客户端接入..." + "\n");
            client = server.accept();
            msgRecArea.append("客户端"
+ client.getInetAddress().toString() + ":" + client.getPort() + "已接入" + "\n");
            while(true){
                
                SimpleDateFormat sm = new SimpleDateFormat("HH:mm:ss");
                //接受消息
                br = new BufferedReader(new
InputStreamReader(client.getInputStream()));
                String info = br.readLine();
                
        if(info.equals("我已断开连接,下次再聊,拜拜!")){
                    msgRecArea.append(sm.format(new Date()) + "    客户端说:" +
info + "\n");
                    info = "又再次连接上" + "\n";
            msgRecArea.append("服务器已启动,等待客户端的再次连接!" + "\n");
            client.close();
            client = null;
            client = server.accept();
                }
                
                if(!info.equals("")){
                    msgRecArea.append(sm.format(new Date()) + "    客户端说:" +
info + "\n");
                }
            }
        }catch(Exception e){
            e.printStackTrace();
        }
    }
    
    
    class ButtonListener implements ActionListener{
        public void actionPerformed(ActionEvent e){
            String cmd = e.getActionCommand();
            if(cmd.equals("发送")){
```

```java
                    try{
                        SimpleDateFormat sm = new SimpleDateFormat("HH:mm:ss");
                            String infotokehu = msgField.getText();
                            pw = new PrintWriter(client.getOutputStream(),true);
                            pw.println(infotokehu);
                            msgRecArea.append(sm.format(new Date()) + "    我说:" + infotokehu + "\n");
                            msgField.setText("");
                    }
                        catch(IOException ex){
                            JOptionPane.showMessageDialog(null,ex.getMessage(),"提示",JOptionPane.ERROR_MESSAGE);
                        }
                    }
                }
            }
        public static void main(String[ ]args){
            ChatServer chatServer = new ChatServer();
        }
    }
```

2. 客户端程序设计实现

```java
import javax.swing.*;
import java.awt.*;
import java.awt.event.*;
import java.io.*;
import java.net.*;
import java.text.SimpleDateFormat;
import java.util.Date;
public class ChatClient extends JFrame
{
    JLabel ipJLabel = new JLabel("IP:");
    JLabel portJLabel = new JLabel("端口:");
    JLabel userNameJLabel = new JLabel("用户名:");
    JLabel sendJLabel = new JLabel("发送内容:");
    JTextField ipField    = new JTextField();
```

```java
        JTextField portField = new JTextField();
        JTextField userNameField = new JTextField();
        JTextField msgField    = new JTextField();
        JButton connectBtn    = new JButton("连接");
        JButton disConBtn     = new JButton("断开");
        JButton sendBtn       = new JButton("发送");
        JTextArea msgRecArea = new JTextArea();
        JScrollPane scrollPanel = new JScrollPane(msgRecArea);
        Socket client;
        BufferedReader br;
        PrintWriter pw;
        BufferedReader in;

        public ChatClient(){
            super("客户端");
            this.setSize(490,480);
            this.setLayout(null);
    this.setResizable(false);
            this.setVisible(true);
            ipJLabel.setBounds(20,30,20,30);
            ipField.setBounds(50,30,150,30);
            portJLabel.setBounds(210,30,50,30);
            portField.setBounds(260,30,60,30);
            connectBtn.setBounds(330,30,60,30);
            disConBtn.setBounds(400,30,60,30);
            this.add(ipJLabel);
            this.add(ipField);
            this.add(portJLabel);
            this.add(portField);
            this.add(connectBtn);
            this.add(disConBtn);
    userNameJLabel.setBounds(20,70,50,30);
    userNameField.setBounds(70,70,65,30);
    sendJLabel.setBounds(140,70,60,30);
            msgField.setBounds(200,70,200,30);
            sendBtn.setBounds(400,70,60,30);
    this.add(userNameJLabel);
    this.add(userNameField);
```

```java
    this.add(sendJLabel);
        this.add(msgField);
        this.add(sendBtn);
        scrollPanel.setBounds(20,120,440,300);
        this.add(scrollPanel);
        msgRecArea.setEditable(false);
        connectBtn.setEnabled(false);
disConBtn.setEnabled(false);;

        InetAddress inetAddr;
try{
        inetAddr = InetAddress.getLocalHost();
        ipField.setText(inetAddr.getHostAddress());
        portField.setText("8888");
        userNameField.setText("客户端");
}catch(UnknownHostException e){
        e.printStackTrace();
        }

        ButtonListener listener = new ButtonListener();
        sendBtn.addActionListener(listener);
        this.addWindowListener(new WindowAdapter(){
            public void windowClosing(WindowEvent e){
        try{
        pw = new PrintWriter(client.getOutputStream(),true);
            pw.println("我已断开连接,下次再聊,拜拜!");
            System.exit(0);
        client.close();
        }catch(Exception e1){
        e1.printStackTrace();
        }

            }
        });

        //创建客户端+接受消息
        try{
            int port = Integer.parseInt(portField.getText());
```

```java
                    String ip = ipField.getText();
                    client = new Socket(ip,port);
                    msgRecArea.append("已连接到服务器:" +
client.getInetAddress().getHostAddress() + ":" + client.getPort() + "\n");
                    while(true){
                    //获取时间
                    SimpleDateFormat sm = new SimpleDateFormat("HH:mm:ss");
                    //接受消息
                    br = new BufferedReader(new
InputStreamReader(client.getInputStream()));
                    String info = br.readLine();

              if(!info.equals(""))
                     msgRecArea.append(sm.format(new Date()) + "       服务器说:" +
info + "\n");
                    }
            }catch(Exception e){
                    e.printStackTrace();
                }
            }


         class ButtonListener implements ActionListener{
             public void actionPerformed(ActionEvent e){
                    String cmd = e.getActionCommand();
                if(cmd.equals("发送")){
                        try{
                        //获取时间
                        SimpleDateFormat sm = new SimpleDateFormat("HH:mm:ss");

                        //发消息
                        String infotofuwu = msgField.getText();
                        pw = new PrintWriter(client.getOutputStream(),true);
                        pw.println(infotofuwu);
                         msgRecArea.append(sm.format(new Date()) + "       我说:" +
infotofuwu + "\n");
                        msgField.setText("");
                        }
```

```
                catch(IOException ex){
                    JOptionPane.showMessageDialog(null,ex.getMessage(),
                        "提示",JOptionPane.ERROR_MESSAGE);
                }
            }
        }
    }
    public static void main(String[]args){
        ChatClient chatClient = new ChatClient();
    }
}
```

如图 8-13、图 8-14 所示。

注意:

测试运行时,应先运行服务器端,然后再运行客户端。退出时,应先退出客户端,然后再退出服务器端。

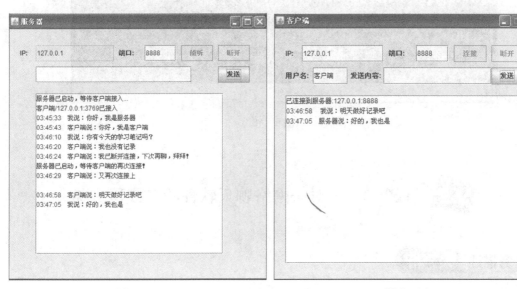

图 8-13   图 8-14

任务 拓展

模拟实现一对一的网络文件传输,要求如下:

(1) 使用客户端程序,将本机 E 盘根目录下名称为"测试.txt"的记事本文件,上传到服务器的 F 盘名称为"已收到"的文件夹。此外,客户端程序在发送文件时,应提示"开始传输文件",接收过程中实时显示发送进度,发送完毕后提示"文件传输成功"。

(2) 服务器端程序在接收文件完成后,应提示"文件接收成功",并输出收到的文件名及大小等信息,可以不使用图形界面实现,效果如图 8-15、图 8-16 所示:

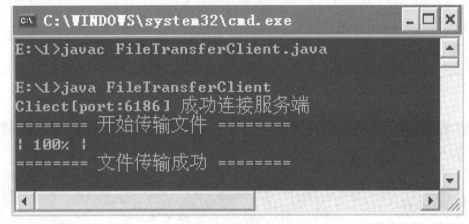

图 8-15

图 8-16

 任务 2　一对一网络聊天软件的多线程实现

任务描述 及分析

小明接到新任务,该任务要求进一步使用多线程实现一对一的网络聊天功能。要完成这个工作任务,第一步,我们需要掌握 Java 多线程程序设计相关知识;第二步,利用多线程技术结合 Java 网络程序设计相关知识,设计实现网络聊天功能。

相关 知识

8.2.1　进程与线程

随着计算机技术的飞速发展,当今的计算机操作系统已经普遍采用多任务和分时设计,可以在同一时间内执行多个程序。而所谓的一个进程就是一个执行中的程序,每一个进程都有自己独立的一块内存空间、一组系统资源。因此,进程是对操作系统而言的,即多个进程(程序)几乎可以在同一时间执行多个任务。例如,我们在打开 Outlook 程序收取邮件的同时,又打开 Word 程序编辑文档,此时,我们可以认为系统内有两个进程在运行,即两个程序几乎在同一时间同时执行两个任务(一个在接收邮件,一个在编辑文档)。

线程一般是对某一进程(程序)而言的,线程是比进程更小的执行单位,是在进程的基础之上进行的进一步划分,是实现并发机制的一种有效手段,即每一个进程(程序)在同一时间内可以执行多个线程(任务)。如上例,在你打开 Outlook 程序收新邮件的同时,还可以阅读已下载的邮件,这两件事没有相互干扰,那么我们就说此时这一进程(程序)中至少有两个线程在运行。由此可知,多线程指的是在单个进程(程序)中可以同时运行多个不同的线程,执行不同的任务。

综上可知,线程与进程相似,是一段完成某个特定功能的代码,是程序中单个顺序的运行流程,而多线程机制就是指同时运行多段完成不同特定功能的代码(其各自运行流程互不相同),完成不同任务;但与进程不同的是,同类的多个线程共享一块内存空间和一组系统资源,系统在产生一个线程,或者在各个线程之间切换时,负担要比进程小得多,正因如此,线程又被称为轻负荷进程(light-weight process)。此外,所有的线程都是在某个进程的基础之上产生和并发运行的(即进程在执行过程中产生多个线程,这些线程同时存在、同时执行)。因此,如果一个进程没有了,则线程肯定会消失;但是线程消失了,进程未必会消失。

使用多线程进行程序设计具有如下优点:
(1) 多线程编程简单,效率高(能直接共享数据和资源,多进程不能)。
(2) 适合于开发服务程序(如 Web 服务、聊天服务等)。
(3) 适合于开发有多种交互接口的程序(如聊天程序的客户端、网络下载工具)。
(4) 减轻编写交互频繁、涉及面多的程序的困难(如监听网络端口)。
(5) 程序的吞吐量会得到改善(同时监听多种设备,如网络端口、串口、并口以及其他外设)。
(6) 对于多处理器系统,可以真正实现并发运行不同的线程,极大提高运行效率。

8.2.2 线程的创建

Java 语言创建线程有两种主要的方法,一种是继承 Thread 类,另一种是实现 Runnable 接口。

1. 继承 Thread 类创建线程

Thread 类是在 java.lang 包中定义的,一个类只要继承了 Thread 类,此类就称为多线程操作类,然后只需要覆盖 Thread 类的 run()方法,就可完成线程的创建。线程定义的语法格式如下:

```
class 类名称 extends Thread{//继承 Thread 类
    属性;//类中定义属性
    方法;//类中定义方法
```

```
public void run(){//覆写 Thread 类中的 run()方法,此方法是线程的主体
   线程主体;
   }
}
```

注意:

(1) Thread 子类的 run()方法覆写后,只是能够创建线程,线程的启动(执行)必须使用 start()方法,线程启动后才可以执行,执行时调用 run()方法。

(2) java.lang 包会在程序运行时自动导入,无须手工编写 import 语句。

【例 8.9】使用继承 Thread 类方法创建线程。

```
class MyThread extends Thread{//继承 Thread 类,作为线程的实现类
   private String name;//表示线程的名称
   public MyThread(String name){
      this.name = name;//通过构造方法配置 name 属性
   }
   public void run(){//覆写 run()方法,作为线程的操作主体
      for(int i = 0;i<3;i++){
         System.out.println(name + "运行,第" + i + "次.");
      }
   }
}
public class ThreadTest1{
   public static void main(String args[]){
      MyThread mt1 = new MyThread("线程 1");//实例化对象
      MyThread mt2 = new MyThread("线程 2");//实例化对象
      mt1.start();//调用线程主体
      mt2.start();//调用线程主体
   }
}
```

如图 8-17 所示。

2. 实现 Runnable 接口创建线程

Java 语言还可以通过实现 Runnable 接口创建线程,该接口中只定义了一个抽象方法 public void run(),该方法定义线程的语法格式如下:

```
class 类名称 implement Runnable{//实现 Runnable 接口
   属性;//类中定义属性
   方法;//类中定义方法
   public void run(){//覆写 Runnable 接口中的 run()方法
```

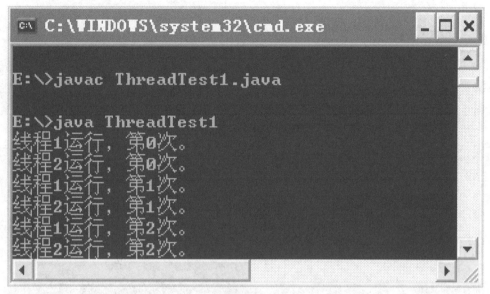

图8-17

```
        线程主体;
    }
}
```

注意:

(1) Runnable 接口只有 run()方法,但线程需要 start()方法启动。因此,必须通过 Thread 类的构造方法 public Thread(Runnable target)接收 Runnable 子类的实例对象生成一个 Thread 类的对象,然后通过 Thread 类对象调用 start()方法启动线程。

(2) 使用继承 Thread 类方法创建的线程无法实现资源共享,而使用实现 Runnable 接口方法创建的线程可以实现资源共享,因此推荐使用实现 Runnable 接口方法创建线程。此外,在后面的内容中,使用多线程时都将以 Runnable 接口的实现为操作的重点。

【例 8.10】 使用实现 Runnable 接口方法创建线程。

```java
class MyThread implements Runnable{//实现 Runnable 接口,作为线程的实现类
    private String name;//表示线程的名称
    public MyThread(String name){
        this.name = name;//通过构造方法配置 name 属性
    }
    public void run(){//覆写 run()方法,作为线程的操作主体
        for(int i = 0;i<3;i++){
            System.out.println(name + "运行第" + i + "次。");
        }
    }
```

```
}
public class RunnableTest1{
public static void main(String args[]){
MyThread mt1 = new MyThread("线程1");//实例化对象
MyThread mt2 = new MyThread("线程2");//实例化对象
Thread t1 = new Thread(mt1);//实例化 Thread 类对象
Thread t2 = new Thread(mt2);//实例化 Thread 类对象
t1.start();//启动多线程
t2.start();//启动多线程
}
}
```

如图 8-18 所示。

图 8-18

3. 线程的常用方法

虽然我们推荐以 Runnable 接口实现 Java 多线程程序设计,但是,操作线程的主要方法并不在 Runnable 接口中,而是在 Thread 类中。Thread 类中线程的常用操作方法如表 8-9 所示。

表 8-9

方法名称	描述	类型
public Thread(Runnable target)	接收 Runnable 接口子类对象,实例化 Thread 对象	构造

续 表

方法名称	描 述	类型
public Thread (Runnable target, String name)	接收 Runnable 接口子类对象,实例化 Thread 对象,并设置线程名称	构造
public Thread(String name)	实例化 Thread 对象,并设置线程名称	构造
public static Thread currentThread()	返回目前正在执行的线程	普通
public final void setName (String name)	改变线程名称,使之与参数 name 相同	普通
public final String getName()	返回线程的名称	普通
public final void setPriority (int newPriority)	设定线程的优先值(案例见 8.2.3)	普通
public final int getPriority()	返回线程的优先级(案例见 8.2.3)	普通
public void run()	执行线程	普通
public void start()	开始执行线程	普通
public static void sleep(long millis) throws InterruptedException	使目前正在执行的线程休眠 millis 毫秒(案例见 8.2.3)	普通
public static void yield()	将目前正在执行的线程暂停,允许其他线程执行(案例见 8.2.3)	普通
public final boolean isAlive()	判断线程是否在活动:返回 true 或 false	普通
public final void join ( ) throws InterruptedException	等待线程死亡(案例见 8.2.3)	普通
public final void join(long millis) throws InterruptedException	等待 millis 毫秒后,线程死亡	普通
public final void setDaemon(boolean on)	将一个线程设置成为后台运行(案例见 8.2.3)	普通

【例 8.11】取得和设置线程名称。

```
class MyThread implements Runnable{//实现 Runnable 接口
public void run(){//覆写 run()方法
for(int i = 0;i<3;i++){
//currentThread()取得当前线程,getName()获取线程的名字
System.out.println(Thread.currentThread().getName()+"运行第"+i+"次。");
```

```
    }
  }
}
public class ThreadNameTest{
public static void main(String args[]){
MyThread mt = new MyThread();//实例化 Runnable 子类对象
Thread th1 = new Thread(mt);//系统自动设置 th1 线程名称
Thread th2 = new Thread(mt);//系统自动设置 th2 线程名称
th2.setName("线程 2");//使用 setName()方法将 th2 的名称设置为"线程 2"
Thread th3 = new Thread(mt,"线程 3");//手工设置线程名称
th1.start();
th2.start();
th3.start();
  }
}
```

如图 8-19 所示。

图 8-19

【例 8.12】判断线程是否启动。

```
class MyThread implements Runnable{//实现 Runnable 接口
public void run(){//覆写 run()方法
```

```
for(int i = 0;i<3;i++){
System.out.println(Thread.currentThread().getName()+"运行第"+i+"次。");
//取得当前线程的名字
}
}
};
public class ThreadAliveTest{
public static void main(String args[]){
MyThread mt = new MyThread();//实例化 Runnable 子类对象
Thread t = new Thread(mt,"线程");//实例化 Thread 对象
System.out.println("线程开始执行之前 - - >"+t.isAlive());//判断是否启动
t.start();//启动线程
System.out.println("线程开始执行之后 - - >"+t.isAlive());//判断是否启动
}
}
```

如图 8-20 所示。

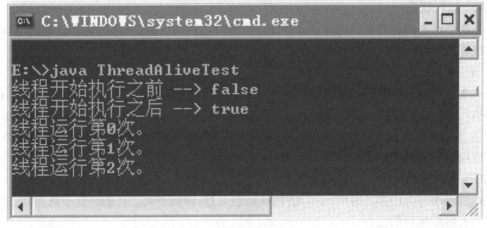

图 8-20

8.2.3 线程的生命周期及调度

1. 线程的状态

线程是动态的,具有一定的生命周期,分别经历从创建、执行、阻塞直到消亡的过程。在每个线程类中都定义了用于完成实际功能的 run() 方法,这个 run() 方法称为线程体 (Thread Body)。按照线程体在计算机系统内存中的状态不同,可以将线程分为创建 (new)、就绪(runnable)、运行、阻塞(blocked)和死亡(dead)5 个状态,如图 8-21 所示。

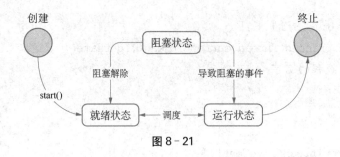

图 8-21

注意：
每个 Java 程序都有一个缺省的主线程,对于 Java 应用程序,主线程是 main()方法。

(1) 创建状态。

在程序中用构造方法创建了一个线程对象后,新的线程对象便处于新建状态,此时,它已经有了相应的内存空间和其他资源,但还处于不可运行状态。创建一个线程对象可采用线程构造方法来实现。例如：

Thread thread＝new Thread();

(2) 就绪状态。

创建线程对象后,调用该线程的 start()方法就可以启动线程。当线程启动时,线程进入就绪状态。此时,线程将进入线程队列排队,等待 CPU 服务,这表明它已经具备了运行条件。

(3) 运行状态。

当就绪状态的线程被调用并获得处理器资源时,线程就进入了运行状态。此时,自动调用该线程对象的 run()方法,run()方法定义了该线程的操作和功能。

(4) 堵塞状态。

一个正在执行的线程在某些特殊情况下,如被人为挂起时,将让出 CPU 并暂时中止自己的执行,进入堵塞状态。在可执行状态下,如果调用 sleep()、suspend()、wait()等方法,线程都将进入堵塞状态。堵塞时,线程不能进入排队队列,只有当引起堵塞的原因被消除后,线程才可以转入就绪状态。

(5) 死亡状态。

线程调用 stop()方法时或 run()方法执行结束后,即处于死亡状态。处于死亡状态的线程不具有继续运行的能力。

线程的上述 5 种状态在转换过程中用到的各种方法如图 8-22 所示。

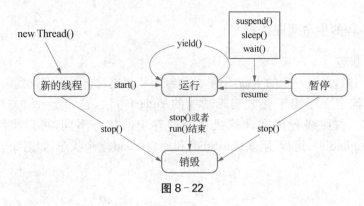

图 8-22

图中，在使用 new Thread()创建一个新的线程之后，就可以通过 start()方法进入运行状态，在运行状态中可以使用 yield()方法进行礼让(即让其他同级别优先级的线程先运行)，但是该线程仍然可以进行；如果一个线程需要暂停的话，可以使用 suspend()、sleep()和 wait()方法，当暂停因素取消后(suspend()方法产生的暂停，必须使用对应的 resume()取消暂停)，又可进入运行；如果线程不需要再执行，则可以通过 stop()方法结束(如果 run()方法执行完毕也表示结束)，此外，一个新的线程在第一次执行前(即一次也没有执行)也可以直接调用 stop()进行结束。

注意：

① 在实际开发中 suspend()方法、resume()方法和 stop()方法不推荐使用，因为这3个方法会产生死锁、数据不完整等问题。

② 如果不使用 stop()方法，又如何停止线程呢？实际开发过程中，一般通过控制 run()方法中循环条件的方式来结束一个线程。

【例8.13】使用设置标志位(循环控制条件)的方法停止线程的运行。

```java
class MyThread implements Runnable{
private boolean flag = true;//定义标志位
public void run(){
while(this.flag){
System.out.println(Thread.currentThread().getName() + "在运行");
}
}
public void stopMe(){
this.flag = false;//修改标志位
}
}
public class StopTest{
public static void main(String args[]){
MyThread my = new MyThread();
Thread t1 = new Thread(my,"线程1");//建立线程对象 t1
Thread t2 = new Thread(my,"线程2");//建立线程对象 t2
t1.start();//启动线程
t2.start();//启动线程
for(int i = 0;i<6;i++){
if(i = = 2)
my.stopMe();//修改标志位,停止运行
System.out.println("Main 线程在运行");
}
}
}
```

如图 8-23 所示。

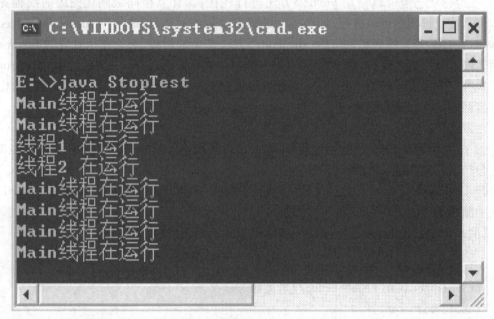

图 8-23

在状态转换的各个过程中,最关键也是最复杂的就是就绪状态和阻塞状态转换的过程。Java 提供了大量方法来支持阻塞,下面让我们逐一分析:

(1) sleep()方法。sleep()允许指定以毫秒为单位的一段时间作为参数,它使得线程在指定的时间内进入阻塞状态,不能得到 CPU 时间;指定的时间一过,线程重新进入可执行状态。典型地,sleep()被用在等待某个资源就绪的情形,例如测试发现条件不满足后,让线程阻塞一段时间后重新测试,直到条件满足为止。

注意:

由于使用 sleep()方法会抛出一个 InterruptedException,所以在程序中需要用 try...catch()捕获。

(2) join()和 join(long timeout)方法。join()方法等待指定的线程运行结束,当线程不活动时,join()才会返回(即可以使用 join()方法让一个线程强制运行,线程强制运行期间,其他线程无法运行,必须等待此线程完成之后才可以继续执行)。而 join(long timeout)方法则是等待指定的线程运行结束,但不超过指定的超时值。

(3) yield()方法。yield()可以用来使具有相同优先级的线程获得执行的机会。如果具有相同优先级的其他线程是可运行的,yield()将把调用线程放到可运行池中并使另一个线程运行。如果没有相同优先级的可运行进程,yield()什么都不做。

(4) wait()和 notify()方法。这两个方法配套使用,wait()使得线程进入阻塞状态,它有两种形式,一种允许指定以毫秒为单位的一段时间作为参数,另一种没有参数。前者当对应的 notify()被调用或者超出指定时间时线程重新进入就绪,后者则必须对应的 notify()被调用。

【例8.14】线程的休眠。

```
class MyThread implements Runnable{//实现Runnable接口
public void run(){//覆写run()方法
for(int i=0;i<3;i++){
try{
Thread.sleep(500);//线程休眠
}catch(InterruptedException e){}
System.out.println(Thread.currentThread().getName()+"运行第"+i+"次。");
}
}
}
public class ThreadSleepTest{
public static void main(String args[]){
MyThread mt=new MyThread();//实例化Runnable子类对象
Thread t=new Thread(mt,"线程");//实例化Thread对象
t.start();//启动线程
}
}
```

如图8-24所示。

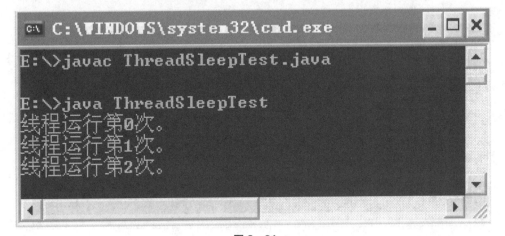

图8-24

【例8.15】线程的强制运行。

```
class MyThread implements Runnable{//实现Runnable接口
public void run(){//覆写run()方法
for(int i=0;i<5;i++){
```

```java
System.out.println(Thread.currentThread().getName() + "运行第" + i + "次。");
            }
        }
    }
public class ThreadJoinTest{
    public static void main(String args[]){
        MyThread mt = new MyThread();//实例化 Runnable 子类对象
        Thread t1 = new Thread(mt,"线程 1");//实例化 Thread 对象
        t1.start();
        for(int i = 0;i<5;i++){
            if(i>1){
                try{
                    t1.join();//线程强制运行
                }catch(InterruptedException e){}
            }
            System.out.println("Main 线程运行 - ->" + i);
        }
    }
}
```

如图 8-25 所示。

图 8-25

【例 8.16】线程的礼让。

```java
class MyThread implements Runnable{//实现 Runnable 接口
    public void run(){//覆写 run()方法
        for(int i = 0;i<4;i++){
```

```
try{
Thread.sleep(1000);
}catch(Exception e){}
System.out.println(Thread.currentThread().getName()+"运行第"+i+"次。");
if(i= =1){
System.out.println(Thread.currentThread().getName()+"线程礼让.");
Thread.currentThread().yield();//线程礼让
    }
   }
  }
 }
public class ThreadYieldTest{
public static void main(String args[]){
MyThread mt = new MyThread();//实例化 MyThread 对象
Thread t1 = new Thread(mt,"线程 1");
Thread t2 = new Thread(mt,"线程 2");
t1.start();
t2.start();
  }
 }
```

如图 8-26 所示。

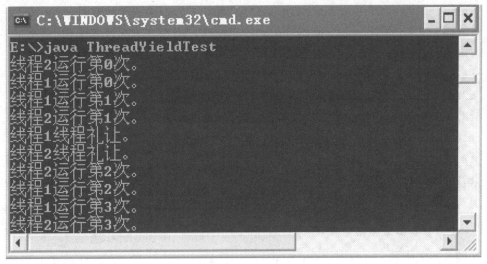

图 8-26

2. 线程的调度

在 Java 的线程操作中,所有的线程在运行前都会保持在就绪状态,线程调度器按线程的优先级高低选择高优先级线程(进入运行中状态)执行;同时线程调度是抢先式调度,即如果在当前线程执行过程中,一个更高优先级的线程进入可运行状态,则这个线程立即被调度执行。

Java 将线程的优先级分为 10 个等级,分别用 1～10 之间的数字表示。数字越大表明线程的级别越高。相应地,在 Thread 类中定义了表示线程最低、最高和普通优先级的成员变量 MIN_PRIORITY、MAX_PRIORITY 和 NORMAL_PRIORITY,代表的优先级等级分别为 1、10 和 5。当一个线程对象被创建时,其默认的线程优先级是 5。此外,在 Java 程序中可以使用 setPriority()方法设置线程的优先级,getPriority()方法返回线程的优先级。

【例 8.17】使用 setPriority()、getPriority()方法设置、返回线程优先级。

```
class MyThread implements Runnable{//实现 Runnable 接口
public void run(){//覆写 run()方法
for(int i = 0;i<3;i++){
try{
Thread.sleep(500);//线程休眠
}catch(InterruptedException e){}
System.out.println(Thread.currentThread().getName()+"运行第"+i+"次,该线程优先级为"+Thread.currentThread().getPriority());//取得当前线程的名字
}
}
}
public class ThreadPriorityTest{
public static void main(String args[]){
Thread t1 = new Thread(new MyThread(),"线程 1");//实例化线程对象
Thread t2 = new Thread(new MyThread(),"线程 2");//实例化线程对象
Thread t3 = new Thread(new MyThread(),"线程 3");//实例化线程对象
t1.setPriority(Thread.MIN_PRIORITY);//优先级最低
t2.setPriority(Thread.MAX_PRIORITY);//优先级最高
t3.setPriority(Thread.NORM_PRIORITY);//优先级为默认
t1.start();//启动线程
t2.start();//启动线程
t3.start();//启动线程
}
}
```

如图 8-27 所示。

在 Java 中有一类比较特殊的线程是被称为守护(Daemon)线程的低级别线程,这类线程具有最低的优先级,用于为系统中的其他对象和线程提供服务。将一个用户线程设置为守护线程的方式是在线程对象创建之前调用线程对象的 setDaemon 方法。典型的守护线程

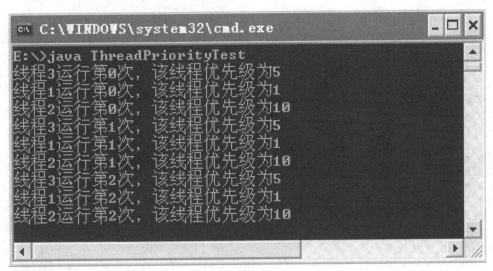

图 8-27

例子是 JVM 中的系统资源自动回收线程,它始终在低级别的状态中运行,用于实时监控和管理系统中的可回收资源。

【例 8.18】使用 setDaemon( )方法设置后台线程。

```
class MyThread implements Runnable{//实现 Runnable 接口
public void run(){//覆写 run()方法
while(true){
System.out.println(Thread.currentThread().getName()+"在运行。");
}
}
}
public class ThreadDaemonTest{
public static void main(String args[]){
MyThread mt = new MyThread();//实例化 Runnable 子类对象
Thread t = new Thread(mt,"线程 1");//实例化 Thread 对象
t.setDaemon(true);//此线程在后台运行
t.start();//启动线程
for(int i = 0;i<3;i++){
System.out.println(Thread.currentThread().getName()+"在运行。");
}
}
}
```

如图 8-28 所示。

该案例中虽然线程的 run()方法中是死循环,但程序依然可以执行完,这是因为 run()

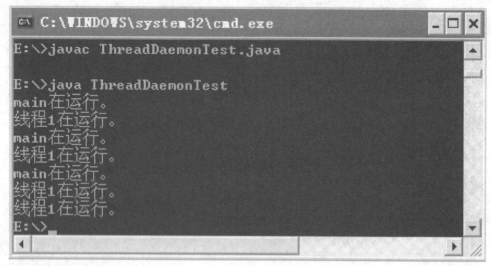

图 8-28

方法中的死循环已经设置为后台运行。

8.2.4 线程的同步与死锁

1. 问题的提出

在 Java 语言中使用 Runnable 接口创建的多个线程可以实现资源共享,但如果不采取相关措施,那么在多个线程对同一资源进行操作时会产生异常。我们以购买火车票为例,如果要购买火车票,可以选择去火车站或者售票点,但不管售票点(或火车站)有多少个,最终某一趟列车的车票数量是固定的。如果把各个售票点(或火车站)理解为线程的话,所有的线程应该共同拥有同一列车的票数(即共享同一列车的票数,售票时都对该车的车票数量进行修改)。

【例8.19】假设某一班列车共有6张票,4个售票点同时售票,要求每个站点每人一次只能购买一张票并且每销售一张车票就实时显示该车票的编号。

```
class MyThread implements Runnable{
private int ticket = 6;//假设一共有6张票
public void run(){
for(int i = 0;i<100;i++){
if(ticket>0){//还有票
try{
Thread.sleep(500);//加入延迟
}catch(InterruptedException e){
e.printStackTrace();
}
```

```
            System.out.println(Thread.currentThread().getName() + "本次出售的车票号
为:" + ticket - -);
        }
    }
}
public class SyncTest{
    public static void main(String args[]){
        MyThread mt = new MyThread();//定义线程对象
        Thread t1 = new Thread(mt,"线程1");//定义 Thread 对象 t1,站点1
        Thread t2 = new Thread(mt,"线程2");//定义 Thread 对象 t2,站点2
        Thread t3 = new Thread(mt,"线程3");//定义 Thread 对象 t3,站点3
        Thread t4 = new Thread(mt,"线程4");//定义 Thread 对象 t4,站点4
        t1.start();
        t2.start();
        t3.start();
        t4.start();
    }
}
```

如图 8-29 所示。

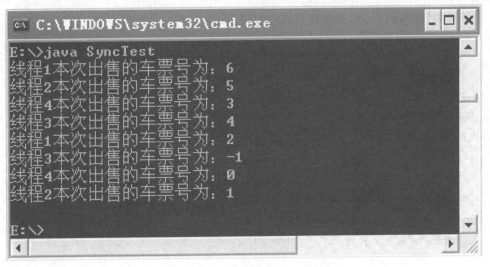

图 8-29

从执行结果可以发现,车票的编号成为负数了,程序代码出现了问题。为什么会这样呢?

在上例中程序对车票的操作步骤如下:

(1) 判断票数是否大于0,大于0则表示还有票可以卖。

(2) 如果票数大于0,则将票卖出去(执行票数减一操作,就得到剩余车票数及车票号),并输出车票编号。

但是,在上面的操作代码中,在第(1)步和第(2)步之间加入了延迟操作,那么一个线程就有可能在还没有对票数进行减操作之前,其他线程就已经将票数减少了,这样就会出现程序中车票标号为负数的情况。

如果想解决这样的问题,就必须使用同步。所谓同步是指多个线程之间互相协作,按照某种既定的步骤来共同完成任务,即多个操作在同一个时间段内只能有一个线程进行,其他线程要等待此线程完成之后才可以继续执行。

2. 同步的实现

线程的同步有两种方法:同步代码块、同步方法。

(1) 同步代码块。

在代码块上加上"synchronized"关键字,此代码块就称为同步代码块,其语法格式如下:

```
synchronized(同步对象){
需要同步的代码;
}
```

注意:

① 同步代码的时候必须指明同步的对象,一般情况下,会将当前对象作为同步的对象,使用 this 表示。

② 在同一时刻只能有一个线程可以进入同步代码块内运行,只有当该线程离开同步代码块后,其他线程才能进入同步代码块内运行。

【例 8.20】修改例 8.19,实现售票代码的同步。

```
class MyThread implements Runnable{
private int ticket = 6;//假设一共有6张票
public void run(){
for(int i = 0;i<100;i++){
synchronized(this){
if(ticket>0){//还有票
try{
Thread.sleep(300);//加入延迟
}catch(InterruptedException e){
e.printStackTrace();
}
System.out.println(Thread.currentThread().getName() + "本次出售的车票号为:" + ticket--);
}
}
```

```
            }
        }
    }
public class SyncTest1{
    public static void main(String args[]){
        MyThread mt = new MyThread();//定义线程对象
        Thread t1 = new Thread(mt,"线程1");//定义Thread对象t1,站点1
        Thread t2 = new Thread(mt,"线程2");//定义Thread对象t2,站点2
        Thread t3 = new Thread(mt,"线程3");//定义Thread对象t3,站点3
        Thread t4 = new Thread(mt,"线程4");//定义Thread对象t4,站点4
        t1.start();
        t2.start();
        t3.start();
        t4.start();
    }
}
```

如图8-30所示。

图8-30

(2) 同步方法。

除了可以将需要的代码设置成同步代码块之外,还可以使用synchronized关键字将一个方法声明成同步方法,其语法格式如下:

```
访问控制符 synchronized 方法返回值 方法名称(参数列表){
    代码;
}
```

注意：

在同一类中，使用 synchronized 关键字定义的若干方法，可以在多个线程之间同步。当有一个线程进入了有 synchronized 修饰的方法时，其他线程就不能进入同一个对象使用 synchronized 来修饰的所有方法，直到第一个线程执行完它所进入的 synchronized 修饰的方法为止。

【例8.21】使用同步方法修改例8.19程序，完善售票功能。

```java
class MyThread implements Runnable{
private int ticket = 6;//假设一共有6张票
public void run(){
for(int i = 0;i<100;i++){
this.sale();//调用同步方法
}
}
public synchronized void sale(){//声明同步方法
if(ticket>0){//还有票
try{
Thread.sleep(500);//加入延迟
}catch(InterruptedException e){
e.printStackTrace();
}
System.out.println(Thread.currentThread().getName() + "本次出售的车票号为:" + ticket--);
}
}
}
public class SyncTest2{
public static void main(String args[]){
MyThread mt = new MyThread();//定义线程对象
Thread t1 = new Thread(mt,"线程1");//定义 Thread 对象t1,站点1
Thread t2 = new Thread(mt,"线程2");//定义 Thread 对象t2,站点2
Thread t3 = new Thread(mt,"线程3");//定义 Thread 对象t3,站点3
Thread t4 = new Thread(mt,"线程4");//定义 Thread 对象t4,站点4
t1.start();
t2.start();
t3.start();
t4.start();
}
}
```

如图 8-31 所示。

图 8-31

3. 线程的死锁

张三想要李四的画,李四想要张三的书。张三对李四说:"把你的画给我,我就给你书。"李四也对张三说:"把你的书给我,我才会给你画。"此时,张三在等着李四的答复,而李四也在等着张三的答复,这样下去,最终结果是张三得不到李四的画,李四也得不到张三的书,这就是死锁。线程死锁是在并发程序设计中经常遇到的问题,它是指程序运行中,多个线程竞争共享资源时可能出现的一种系统状态:线程 1 拥有资源 1,并等待资源 2,而线程 2 拥有资源 2,并等待资源 3……依此类推,线程 n 拥有资源 n-1,并等待资源 1。在这种状态下,各个线程互不相让,永远进入一种等待状态。

注意:

线程死锁并不是必然会发生的,在某些情况下,可能会非常偶然地产生,该状态出现的机会可能会非常小,因此简单的测试往往无法发现。遗憾的是 Java 语言也没有有效的方法可以避免或检测死锁,因此我们只能在程序设计中尽力去减少这种情况的出现。

导致死锁的根源在于不适当地运用"synchronized"关键词来管理线程对特定对象的访问。"synchronized"关键词的作用是,确保在某个时刻只有一个线程被允许执行特定的代码块,因此,被允许执行的线程首先必须拥有对变量或对象的排他性访问权。当线程访问对象时,线程会给对象加锁,而这个锁导致其他也想访问同一对象的线程被阻塞,直至第一个线程释放它加在对象上的锁。

同步锁导致的死锁虽然具有偶然性,但死锁的产生是有规律可循的。一般来说,要出现死锁必须同时具备 4 个条件。因此,如果能够尽可能地破坏这 4 个条件中的任意一个,就可以避免死锁的出现。死锁产生必需的 4 个条件如下:

(1) 互斥条件:一个资源每次只能被一个进程使用。

(2) 请求与保持条件:一个进程因请求资源而阻塞时,对已获得的资源保持不放。

(3) 不剥夺条件:进程已获得的资源,在未使用完之前,不能强行剥夺。

（4）循环等待条件：若干进程之间形成一种头尾相接的循环等待资源关系。

任务 实施

1. 服务器端程序设计

```java
import javax.swing.*;
import java.awt.*;
import java.awt.event.*;
import java.io.*;
import java.net.*;

public class ChatServer extends JFrame{
    JLabel ipJLabel = new JLabel("IP:");
    JLabel portJLabel = new JLabel("端口:");
    JTextField ipField    = new JTextField();
    JTextField portField  = new JTextField();
    JTextField msgField   = new JTextField();
    JButton listenBtn     = new JButton("侦听");
    JButton disConBtn     = new JButton("断开");
    JButton sendBtn       = new JButton("发送");
    JTextArea msgRecArea = new JTextArea();
    JScrollPane scrollPanel = new JScrollPane(msgRecArea);
    ServerSocket server;
    Socket client;
    DataInputStream dis = null;
    DataOutputStream dos = null;
    boolean isConnected = false;
    ListenClient lisClient = null;

    public ChatServer(){
        super("服务器");
        this.setSize(500,500);
        this.setLayout(null);
        this.setVisible(true);
        ipJLabel.setBounds(20,30,20,30);
        ipField.setBounds(50,30,150,30);
        portJLabel.setBounds(210,30,50,30);
```

```java
            portField.setBounds(260,30,60,30);
            listenBtn.setBounds(330,30,60,30);
            disConBtn.setBounds(400,30,60,30);
            this.add(ipJLabel);
            this.add(ipField);
            this.add(portJLabel);
            this.add(portField);
            this.add(listenBtn);
            this.add(disConBtn);

            msgField.setBounds(50,70,300,30);
            sendBtn.setBounds(400,70,60,30);
            this.add(msgField);
            this.add(sendBtn);
            scrollPanel.setBounds(50,120,360,300);
            this.add(scrollPanel);
            msgRecArea.setEditable(false);
            disConBtn.setEnabled(false);
            ipField.setEditable(false);

            InetAddress inetAddr;
            try{
                inetAddr = InetAddress.getLocalHost();
                ipField.setText(inetAddr.getHostAddress());
            }catch(UnknownHostException e){
                e.printStackTrace();
            }
            ButtonListener listener = new ButtonListener();
            listenBtn.addActionListener(listener);
            disConBtn.addActionListener(listener);
            sendBtn.addActionListener(listener);
            this.addWindowListener(new WindowAdapter(){
                public void windowClosing(WindowEvent e){
                    System.exit(0);
                }
            });
        }
        private class ListenClient implements Runnable{
```

```java
            public void run(){
                try{
                    while(isConnected){
                        String line = dis.readUTF();
                        msgRecArea.append("客户端说:" + line + "\n");
                    }
                }
                catch(IOException ex){
                }
            }
        }

        class ButtonListener implements ActionListener{
            public void actionPerformed(ActionEvent e){
                String cmd = e.getActionCommand();
                if(cmd.equals("侦听")){
                    if(portField.getText().equals("")){
                        JOptionPane.showMessageDialog(null,"请输入端口号",
                            "提示",JOptionPane.ERROR_MESSAGE);
                        return;
                    }
                    int port = Integer.parseInt(portField.getText());
                    listenBtn.setEnabled(false);
                    disConBtn.setEnabled(true);
                    try{
                        server = new ServerSocket(port);
                        client = server.accept();
                        msgRecArea.append("已连接客户机:" +
client.getInetAddress().toString() + ":" + client.getPort() + "\n");
                        dis = new DataInputStream(client.getInputStream());
                        dos = new DataOutputStream(client.getOutputStream());
                        isConnected = true;
                        lisClient = new ListenClient();
                        new Thread(lisClient).start();
                    }
                    catch(IOException ex){
                        JOptionPane.showMessageDialog(null,ex.getMessage(),
                            "提示",JOptionPane.ERROR_MESSAGE);
```

```java
                }
            }
            else if(cmd.equals("断开")){
                try{
                    server.close();
                }catch(IOException ex){
                    JOptionPane.showMessageDialog(null,ex.getMessage(),
                        "提示",JOptionPane.ERROR_MESSAGE);
                }
                listenBtn.setEnabled(true);
                disConBtn.setEnabled(false);
            }
            else if(cmd.equals("发送")){
                try{
                    String content = msgField.getText();
                    msgField.setText(null);
                    dos.writeUTF(content);
                    msgRecArea.append("我说:" + content + "\n");
                }
                catch(IOException ex){
                    JOptionPane.showMessageDialog(null,ex.getMessage(),
                        "提示",JOptionPane.ERROR_MESSAGE);
                }
            }
        }
    }
    public static void main(String[]args){
        ChatServer chatServer = new ChatServer();
    }
}
```

2. 客户端程序设计

```java
import javax.swing.*;
import java.awt.*;
import java.awt.event.*;
import java.io.*;
import java.net.*;
```

```java
public class ChatClient extends JFrame
{
    JLabel ipJLabel = new JLabel("IP:");
    JLabel portJLabel = new JLabel("端口:");
    JTextField ipField    = new JTextField();
    JTextField portField = new JTextField();
    JTextField msgField   = new JTextField();
    JButton connectBtn   = new JButton("连接");
    JButton disConBtn    = new JButton("断开");
    JButton sendBtn      = new JButton("发送");
    JTextArea msgRecArea = new JTextArea();
    JScrollPane scrollPanel = new JScrollPane(msgRecArea);
    Socket client;
    DataInputStream dis = null;
    DataOutputStream dos = null;
    ListenClient lisClient = null;
    boolean isConnected = false;

    public ChatClient(){
        super("客户端");
        this.setSize(500,500);
        this.setLayout(null);
        this.setVisible(true);
        ipJLabel.setBounds(20,30,20,30);
        ipField.setBounds(50,30,150,30);
        portJLabel.setBounds(210,30,50,30);
        portField.setBounds(260,30,60,30);
        connectBtn.setBounds(330,30,60,30);
        disConBtn.setBounds(400,30,60,30);
        this.add(ipJLabel);
        this.add(ipField);
        this.add(portJLabel);
        this.add(portField);
        this.add(connectBtn);
        this.add(disConBtn);
        msgField.setBounds(50,70,300,30);
        sendBtn.setBounds(400,70,60,30);
        this.add(msgField);
```

```java
            this.add(sendBtn);
            scrollPanel.setBounds(50,120,360,300);
            this.add(scrollPanel);
            msgRecArea.setEditable(false);
            disConBtn.setEnabled(false);
            ButtonListener listener = new ButtonListener();
            connectBtn.addActionListener(listener);
            disConBtn.addActionListener(listener);
            sendBtn.addActionListener(listener);
            this.addWindowListener(new WindowAdapter(){
                public void windowClosing(WindowEvent e){
                    System.exit(0);
                }
            });
        }
        private class ListenClient implements Runnable{
            public void run(){
                try{
                    while(isConnected){
                        String line = dis.readUTF();
                        msgRecArea.append("服务器说:" + line + "\n");
                    }
                }
                catch(IOException ex){
                }
            }
        }
        class ButtonListener implements ActionListener{
            public void actionPerformed(ActionEvent e){
                String cmd = e.getActionCommand();
                if(cmd.equals("连接")){
                    if(ipField.getText().equals("")){
                        JOptionPane.showMessageDialog(null,"请输入 IP",
                            "提示",JOptionPane.ERROR_MESSAGE);
                        return;
                    }
                    if(portField.getText().equals("")){
                        JOptionPane.showMessageDialog(null,"请输入端口号",
```

```java
                            "提示",JOptionPane.ERROR_MESSAGE);
                        return;
                    }
                    int port = Integer.parseInt(portField.getText());
                    String ip = ipField.getText();
                    connectBtn.setEnabled(false);
                    disConBtn.setEnabled(true);
                    try{
                        client = new Socket(ip,port);
                        msgRecArea.append("已连接到服务器:" +
client.getInetAddress().getHostAddress() + ":" + client.getPort() + "\n");
                        dis = new DataInputStream(client.getInputStream());
                        dos = new DataOutputStream(client.getOutputStream());
                        isConnected = true;
                        lisClient = new ListenClient();
                        new Thread(lisClient).start();
                    }
                    catch(IOException ex){
                        JOptionPane.showMessageDialog(null,ex.getMessage(),
                            "提示",JOptionPane.ERROR_MESSAGE);
                    }
                }
                else if(cmd.equals("断开")){
                    try{
                        client.close();
                    }catch(IOException ex){
                        JOptionPane.showMessageDialog(null,ex.getMessage(),
                            "提示",JOptionPane.ERROR_MESSAGE);
                    }
                    connectBtn.setEnabled(true);
                    disConBtn.setEnabled(false);
                }
                else if(cmd.equals("发送")){
                    try{
                        String content = msgField.getText();
                        msgField.setText(null);
                        dos.writeUTF(content);
                        msgRecArea.append("我说:" + content + "\n");
```

```
            }
            catch(IOException ex){
                JOptionPane.showMessageDialog(null,ex.getMessage(),
                    "提示",JOptionPane.ERROR_MESSAGE);
            }
        }
    }
}
public static void main(String[]args){
    ChatClient chatClient = new ChatClient();
}
}
```

如图 8-32、图 8-33 所示。

注意：

程序调试运行时，应先运行服务器端程序并点击【侦听】按钮，开始等待客户端连接后，再点击客户端程序的【连接】按钮（连接之前请输入 IP 地址和端口号），完成连接，然后就可以输入内容并点击【发送】按钮开始聊天。结束聊天时，服务器和客户端都应先点击各自的【断开】按钮，再关闭程序。

图 8-32

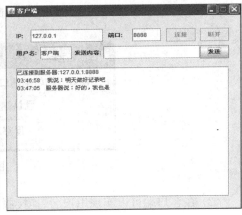

图 8-33

任务 拓展

使用 TCP/IP 协议（SocketServer、Socket 类）和多线程设计实现具有多人聊天功能的聊天室，要求如下：

（1）在服务器端可以实现端口设置、启动服务、停止服务、发送信息、发送信息的用户、退出程序等操作，如图 8-34、图 8-35 所示：

图 8-34

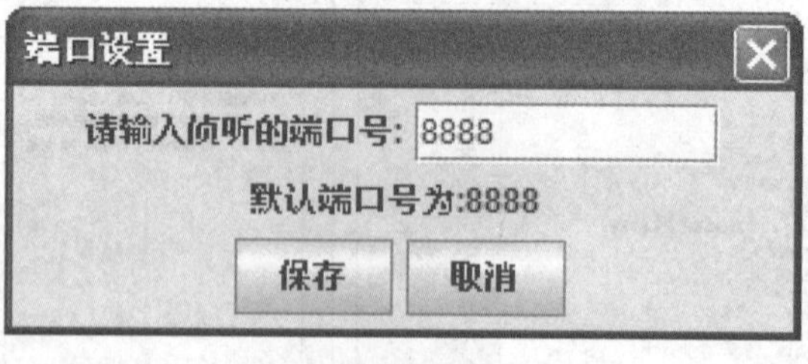

图 8-35

(2) 在客户端可以实现用户名、连接服务器的 IP 地址、端口等信息的设置,以及用户的登录、注销、消息发送、程序退出等操作,如图 8-36、图 8-37、图 8-38 所示:

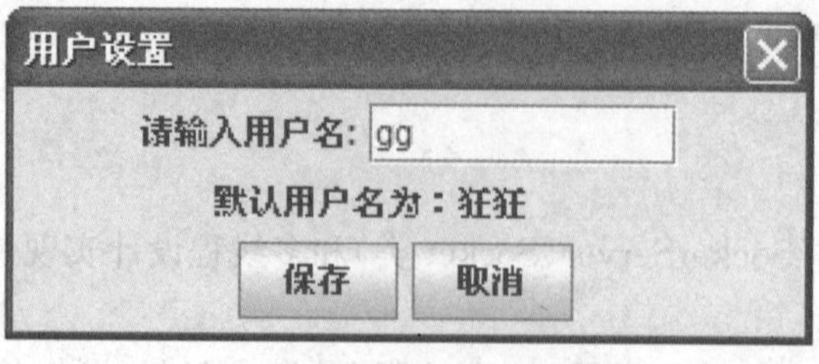

图 8-36

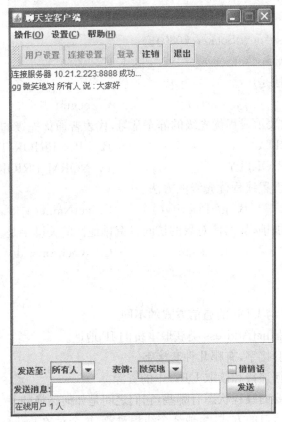

图 8-37

图 8-38

 习　　题

一、单选题

1. 请选择您认为正确的 Socket 工作流程。（　　）
① 打开连接到 Socket 的输入/输出
② 按照某个协议对 Socket 进行的读/写操作
③ 创建 Socket

④ 关闭 Socket
   A. ①③②④   B. ②①③④   C. ③①②④   D. ①②③④
2. 已知创建 URL 对象的 Java 语句为：URL u＝new URL("http://www.cq113.com");。如果 www.cq113.com 不存在，则返回（   ）。
   A. http://www.cq113.com   B. ""
   C. null   D. 抛出异常
3. 请选出能够正确创建 socket 对象的 Java 语句。（   ）
   A. Socket s＝new Socket(8080);
   B. Socket s＝new Socket("192.168.1.1","8080")
   C. SocketServer s＝new Socket(8080);
   D. Socket s＝new SocketServer("192.168.1.1","8080")
4. 线程的启动方法是（   ）。
   A. run()   B. start()   C. begin()   D. accept()
5. Thread 类提供表示线程优先级的静态常量，代表普通优先级的静态常量是（   ）。
   A. MAX_PRIORITY   B. MIN_PRIORITY
   C. NORMAL_PRIORITY   D. NORM_PRIORITY
6. 请选出正确的设置线程优先级的方法。（   ）
   A. setPriority()   B. getPriority()   C. getName()   D. setName()
7. 通常用于对象加锁，从而使对象的访问具有排他性的关键字是（   ）。
   A. serialize   B. transient   C. synchronized   D. static

二、简答题
   1. 简要阐述 TCP 与 UDP 的通信方式的不同。
   2. 编写程序，使用 InetAddress 类获取本机的 IP 地址。
   3. 请简述什么是套接字，有哪几种套接字。
   4. 请简要描述如何连接和读取 URL 中的资源。
   5. 请简要描述主要的线程状态有哪些，它们之间是如何转换的。
   6. 分别用继承和接口的方式，生成 3 个线程对象，每个对象循环打印 10 次，要求从结果中证明线程的无序性。
   7. 创建两个线程对象，要求用同步块的方式使第一个线程运行 10 次，然后将自己阻塞起来，唤醒第二个线程，第二个线程再运行 10 次，然后将自己阻塞起来，唤醒第一个线程……两个线程交替执行。

参考文献

[1] 刘宝林. Java 程序设计与案例[M]. 北京:高等教育出版社,2004.
[2] 洪维恩. Java 2 面向对象程序设计[M]. 北京:中国铁道出版社,2002.
[3] 李钟尉,马文强,陈丹丹. Java 从入门到精通[M]. 北京:清华大学出版社,2008.
[4] 李兴华. Java 核心技术精讲[M]. 北京:清华大学出版社,2013.
[5] 李兴华. Java 开发实战经典[M]. 北京:清华大学出版社,2009.
[6] 朱喜福. Java 程序设计(第 2 版)[M]. 北京:人民邮电出版社,2007.
[7] https://baike.baidu.com[EB/OL].
[8] http://image.baidu.com[EB/OL].

图书在版编目(CIP)数据

Java 程序设计项目化教程/范凌云,兰伟,杨东主编. —上海:复旦大学出版社,2020.8
ISBN 978-7-309-14646-2

Ⅰ.①J… Ⅱ.①范… ②兰… ③杨… Ⅲ.①JAVA 语言-程序设计-高等职业教育-教材
Ⅳ.①TP312.8

中国版本图书馆 CIP 数据核字(2019)第 220559 号

Java 程序设计项目化教程
范凌云 兰 伟 杨 东 主编
责任编辑/陆俊杰

复旦大学出版社有限公司出版发行
上海市国权路 579 号 邮编:200433
网址:fupnet@fudanpress.com http://www.fudanpress.com
门市零售:86-21-65102580   团体订购:86-21-65104505
外埠邮购:86-21-65642846   出版部电话:86-21-65642845
上海春秋印刷厂

开本 787×1092 1/16 印张 36 字数 832 千
2020 年 8 月第 1 版第 1 次印刷

ISBN 978-7-309-14646-2/T·654
定价:79.00 元

如有印装质量问题,请向复旦大学出版社有限公司出版部调换。
版权所有 侵权必究